Statics
and Strength
of Materials

Statics
and Strength
of Materials

H. W. MORROW

Professor
Nassau Community College
Garden City, New York

PRENTICE-HALL, INC., Englewood Cliffs, New Jersey 07632

Library of Congress Cataloging in Publication Data

Morrow, Harold W
   Statics and strength of materials.

   Includes index.
   1. Strength of materials. 2. Statics.
3. Strength of materials—Problems, exercises, etc.
4. Statics—Problems, exercises, etc.
5. Trigonometry. I. Title.
TA405.M877   620.1'053   80-24085
ISBN 0-13-844720-9

Editorial/production supervision and interior design
  by Barbara A. Cassel
Cover design by Miriam Recio
Manufacturing buyer: Anthony Caruso

Printed in the United States of America

10 9 8 7 6 5 4 3 2 1

PRENTICE-HALL INTERNATIONAL, INC., *London*
PRENTICE-HALL OF AUSTRALIA PTY. LIMITED, *Sydney*
PRENTICE-HALL OF CANADA, LTD., *Toronto*
PRENTICE-HALL OF INDIA PRIVATE LIMITED, *New Delhi*
PRENTICE-HALL OF JAPAN, INC., *Tokyo*
PRENTICE-HALL OF SOUTHEAST ASIA PTE. LTD., *Singapore*
WHITEHALL BOOKS LIMITED, *Wellington, New Zealand*

To my mother and father,

Hazel Nicholson Morrow
Harold S. Morrow

Contents

⑥   FORCE ANALYSIS OF STRUCTURES AND MACHINES   112

⑦   FRICTION   145

⑧   CENTER OF GRAVITY, CENTROIDS, AND MOMENTS OF INERTIA OF AREAS   169

13 BENDING AND SHEARING STRESSES IN BEAMS   318

14 DEFLECTION OF BEAMS DUE TO BENDING   358

15 COMBINED STRESSES—MOHR'S CIRCLE   400

16 COLUMNS   447

APPENDIX   467

INDEX   498

ANSWERS TO EVEN-NUMBERED PROBLEMS   505

Preface

This textbook is designed to cover the topics of statics and strength of materials at an elementary level not requiring calculus. It is appropriate for students enrolled in an engineering technology curricula and for university-level courses for nonengineering majors, such as architecture. It can be used for self-study and may also be used as a reference for courses in materials testing, machine design, and structural design.

Chapters 1 through 8 provide coverage of statics and include a review of the basic principles of trigonometry. Graphical methods, together with trigonometric formulas and the component method, are used to solve concurrent force problems and to encourage visualization of the geometric relationships involved. With nonconcurrent forces the concept of the moment of a force is introduced, with special emphasis on the theorem of moments. The study of the equilibrium of forces includes the analysis of trusses, frames, and machines. The coverage of statics concludes with the study of friction, centers of gravity, centroids, and moments of inertia of areas.

Chapters 9 through 16 are devoted to the study of strength of materials. The study begins with the concepts of stress and strain, together with the stress and strain of axially loaded members. These concepts are followed by the torsion of circular bars, bending and deflection of beams, combined stresses using Mohr's circle, and columns.

More than seven hundred exercise problems on various levels of difficulty are provided. The book can be used for several years without repeating assignments and can be used for classes of varying abilities.

The United States is in the process of converting from the U.S. common system of units to the international system of units (SI metric units). Consequently, both systems are introduced and used equally in example and practice problems. In problems using SI units, as a concession to previous U.S. practice, some bodies are described in terms of weight rather than mass. Exercises are provided to aid the student in converting between the two systems but, in general, conversions have not been emphasized.

The original draft of the text was written during my 1975–76 sabbatical leave and was used in photo-offset form at Nassau Community College during the school years 1976–77, 1977–78, and 1978–79.

I wish to acknowledge the helpful suggestions and criticisms made by my colleagues Professors William Herschcopf and Joseph F. Keuler while teaching from the text in photo-offset form. I would also like to acknowledge the assistance of Professor Andrew C. Kowalik in arranging for classroom testing of the text.

Finally, I would like to thank my good friend, Charles Guy Williams, for his assistance and encouragement during the preparation of the many drafts of this text.

HAROLD W. MORROW

New York, NY
Ft. Lauderdale, FL

1

Basic Concepts _____

1.1 INTRODUCTION

Statics is that branch of mechanics which involves the study of forces and the effect of forces acting on bodies in equilibrium without motion. From our study of statics we will be able to answer questions concerning the loads that such bodies in equilibrium support.

Strength of materials is concerned with the stresses and deflections or deformations in a body that are produced by loads on the body and by the dimensions of the body. We will be able to determine from our study of strength of materials if a body fulfills its intended purpose and to design the body so that it can safely support a given loading.

Statics is considered in Chapters 1 through 8 and strength of materials in Chapters 9 through 16.

Statics is one of the oldest branches of science. Its origins date back to the Egyptians and Babylonians, who used statics in the building of pyramids and temples. Among the earliest written records are the theories developed by Archimedes (287–212 B.C.), who explained the equilibrium of the lever and the law of buoyancy in hydrostatics. Modern statics date from about A.D. 1600 with the use by Simon Stevinus of the parallelogram law to combine forces vectorally.

1.2 FUNDAMENTAL QUANTITIES—UNITS

The two fundamental quantities used in the solution of statics problems are force and length.

A *force* may be defined as the action of one body on another which tends to change the size or shape and/or state of motion of the body. Aside from a simple push or pull which we can exert on a body with our hands, the force we are most familiar with is gravitational force. The gravitational force or attraction exerted by the earth on a body is called the weight of the body. Other familiar forces are magnetic attraction, wind force, automobile tire traction on pavement, and water pressure.

The concept of *length* or distance is required to describe the size of the bodies in equilibrium without motion and to locate the position of the forces that act on these bodies.

Force is described by its magnitude, direction, and point of application. The magnitude of a force is given by a certain number of force units. The direction may be given by the angle the force makes with a selected reference axis. The point of application is the point at which the force may be considered to be concentrated.

Length is described by magnitude only. It is given by a certain number of length units. The magnitude of both force and length are defined on the basis of arbitrarily chosen units.

U.S. Common System of Units

Engineers in the United States have commonly used the pound (lb) as the unit of force and the foot (ft) as the unit of length. The pound force represents the weight of a certain platinum cylinder placed at sea level and at a latitude of 45°. Other commonly used units of force are the kilopound (kip), equal to 1000 lb, and the ton, equal to 2000 lb. The foot is defined as 0.3048 meter. (See the following paragraph for the definition of meter.) Other units of length based on the foot are the mile (mi), equal to 5280 ft; the inch (in.), equal to $\frac{1}{12}$ ft; and the yard (yd), equal to 3 ft.

International System of Units (SI)

The United States is in the process of converting from the common units to a modernized version of the metric system called the *International*

System of Units, abbreviated SI. The meter or metre (m) is the unit of length and the newton (N) is the unit of force. The meter was at one time defined as the distance between parallel lines marked on a bar kept at standard temperature and pressure and located near Paris, France. The latest definition is based on the wavelength of a color line in the spectrum of the gas krypton. The advantage of such a definition is that it can be reproduced at any location. The meter is 3.2808 ft. The unit of force, the newton, is a derived unit. It is based on the change in the state of motion of a standard body on which the force acts. The newton (N) is approximately two-tenths of a pound of force or, more precisely, 1 N = 0.2248 lb. (The pound force is approximately $4\frac{1}{2}$ N or, more precisely, 1 lb = 4.4482 N.)

Weight and mass

The *weight* or *force of gravity* on a body is determined from the *mass* of the body. The unit of mass is called a kilogram (approximately equal to a volume of 0.001 m^3 of water) and is defined as the mass of a certain platinum cylinder kept near Paris, France.

To find the weight W of a body in newtons (N) from the mass m in kilograms (kg), we use the equation

$$W = mg \qquad\qquad (1.1)$$

where g is the acceleration due to gravity in meters per second squared (m/s^2). The acceleration varies only slightly from place to place on earth. We will use the approximation 9.81 m/s^2. For example, to find the weight of a body with a mass of 2.5 kg, we multiply by 9.81. $W = 2.5(9.81) = 24.5$ N.

Multiples and submultiples of length and force commonly used are the kilometer (km), equal to 1000 m; the millimeter (mm), equal to 0.001 m; the kilonewton (kN), equal to 1000 N; and the meganewton (MN), equal to 1 000 000 N. Thus we see that the prefix milli means 0.001 or 10^{-3}, the prefix kilo means 1000 or 10^3, and the prefix mega means 1 000 000 or 10^6. See Table A.1 of the Appendix for other common prefixes.

1.3 SI STYLE AND USAGE

Precise rules of style and usage have been established in the SI system. Several of the rules follow.

A dot should be used to separate units that are multiplied together. For example, for a newton·meter, the moment of force, we write N·m. This helps avoid confusion with the millinewton, which would be written mN.

Except for the prefix kilo (k) in kilograms, prefixes should be avoided in the denominator of compound units. For example, as a measure of stress we use N/m^2 rather than N/mm^2, to avoid having the prefix milli (m) appear in the denominator.

Numbers having five or more digits on either or both sides of the decimal point should be placed in groups of three separated by spaces rather than commas both to the right and left of the decimal point. For example, we write 12 345.543 21 rather than 12,345.54321.

Numbers having four digits on either or both sides of the decimal point are not required to be placed in groups of three unless the numbers are part of a column of numbers with five or more digits. For example, we write 1234.4321 rather than 1 234.432 1.

1.4 CONVERSION OF UNITS

In general, we will solve problems in the same units as those used in giving information for the problem. In special cases it may be necessary to convert from one system of units to the other.

To convert a length or force in one set of units to another, we multiply by the appropriate conversion factor. For example, to convert a length L of 12.5 in. into m, we must replace inches by meters. From Table 1.1 we see that

$$1 \text{ m} = 39.37 \text{ in.} \quad \text{or} \quad \frac{1 \text{ m}}{39.37 \text{ in.}} = 1$$

Since this ratio has the value of unity, the value of L will not be changed if we write

$$L = 12.5 \text{ in.} \left(\frac{1 \text{ m}}{39.37 \text{ in.}} \right)$$

TABLE 1.1

Equivalents of Common and SI Units

U.S. common units	SI units
Length	
1 in. = 25.40 mm	1 mm = 0.03937 in.
1 in. = 0.02540 m	1 m = 39.37 in.
1 ft = 0.3048 m	1 m = 3.281 ft
Force	
1 lb = 4.448 N	1 N = 0.2248 lb
1 kip = 4.448 kN	1 kN = 0.2248 kip

Performing the numerical calculations and canceling units that appear in the numerator and denominator, we obtain the desired result:

$$L = 12.5\,(0.0254) = 0.318 \text{ m}$$

The number 0.0254, which is used to convert length in inches to length in meters, is called a *conversion factor*.

To convert a stress of 18,540 lb/in.^{2} to units of N/m^2, we replace pounds by newtons and inches by meters. From Table 1.1 we have

$$1 \text{ N} = 0.2248 \text{ lb} \quad \text{or} \quad \frac{1 \text{ N}}{0.2248 \text{ lb}} = 1$$

and

$$1 \text{ m} = 39.37 \text{ in.} \quad \text{or} \quad \frac{39.37 \text{ in.}}{1 \text{ m}} = 1$$

Since each ratio is equal to unity, the value of stress is not changed if we write

$$\text{Stress} = 18{,}540\,\frac{\text{lb}}{\text{in.}^2}\left(\frac{1 \text{ N}}{0.2248 \text{ lb}}\right)\left(\frac{39.37 \text{ in.}}{1 \text{ m}}\right)\left(\frac{39.37 \text{ in.}}{1 \text{ m}}\right)$$

Calculating and canceling units, we have

$$\text{Stress} = 18{,}540\,(6895) = 127.8 \times 10^6\,\frac{\text{N}}{\text{m}^2}$$

The number 6895 is a conversion factor from lb/in.^{2} to N/m^2. From Table A.1 of the Appendix we see that 10^6 newtons is equal to a meganewton (MN); therefore,

$$\text{Stress} = 127.8\,\frac{\text{MN}}{\text{m}^2}$$

Conversion factors for several quantities are given in Table A.2 of the Appendix.

1.5 NUMERICAL COMPUTATIONS

Accuracy

The accuracy of a solution can be no greater than the accuracy of the data on which the solution is based. For example, the length of one side of a right triangle may be given as 20 ft. Without knowing the possible error in the length measurement, it is impossible to determine the error in the answer obtained from it. We will usually assume that the data are known with an accuracy of 0.2 percent. The possible error in the 20-ft length would therefore be 0.04 ft.

To maintain an accuracy of approximately 0.2 percent in our calculations, we will use the following practical rule: use four digits to record numbers beginning with 1 and three digits to record numbers beginning with 2 through 9. Thus a length of 19 ft becomes 19.00 ft, while a length of 20 ft becomes 20.0 ft and a length of 43 ft becomes 43.0 ft.

Rounding Off Numbers

If the data are given with greater accuracy than we wish to maintain, the following rules may be used to round off their values:

1. When the digit to be dropped is greater than 5: increase the digit to the left by 1. *Example:* 23.56 ft becomes 23.6 ft.
2. When the digit to be dropped is 5: increase the digit to the left by 1 only if it becomes even. If the digit to the left becomes odd, drop the 5 without changing the digit to the left. *Example:* 23.55 ft becomes 23.6 ft while 23.45 ft becomes 23.4 ft.
3. When the digit to be dropped is less than 5: drop it without changing the digit to the left. *Example:* 23.34 ft becomes 23.3 ft.

Calculators

Pocket electronic calculators are widely available for use in engineering. Their speed and accuracy make it possible to perform difficult numerical computations in a routine manner. Because of the large number of digits displayed by the calculator, however, the accuracy of the answer is frequently misleading. As pointed out previously, the accuracy of the solution can be no greater than the accuracy of the data on which the solution is based.

PROBLEMS

1.1 Convert the following lengths to millimeters and meters: (a) 16.8 in., (b) 5.8 ft, (c) 12.5 in., and (d) 93.4 ft.

1.2 Convert the following lengths to inches and feet: (a) 10.2 m, (b) 45.0 m, (c) 204 mm, and (d) 4600 mm.

1.3 Convert the following forces to newtons and kilonewtons: (a) 23.5 lb, (b) 5.8 kips, (c) 250 lb, and (d) 15.9 kips.

1.4 Convert the following forces to pounds and kilopounds: (a) 52.9 N, (b) 6.85 kN, (c) 1200 N, and (d) 20.8 kN.

See Table A.3 of the Appendix for the areas of rectangles, triangles, and circles. To avoid very large or small numerical values in the computation of areas or volumes, the decimeter (dm) may be used.

1.5 through 1.16 For the figure shown: (a) Determine the area in the units given. (b) Convert the dimensions to decimeters (dm) and determine the area in units of dm². (10 decimeters = 1 meter.)

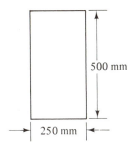

PROB. 1.8

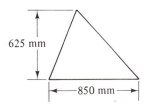

PROB. 1.9

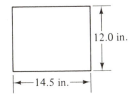

PROB. 1.5

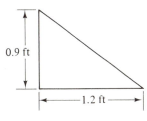

PROB. 1.10

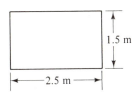

PROB. 1.6

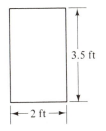

PROB. 1.7

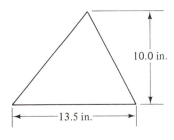

PROB. 1.11

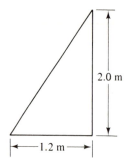

PROB. 1.12

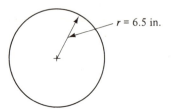

PROB. 1.16

1.17 through 1.20 The volume of a rectangular box is equal to the product of
width, length, and height. (a) Determine the volume of the box in the
units given on the figure. (b) Convert
the dimensions to decimeters (dm)
and determine the volume in units of
dm^3 and liters. (10 decimeters = 1
meter; 1 dm^3 = 1 liter.)

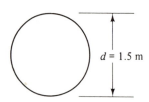

PROB. 1.13

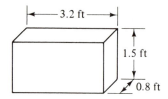

PROB. 1.17

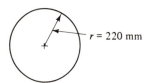

PROB. 1.14

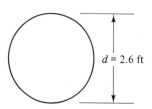

PROB. 1.15

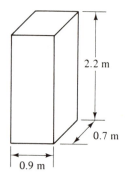

PROB. 1.18

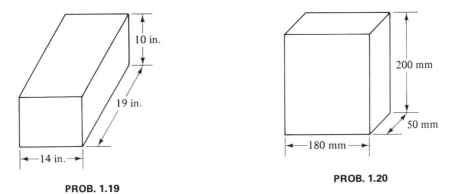

PROB. 1.19

PROB. 1.20

1.6 GRAPHICAL REPRESENTATION OF FORCES—VECTORS

Physical quantities such as length and temperature require a magnitude for their complete description and are called *scalar* quantities. Forces require both magnitude and direction and are called *vector* quantities.

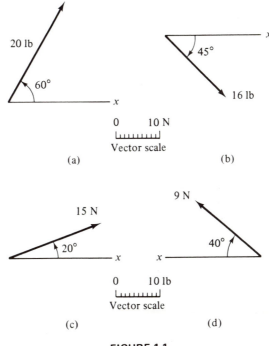

FIGURE 1.1

Vector quantities such as forces can be represented graphically by arrows drawn to an appropriate scale. The length of the arrow represents the magnitude and the direction in which the arrow points represents the direction. For example, four forces, of 20 lb $\angle 60°$, 16 lb $\angle 45°$, 15 N $\angle 20°$, and 9 N $\angle 40°$, are illustrated graphically in Fig. 1.1.

In print a vector is commonly represented in boldface (heavy) type (**F**). The same letter in lightface italic type (*F*) represents the magnitude of the same vector. In longhand or typewritten copy a vector can be indicated by an arrow over the letter ($\vec{F}$) or by underlining the letter (F). Underlining the letter has the advantage that it can be done with the typewriter.

PROBLEMS

1.21 Using a scale of 1 in. = 10 lb or 10 N, represent the following forces graphically: (a) 12.5 lb $\angle 20°$, (b) 15 lb $\angle 80°$, (c) 9.0 N $\angle 55°$, and (d) 17.5 N $\angle 22°$.

1.22 Using a scale of 1 in. = 10 N or 10 lb, represent the following forces graphically: (a) 20.5 N $\angle 32°$, (b) 8.5 N $\angle 35°$, (c) 14 lb $\angle 15°$, and (d) 17.5 lb $\angle 60°$.

1.23 Using a scale of 10 mm = 10 N or 10 lb, represent the following forces graphically: (a) 45 lb $\angle 20°$, (b) 68 lb $\angle 40°$, (c) 42 N $\angle 30°$, and (d) 52 N $\angle 45°$.

1.24 Using a scale of 10 mm = 10 lb or 10 N, represent the following forces graphically: (a) 50 N $\angle 34°$, (b) 44 lb $\angle 70°$, (c) 71 lb $\angle 46°$, and (d) 38 N $\angle 80°$.

1.7 TRIGONOMETRIC FUNCTIONS

In the study of statics we will be concerned with trigonometric angles, as we have already seen in Fig. 1.1. Consider the following right triangle [Fig. 1.2(a)], a three-sided closed figure where one of the angles is a right angle (90°). Sides *b* and *c* of the triangle form an angle *A* and sides *a* and *c* form an angle *B*. The right angle is angle *C*.

The side *c* opposite the right angle is called the *hypotenuse*. The other two sides are named with reference to either angle *A* or angle *B*. When referring to angle *A* the side opposite is *a* and the side adjacent is *b* [Fig. 1.2(b)]. Similarly with angle *B*, the side opposite is *b* and the side adjacent is *a* [Fig. 1.2(c)].

Trigonometric functions are ratios of the length of the sides of a right triangle. The three trigonometric functions of angle *A* that are of interest are known as the tangent, sine, and cosine of the angle *A*. They are abbreviated as tan *A*, sin *A*, and cos *A* but are always read in full. For example, tan *A* is read tangent of *A* or tangent of angle *A*. The functions are defined as follows:

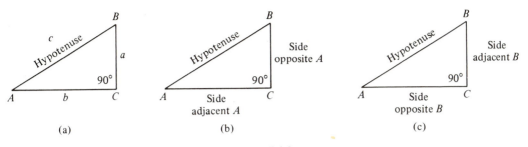

FIGURE 1.2

$$\tan A = \frac{\text{length of side opposite angle}}{\text{length of side adjacent angle}} = \frac{a}{b} \qquad (1.2)$$

$$\sin A = \frac{\text{length of side opposite angle}}{\text{length of the hypotenuse}} = \frac{a}{c} \qquad (1.3)$$

$$\cos A = \frac{\text{length of side adjacent angle}}{\text{length of the hypotenuse}} = \frac{b}{c} \qquad (1.4)$$

They are not independent but are related in the following way:

$$\frac{\sin A}{\cos A} = \frac{a/c}{b/c} = \frac{a}{b} = \tan A$$

For a given angle A the trigonometric functions remain constant. Their values are usually irrational numbers represented by non-repeating, nonterminating decimals. We shall use a pocket electronic calculator for the values of the trigonometric functions. Values of trigonometric functions for some common angles are tabulated in Table 1.2 for illustration purposes only.

TABLE 1.2

Trigonometric Functions

Angle	Sine	Cosine	Tangent
$0°$	0.0	1.000	0.0
$30°$	0.500	0.866	0.577
$45°$	0.707	0.707	1.000
$60°$	0.866	0.500	1.732
$90°$	1.000	0.0	∞

EXAMPLE 1.1

The shadow of a flagpole falls on the ground as shown in Fig. 1.3(a). The length of the shadow on the ground measures 20 ft and the angle that the sun's rays make with the ground is $60°$. Find the length of the flagpole.

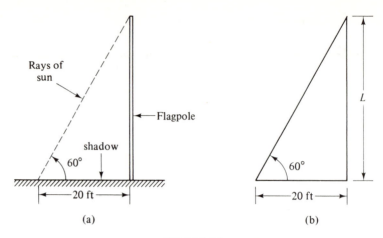

FIGURE 1.3

Solution

In Fig. 1.3(b) we show the right triangle required for solution. The side adjacent to the $60°$ angle is 20 ft and the side opposite the $60°$ angle is the length of the flagpole, L. From Eq. (1.2),

$$\tan 60° = \frac{\text{side opposite } 60° \text{ angle}}{\text{side adjacent } 60° \text{ angle}} = \frac{L}{20}$$

$$L = 20 \tan 60° = 20(1.732) = 34.64$$

Length of flagpole = 34.6 ft   Answer

EXAMPLE 1.2

In Fig. 1.4(a) an 8-m ladder is shown leaning against a house at an angle of $70°$ with the ground. Determine (a) the distance of the bottom of the ladder from the house, and (b) the distance of the top of the ladder from the ground.

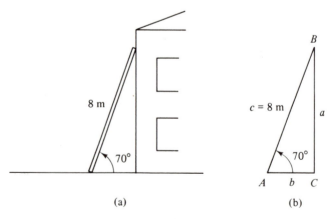

FIGURE 1.4

Solution

Shown in Fig. 1.4(b) is the required right triangle, with a hypotenuse 8 m long, the side b adjacent to the $70°$ angle, and the side a opposite the $70°$ angle.

    (a) Distance of bottom of ladder from house: from Eq. (1.4),

$$\cos 70° = \frac{\text{side adjacent } 70° \text{ angle}}{\text{hypotenuse}} = \frac{b}{8}$$

$$b = 8 \cos 70° = 8(0.3420) = 2.736$$

Bottom of ladder from house = 2.74 m   Answer

    (b) Distance of top of ladder from ground: from Eq. (1.3),

$$\sin 70° = \frac{\text{side opposite } 70° \text{ angle}}{\text{hypotenuse}} = \frac{a}{8}$$

$$a = 8 \sin 70° = 8(0.9397) = 7.517$$

Top of ladder from ground = 7.52 m   Answer

PROBLEMS

1.25 through 1.28 Determine the third side c graphically and mathematically and the $\sin \theta$ and $\cos \theta$ mathematically for the right triangle shown.

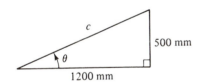

PROB. 1.25

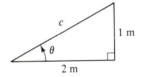

PROB. 1.27

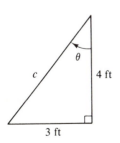

PROB. 1.26

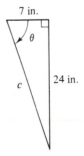

PROB. 1.28

1.29 In the figure, $c = 35$ ft, $\theta_1 = 56°$, and $\theta_2 = 39°$. Determine distances b and d graphically and mathematically.

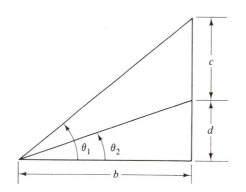

PROB. 1.29 and PROB. 1.30

1.30 In the figure, $d = 300$ m, $\theta_1 = 45°$, and $\theta_2 = 35°$. Determine distances b and c graphically and mathematically.

The angle the line of sight forms with the horizontal is called the *angle of elevation* if the line of sight is above the horizontal and the *angle of depression* if it is below the horizontal.

1.31 A man 5 ft 8 in. tall casts a shadow 10 ft long. Determine the angle of elevation of the sun above the ground graphically and mathematically.

1.32 From a point 30 m away from the base of a flagpole the angle of elevation of the top is $35°$. Determine the height of the flagpole graphically and mathematically.

1.33 The angle of depression of a ship from the top of a 288-ft-high lighthouse is $22°$. Determine graphically and mathematically how far the ship is from the lighthouse.

1.34 A 90-m-long kite string forms an angle of $38°$ with the ground. Determine

graphically and mathematically the distance the kite is flying above the ground. Neglect the sag in the kite string.

1.35 A dock on one side of a canal measures 23 ft 6 in. from B to C. A pile on the other side at A is located so that A, B, and C form a right triangle with angles as shown in the figure. Determine graphically and mathematically the width AB of the canal.

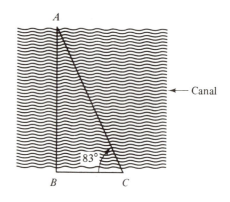

PROB. 1.35

1.36 Two buildings are on opposite sides of a 30-m-wide street. The taller building is known to be 124 m high. If the angle of elevation of the roof of the taller building from the roof of the lower building is $42°$, determine the height of the lower building graphically and mathematically.

1.37 From a point on the ground, a balloon was observed to have an angle of elevation of $40°$. After the balloon ascended 500 m, it was observed to have an angle of elevation of $53°$. Determine graphically and mathematically the height of the balloon above the ground on the first observation.

1.38 Two points M and N on level ground are 4500 ft apart. A balloon over a line from M to N is observed to have an elevation from point M of $42°$ and from point N of $59°$ as shown in the figure. Determine graphically and mathematically the height of the balloon above the ground.

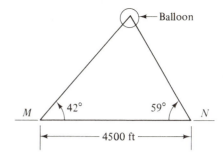

PROB. 1.38

1.8 TRIGONOMETRIC FORMULAS

There are many applications in statics where the triangle to be considered does not involve a right angle. For such applications the cosine law and the sine law may prove useful.

Cosine Law

The square of any side of a triangle is equal to the sum of the squares of the other two sides minus twice the product of these sides and the cosine of their included angle.

The triangle in Fig. 1.5 has sides of length a, b, and c. Angle A is opposite side a, angle B is opposite side b, and angle C is opposite side c. Applying the cosine law, we have formulas for the length of a, b, and c as follows:

$$a^2 = b^2 + c^2 - 2bc \cos A$$
$$b^2 = a^2 + c^2 - 2ac \cos B \qquad (1.5)$$
$$c^2 = a^2 + b^2 - 2ab \cos C$$

Sine Law

The ratio of the side of any triangle to the sine of the opposite angle is a constant. Applying the sine law to the triangle (Fig. 1.5), we have

$$\frac{a}{\sin A} = \frac{b}{\sin B} = \frac{c}{\sin C} \qquad (1.6)$$

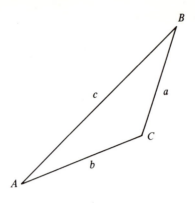

FIGURE 1.5

In the special cases where a right triangle is involved, the Pythagorean theorem may be used.

Pythagorean Theorem

In any right triangle the square of the length of the hypotenuse equals the sum of the squares of the lengths of the other two sides.

This is a special case of the cosine law when one of the angles is equal to 90°. Let angle $C = 90°$. Then $\cos C = \cos 90° = 0$. Equation (1.5) reduces to

$$c^2 = a^2 + b^2 \tag{1.7}$$

which is the Pythagorean theorem.

EXAMPLE 1.3

A triangle [Fig. 1.6(a)] has sides of 3 ft and 4 ft and the included angle equal to 135°. Find the third side and the angle it forms with line AE.

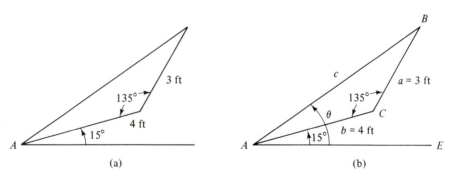

FIGURE 1.6

Solution

In Fig. 1.6(b) sides and angles of the triangle are identified with those shown in Fig. 1.5. Side $a = 3$ ft, side $b = 4$ ft, and angle $C = 135°$. We may solve for side c by applying the cosine law.

$$c^2 = a^2 + b^2 - 2ab \cos C$$

$$\cos 135° = -\cos 45° = -0.707$$

$$c^2 = (3)^2 + (4)^2 - 2(3)(4)(-0.707) = 41.97$$

$$c = 6.48 \text{ ft} \qquad \text{Answer}$$

To find angle A we apply the law of sines.

$$\frac{a}{\sin A} = \frac{c}{\sin C} \qquad \frac{3}{\sin A} = \frac{6.48}{\sin 135°}$$

Solving for $\sin A$, we have

$$\sin A = \frac{3 \sin 135°}{6.48} = \frac{3(0.707)}{6.48}$$

$$= 0.327$$

$$A = \arcsin 0.327 = 19.1°$$

The angle that side c of the triangle forms with line AE is

$$\theta = A + 15°$$

$$= 19.1° + 15° = 34.1° \measuredangle \qquad \text{Answer}$$

EXAMPLE 1.4

A triangle has sides of 6 m and 12 m and one of the angles is $35°$, as shown in Fig. 1.7(a). Find the third side and the other two angles.

Solution

The sides and angles of the triangle are identified in Fig. 1.7(b) with those shown in Fig. 1.5. Since angle C between a and b is unknown, we cannot apply the cosine law directly. Applying the sine law to find angle A, we have

$$\frac{b}{\sin B} = \frac{a}{\sin A} \qquad \frac{12}{\sin 35°} = \frac{6}{\sin A}$$

$$\sin A = \frac{6(\sin 35°)}{12} = \frac{6(0.5736)}{12}$$

$$= 0.287$$

$$A = \arcsin 0.287 = 16.7° \qquad \text{Answer}$$

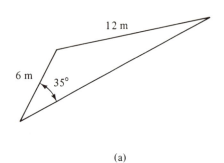

(a)

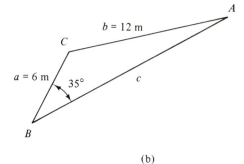

(b)

FIGURE 1.7

Angles $A + B + C = 180°$; therefore, $C = 128.3°$. We may now solve for the third side of the triangle by applying the cosine law.

$$c^2 = a^2 + b^2 - 2ab \cos C$$

$$\cos 128.3° = -\cos 51.7° = -0.620$$

$$c^2 = (6)^2 + (12)^2 - 2(6)(12)(-0.620) = 269$$

$$c = 16.4 \text{ m} \qquad\qquad \text{Answer}$$

PROBLEMS

1.39 Determine the length of the third side of the triangle with the following sides and angles.
(a) $a = 10.3$ in., $b = 14.5$ in., and $C = 54°$
(b) $a = 2.2$ m, $c = 6.8$ m, and $C = 95°$
(c) $b = 12.2$ ft, $c = 25.0$ ft, and $A = 34°$
(d) $b = 4.2$ m, $c = 8.5$ m, and $B = 28°$

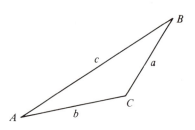

PROB. 1.39 and PROB. 1.40

1.40 Determine, for the triangle shown, the missing sides and angles with the following sides and angles given.

(a) $a = 7$ in., $B = 40°$, and $C = 30°$
(b) $a = 3$ m, $b = 6$ m, and $C = 48°$
(c) $a = 8$ ft, $b = 7$ ft, and $A = 60°$
(d) $a = 4$ m, $b = 7$ m, and $c = 9$ m

1.41 Determine the lengths of the diagonals AC and BD of the parallelogram shown with the following sides and angle.
(a) 8 in., 12.5 in., and $65°$
(b) 550 mm, 320 mm, and $55°$
(c) 10.3 ft, 12.5 ft, and $45°$
(d) 5 m, 12 m, and $125°$

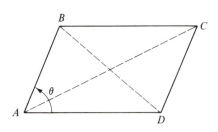

PROB. 1.41

1.42 A boom is supported by a cable as shown. Determine the length of the cable from *A* to *B* and the angle the cable makes with the horizontal.

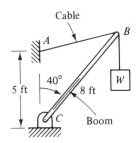

PROB. 1.42

1.43 A boom is supported by two cables as shown. Cable *BD* is 28.3 ft long and cable *CD* is 44.7 ft long. Determine the length of the boom and the angles cable *BD* and cable *CD* make with the horizontal.

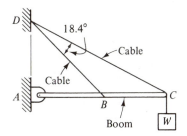

PROB. 1.43

1.44 The truss supports loads as shown. Determine the following angles: (a) $\angle GCB$, (b) $\angle AHJ$, (c) $\angle GFC$.

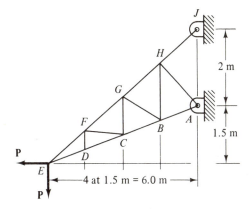

PROB. 1.44

1.45 For the pin-connected tower shown, determine the following angles: (a) $\angle BAH$, (b) $\angle BCG$, (c) $\angle DEF$.

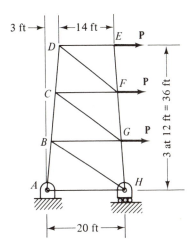

PROB. 1.45

2

Resultant of Concurrent Forces _____

2.1 INTRODUCTION

A force was defined as the action of one body on another which tends to change the size or shape and/or state of motion of a body. Forces were described by their magnitude, direction, and point of application. With previous examples of single forces we were concerned with magnitude and direction only. However, with the study of *concurrent force systems* we must consider the point of application of the forces. A force system is concurrent when all the forces acting on a body have the same point of application or the lines of action of the forces intersect at a common point. Examples of concurrent force systems are shown in Fig. 2.1.

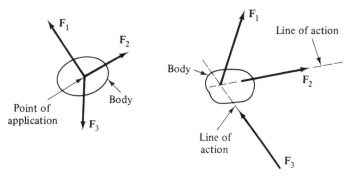

FIGURE 2.1

2.2 RESULTANT OF TWO CONCURRENT FORCES—VECTORS

It has been found by experiment that a concurrent force system can be replaced by a single force or resultant. The resultant has the same physical effect as the force system it replaces.

Parallelogram Method

The resultant of two concurrent forces can be obtained graphically by constructing a parallelogram, as shown in Fig. 2.2. Forces **P** and **Q** acting at O form two sides of the parallelogram. The diagonal that passes through

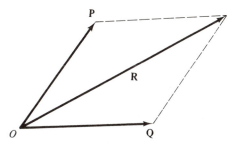

FIGURE 2.2

O is the resultant **R**. This is known as the *parallelogram law*, which states that two forces whose lines of action intersect can be replaced by a single force which is the diagonal of a parallelogram that has sides equal to the two forces. The parallelogram law is based on experimental evidence only. It cannot be proved or derived by mathematics.

As previously indicated, force is a *vector* quantity. Vectors may be defined as mathematical quantities that have magnitude and direction and add according to the parallelogram law.

Triangular Method

The parallelogram construction shown in Fig. 2.2 suggests that forces may also be added by arranging vectors in a tip-to-tail fashion. The method as shown in Fig. 2.3(a) and (b) consists of moving either force parallel to itself until the tail coincides with the tip of the fixed force. The closing side of the triangle forms a resultant with the tail of the resultant at the tail of the fixed force and the tip of the resultant at the tip of the moved force. As we can see from the construction, the order in which the forces are combined does not change the resultant.

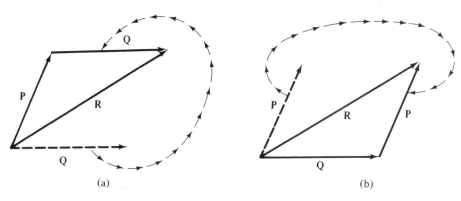

(a)   (b)

FIGURE 2.3

EXAMPLE 2.1

The two forces S and T act as shown in Fig. 2.4(a) at point *O*. Obtain their resultant graphically by the parallelogram method and the triangular method.

Solution

Graphical Parallelogram Method

A parallelogram with sides equal to *S* and *T* is drawn to scale [Fig. 2.4(b)]. The diagonal forms the resultant. The magnitude and direction of the resultant are measured and found to be

$$R = 590 \text{ lb} \diagdown 20°$$ Answer

Graphical Triangular Method

Draw fixed force S at point *O* as shown in Fig. 2.4(c). Move force T parallel to itself until its tail coincides with the tip of force S. The resultant forms the closed side of a triangle. The tail of the resultant is at the tail of the fixed force and the tip of the resultant is at the tip of the moved force. Measurement of the magnitude and direction of the resultant gives

$$R = 590 \text{ lb} \diagdown 20°$$ Answer

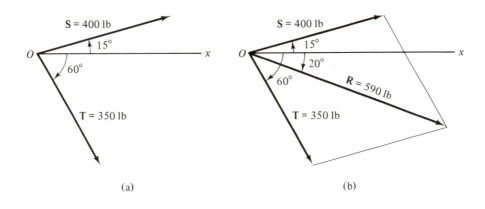

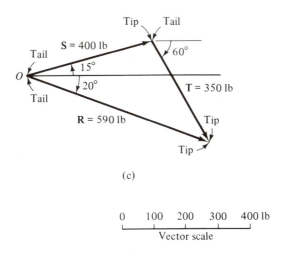

(c)

0   100   200   300   400 lb

Vector scale

FIGURE 2.4

EXAMPLE 2.2

A ship is pulled by two tugboats, as shown in Fig. 2.5(a). The resultant force is 3 kips parallel to the x axis. Find the forces exerted by each of the tugboats graphically by the triangular method.

Solution

The resultant of 3 kips is drawn to scale parallel to the x axis from O to V as shown in Fig. 2.5(b). The line OT is drawn through O at an angle of $25°$ and the line VS through V at an angle of $35°$. The two lines intersect at point U. Force **P** is directed from O to U and force **Q** is directed from U to V. Measuring the magnitude of the forces, we obtain

$$P = 2.0 \text{ kips} \qquad Q = 1.5 \text{ kips} \qquad \text{Answer}$$

23

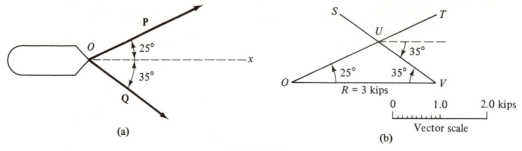

FIGURE 2.5

EXAMPLE 2.3

The two forces P and S act as shown in Fig. 2.6(a) at point A. Obtain their resultant by trigonometry.

Solution

We use the triangular method, arranging P and S in a tip-to-tail fashion. The resultant R forms the third side of the triangle [Fig. 2.6(b)]. From the construction, angle $C = 115°$. Applying the law of cosines yields

$$R^2 = P^2 + S^2 - 2PS \cos C$$

$$= (8.5)^2 + (6.0)^2 - 2(8.5)(6.0) \cos 115°$$

$$R = 12.30 \text{ kN}$$

From the law of sines, we have

$$\frac{\sin A}{S} = \frac{\sin C}{R} \qquad \frac{\sin A}{6.0} = \frac{\sin 115°}{12.3}$$

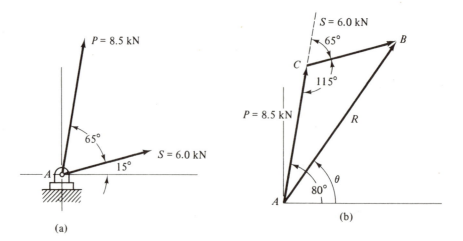

FIGURE 2.6

24

Solving for angle A, we have

$$\sin A = \frac{6.0\,(\sin 115^\circ)}{12.3} = 0.4421$$

$$A = \arcsin 0.4421 = 26.2^\circ$$

The direction of angle $\theta = 80 - A = 53.8^\circ$.

$$R = 12.30 \text{ kN} \angle 53.8^\circ \qquad \qquad \text{Answer}$$

EXAMPLE 2.4

Two cables are used to pull a truck as shown in Fig. 2.7(a). The resultant force lies along the x axis. Determine the magnitude of the force Q and of the resultant R by trigonometry.

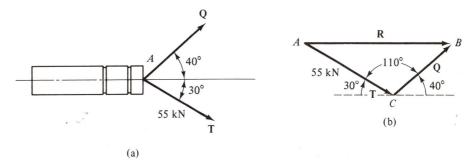

(a)

(b)

FIGURE 2.7

Solution

The force Q at an angle of 40° is added to the force T at an angle of 30° to form the resultant R at an angle of 0° in a tip-to-tail fashion as shown in Fig. 2.7(b). From the construction, angle $B = 40^\circ$ and angle $C = 110^\circ$. Applying the law of sines yields

$$\frac{\sin 30^\circ}{Q} = \frac{\sin 40^\circ}{55} = \frac{\sin 110^\circ}{R}$$

Solving for Q and R, we have

$$Q = \frac{55\,(\sin 30^\circ)}{\sin 40^\circ} = 42.8 \text{ kN} \qquad \qquad \text{Answer}$$

$$R = \frac{55\,(\sin 110^\circ)}{\sin 40^\circ} = 80.4 \text{ kN} \qquad \qquad \text{Answer}$$

PROBLEMS

2.1 and 2.2 The two forces act at point O as shown. Obtain their resultant graphically by using (a) the parallelogram method, and (b) the triangular method.

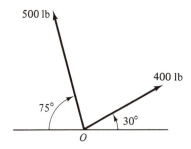

PROB. 2.1

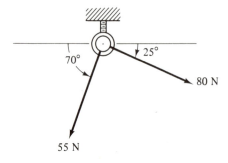

PROB. 2.4

2.5 and 2.6 The bracket supports two forces as shown. Obtain their resultant by trigonometry.

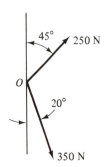

PROB. 2.2

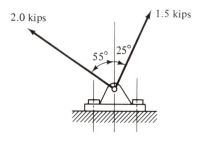

PROB. 2.5

2.3 and 2.4 Determine graphically the resultant force on the screw eye shown. Use (a) the parallelogram method, and (b) the triangular method.

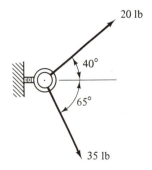

PROB. 2.3

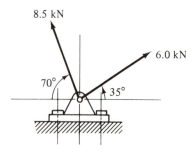

PROB. 2.6

2.7 A barge is pulled by two ropes with tensions P and Q as shown. If $P = 250$ lb and $Q = 500$ lb, determine the resultant force applied on the barge by trigonometry.

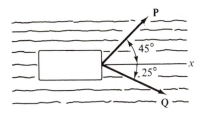

PROB. 2.7 and PROB. 2.8

2.8 A barge is pulled by two ropes with tensions **P** and **Q**. The force $P = 16$ kN and the resultant of the two forces acts along the x axis. Determine the force Q and the resultant force applied on the barge by trigonometry.

2.9 Two forces **P** and **Q** have the resultant **R** as shown. If $R = 18$ kips, $P = 10$ kips, and $\theta_1 = 30°$, determine the force Q

and the angle θ_2 it makes with the vertical by trigonometry.

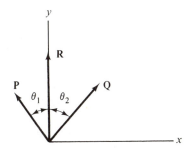

PROB. 2.9 and PROB. 2.10

2.10 Two forces **P** and **Q** have the resultant **R** as shown. The resultant $R = 10$ kN, $\theta_1 = 35°$, and $\theta_2 = 20°$. Determine the forces P and Q by trigonometry.

2.3 RESULTANT OF THREE OR MORE CONCURRENT FORCES

When the resultant of three or more concurrent forces is required, either the parallelogram or triangular method may be used. However, the paral-

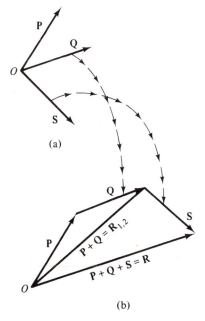

(a)

(b)

FIGURE 2.8

lelogram construction becomes awkward and the triangular method is preferred.

In Fig. 2.8(a) and (b) the resultant of three forces **P**, **Q**, and **S** is determined by the repeated application of the triangular method. First we add **P** and **Q** to find the resultant $R_{1,2}$. Then we add $R_{1,2}$ and **S** to find the resultant **R** of the three forces.

A modification of the triangular method known as the *polygon method* can also be used to find the resultant. In the polygon method the intermediate resultants, in this example $R_{1,2}$, need not be included in the construction.

The polygon method, shown in Fig. 2.9(a), can be described in the following way. Leave the fixed force **P** at point *O*. Move **Q** parallel to itself until its tail coincides with the tip of **P**. Next, move **S** parallel to itself until its tail coincides with the tip of **Q**. The resultant **R** forms the closing side of the polygon with the tail of the resultant at the tail of the fixed force **P** and the tip of the resultant at the tip of the last moved force **S**.

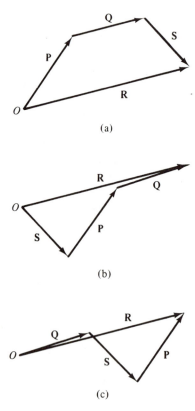

(a)

(b)

(c)

FIGURE 2.9

The order of addition of the forces does not change the resultant, as we can see from the construction shown in Fig. 2.9.

EXAMPLE 2.5

The four forces [Fig. 2.10(a)] act at point P. Obtain their resultant graphically by the polygon method.

Solution

The four forces are drawn to scale tip-to-tail fashion in Fig. 2.10(b). The resultant acts from the tail of the first force to the tip of the last. The magnitude and direction angle are measured and the resultant is found to be

$$R = 4.9 \text{ kN} \angle 24°$$

Answer

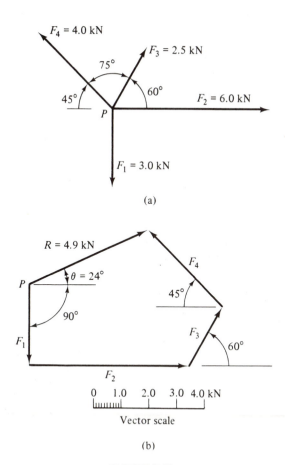

(a)

(b)

FIGURE 2.10

PROBLEMS

2.11 through 2.14   Obtain the resultant of the
concurrent forces shown, graphically by
the polygon method.

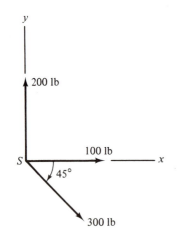

PROB. 2.11

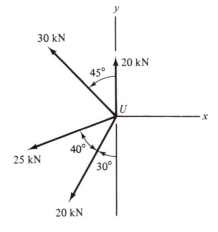

PROB. 2.13

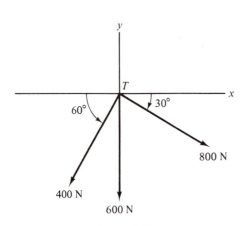

PROB. 2.12

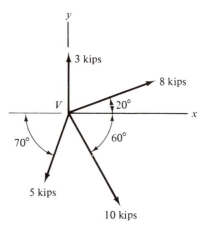

PROB. 2.14

2.15   The gusset plate shown is subjected to
four concurrent forces. If F_1 = 250 lb,
F_2 = 400 lb, F_3 = 300 lb, and F_4 = 500
lb, determine the magnitude and direction of the resultant graphically by the
polygon method.

2.16 The forces on the gusset plate shown in the figure are $F_1 = 1.5$ kips, $F_2 = 1.2$ kips, $F_3 = 1.0$ kips, and $F_4 = 2.0$ kips. Determine the magnitude and direction of the resultant graphically by the polygon method.

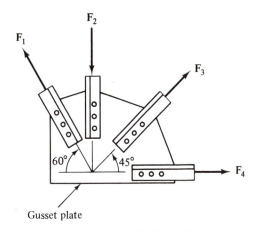

Gusset plate

PROB. 2.15 and PROB. 2.16

2.4 COMPONENTS OF A FORCE VECTOR

Two or more forces may be replaced by their resultant. By reversing the process, any force **P** can be replaced by any number of forces which have as their resultant the original force. The replacement forces are called *components* of the force **P**.

2.5 RECTANGULAR COMPONENTS OF A FORCE VECTOR

Although any force **F** can be replaced by an infinite number of different components, two components at right angles to each other are the most useful. They are called *rectangular components* and are usually determined in the horizontal and vertical direction as shown in Fig. 2.11(a). They may also be found in any two directions at right angles to each other, as shown in Fig. 2.11(b).

A method for finding the rectangular components of a force vector graphically can be described as follows. Let a force **F** be represented by a vector drawn to a convenient scale [Fig. 2.11(a) and (b)]. The tail end of the vector is point O and the arrow end of the vector is point B. Through point O draw any two axes at right angles to each other. The angle that the force **F** makes with the x axis is θ (theta). To find the rectangular component of **F** along the x axis we draw a line parallel to the y axis from the arrow end of **F** at B to point C on the x axis. The vector from O to C represents the component of **F** along the x axis and is represented by $\mathbf{F}_x$. Similarly, by drawing a second line parallel to the x

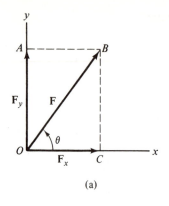
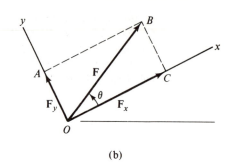

(a)   (b)

FIGURE 2.11

axis from B to point A on the y axis, we find $\mathbf{F}_y$ the component of $\mathbf{F}$ along the y axis.

Rectangular components of a force vector may also be found mathematically by using trigonometric functions. Notice that triangle OBC in Fig. 2.11(a) and (b) is a right triangle and length OA is equal to length CB. From the definition of the cosine of an angle,

$$\cos \theta = \frac{OC}{OB} = \frac{F_x}{F}$$

or

$$F_x = F \cos \theta \qquad (2.1)$$

and from the definition of the sine of an angle

$$\sin \theta = \frac{CB}{OB} = \frac{F_y}{F}$$

or

$$F_y = F \sin \theta \qquad (2.2)$$

The components F_x and F_y are considered either positive or negative, depending on whether they act in the positive or negative direction of the x and y axes.

EXAMPLE 2.6

Find the horizontal and vertical components of the forces shown in Fig. 2.12(a).

Solution

Graphical Method

Select a convenient scale and draw, as in Fig. 2.12(b), the force P = 14 N from the origin to point B at an angle of 55° with the x axis. Draw a line parallel to the y axis from point B until it intersects the x axis at point C and a second line parallel to the x axis from point B until it intersects the y axis at point A. Measure OC to find P_x and OA to find P_y.

$$P_x = 8.0 \text{ N} \qquad\qquad \text{Answer}$$

$$P_y = 11.5 \text{ N} \qquad\qquad \text{Answer}$$

By the same method we draw force Q = 12.5 N at an angle of 65° with the negative x axis and determine the x and y components as shown in Fig. 2.12(b) by measurement.

$$Q_x = -5.3 \text{ N} \qquad\qquad \text{Answer}$$

$$Q_y = -11.3 \text{ N} \qquad\qquad \text{Answer}$$

The components of the force Q are negative because they acted in the negative direction of the x and y axes.

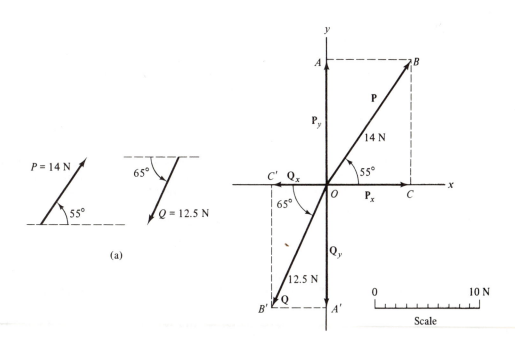

(a)

(b)

FIGURE 2.12

Trigonometric Method

Apply Eqs. (2.1) and (2.2) to force **P** at an angle $\theta = 55°$ with the positive x axis. The components of **P** are both positive because they act in the positive x and y directions.

$$P_x = P \cos \theta = 14 \cos 55°$$

$$= 8.03 \text{ N} \qquad\qquad \text{Answer}$$

$$P_y = P \sin \theta = 14 \sin 55°$$

$$= 11.47 \text{ N} \qquad\qquad \text{Answer}$$

Proceed similarly for force **Q** at an angle $\theta = 65°$ with the negative x axis. The components of **Q** are both negative because they act in the negative x and y direction.

$$Q_x = -Q \cos \theta = -12.5 \cos 65°$$

$$= -5.28 \text{ N} \qquad\qquad \text{Answer}$$

$$Q_y = -Q \sin \theta = -12.5 \sin 65°$$

$$= -11.32 \text{ N} \qquad\qquad \text{Answer}$$

EXAMPLE 2.7

Determine the rectangular components of the forces shown in Fig. 2.12(a) with respect to axes that form angles of $30°$ and $120°$ with the horizontal.

Solution

Graphical Method

Rotate the x and y axes to the new position as in Fig. 2.13. Force $P = 14$ N forms an angle of $25°$ with the new x axis. Draw a line parallel to the y axis until it intersects the x axis at point C and a second line parallel to the x axis from point B until it intersects the y axis at point A. Measure OC to find P_x and OA to find P_y.

$$P_x = 12.7 \text{ N} \qquad\qquad \text{Answer}$$

$$P_y = 5.9 \text{ N} \qquad\qquad \text{Answer}$$

Force $Q = 12.5$ N forms an angle of $35°$ with the new negative x axis. By the same method as for **P**, determine the x and y components of **Q** by measurement.

$$Q_x = -10.2 \text{ N} \qquad\qquad \text{Answer}$$

$$Q_y = -7.2 \text{ N} \qquad\qquad \text{Answer}$$

Trigonometric Method

Applying Eqs. (2.1) and (2.2) to force **P** at an angle of $\theta = 25°$ with the positive x axis, we obtain

$$P_x = P \cos \theta = 14 \cos 25°$$

$$= 12.69 \text{ N} \qquad\qquad \text{Answer}$$

$$P_y = P \sin \theta = 14 \sin 25°$$

$$= 5.92 \text{ N} \qquad\qquad \text{Answer}$$

Proceed similarly for force **Q** at an angle $\theta = 35°$ with the negative x axis.

$$Q_x = -Q \cos \theta = -12.5 \cos 35°$$

$$= -10.24 \text{ N} \qquad\qquad \text{Answer}$$

$$Q_y = -Q \sin \theta = -12.5 \sin 35°$$

$$= -7.17 \text{ N} \qquad\qquad \text{Answer}$$

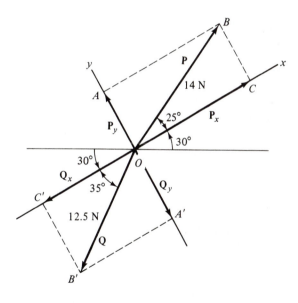

FIGURE 2.13

2.17 Determine components, in the horizontal and vertical direction, of the following forces by graphical and mathematical methods: (a) 1250 lb ∡25°, (b) 5.8 kips ⦡75°, (c) 20 N ⦣12.5°, and (d) 14.5 kN ⦤35°.

2.18 Determine components, in the horizontal and vertical direction, of the following forces by graphical and mathematical methods: (a) 28 kN ⦤35°, (b) 1500 lb ∡15°, (c) 10.2 kips ⦣75°, and (d) 450 N ⦤50°.

2.19 Determine, by graphical and mathematical methods, the rectangular components of the forces given in Prob. 2.17 with respect to the x' and y' axes as shown.

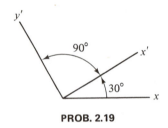

PROB. 2.19

2.20 Determine, by graphical and mathematical methods, the rectangular components of the forces given in Prob. 2.18 with respect to the x' and y' axes as shown.

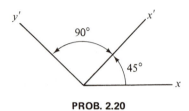

PROB. 2.20

2.21 If $P = 25$ kN and $\theta = 30°$, find the components of **P** parallel and perpendicular to the lines (a) mn, and (b) rs.

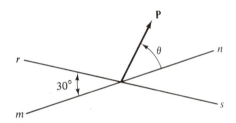

PROB. 2.21 and PROB. 2.22

2.22 If $P = 4000$ lb and $\theta = 45°$, find the components of **P** parallel and perpendicular to the lines (a) mn, and (b) rs.

2.6 RESULTANT OF CONCURRENT FORCES BY RECTANGULAR COMPONENTS

The resultant of three or more concurrent forces may be found graphically by the polygon of forces. The usual mathematical solution, however, is based on the method of rectangular components.

To develop the rectangular component method we consider, in Fig. 2.14(a), two forces **P** and **Q** which act at point A and form angles θ_1 and θ_2 with the x axis. The x and y components of **P** and **Q** can be found graphically by the method described in Sec. 2.5. They may also be found by Eqs. (2.1) and (2.2) as follows:

$$P_x = P \cos \theta_1 \qquad P_y = P \sin \theta_2$$

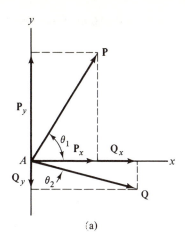 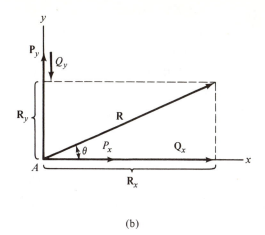

(a)   (b)

FIGURE 2.14

and

$$Q_x = Q \cos \theta_2 \qquad Q_y = -Q \sin \theta_2$$

Since the components P_x and Q_x lie on the same line [Fig. 2.14(a)], they may be added algebraically to find the resultant in the x direction, R_x. That is,

$$R_x = \Sigma F_x = P_x + Q_x$$

where the capital Greek letter Σ (sigma) means the algebraic sum and ΣF_x means the algebraic sum of the x components of the forces.

Similarly, the components P_y and Q_y lie on the same line [Fig. 2.14(a)] and can be added algebraically to find the resultant in the y direction, R_y. That is,

$$R_y = \Sigma F_y = P_y + Q_y$$

where ΣF_y means the algebraic sum of the y components of the forces.

Adding the resultants in the x and y directions, $\mathbf{R}_x$ and $\mathbf{R}_y$ [Fig. 2.14(b)], we find the resultant $\mathbf{R}$ of the force system. The magnitude and direction of the resultant can be found from the Pythagorean theorem and the definition of the tangent of an angle. That is,

$$R = R_x^2 + R_y^2 = (\Sigma F_x)^2 + (\Sigma F_y)^2 \tag{2.3}$$

$$\tan \theta = \frac{R_y}{R_x} = \frac{\Sigma F_y}{\Sigma F_x} \tag{2.4}$$

The rectangular component method can be used to find the resultant of any number of forces.

EXAMPLE 2.8

Find the resultant of the concurrent force system shown in Fig. 2.15(a) by the method of rectangular components.

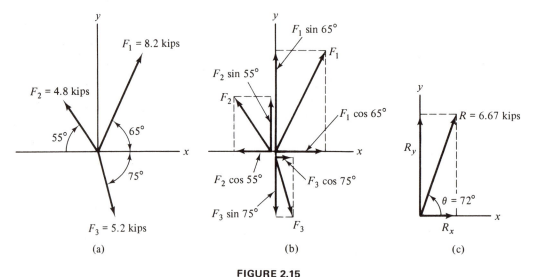

(a)   (b)   (c)

FIGURE 2.15

Solution

The x and y components of each force are determined by trigonometry as shown in Fig. 2.15(b).

Force	Magnitude (kips)	Angle (deg)	x component (kips)	y component (kips)
F_1	8.2	65	3.465	7.432
F_2	4.8	55	-2.753	3.932
F_3	5.2	75	1.346	-5.023
			$\Sigma F_x = 2.058$ kips (R_x)	$\Sigma F_y = 6.341$ kips (R_y)

Calculations for the value of each component can be tabulated as shown. Recall that x components are positive if they act to the right and negative to the left. The y components are positive if they act upward and negative if they act downward.

The resultants in the x and y directions are given by the sums shown in the table.

$$R_x = \Sigma F_x = 2.058 \text{ kips}$$

and

$$R_y = \Sigma F_y = 6.341 \text{ kips}$$

The magnitude and direction of the resultant is determined from the Pythagorean theorem and the definition of the tangent of an angle. From Fig. 2.15(c), we write

$$R^2 = (2.058)^2 + (6.341)^2 = 44.44$$

$$R = 6.67 \text{ kips}$$

and

$$\tan \theta = \frac{R_y}{R_x} = \frac{6.341}{2.058} = 3.081 \qquad \theta = 72.0°$$

$$R = 6.67 \text{ kips} \angle 72.0° \qquad\qquad\qquad \text{Answer}$$

EXAMPLE 2.9

For the concurrent force system shown in Fig. 2.16(a), find the resultant by the rectangular component method.

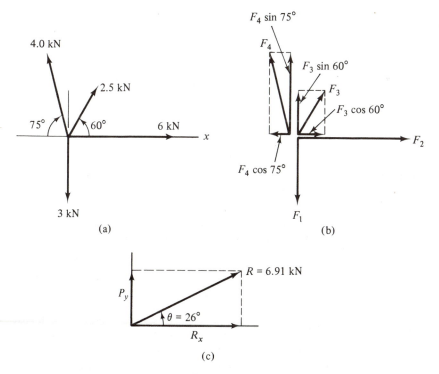

(a)

(b)

(c)

FIGURE 2.16

Solution

The x and y components of each force are determined by trigonometry in Fig. 2.16(b).

Force	Magnitude (kN)	Angle (deg)	x component (kN)	y component (kN)
F_1	3.0	90	0	−3.0
F_2	6.0	0	6.0	0
F_3	2.5	60	1.250	2.165
F_4	4.0	75	−1.035	3.864
			ΣF_x = 6.215 kN (R_x)	ΣF_y = 3.029 kN (R_y)

Calculations for the components are tabulated systematically as shown. The resultants in the x and y directions are given by

$$R_x = \Sigma F_x = 6.215 \text{ kN} \qquad R_y = \Sigma F_y = 3.029 \text{ kN}$$

From the Pythagorean theorem and the definition of the tangent of an angle, we write

$$R^2 = R_x^2 + R_y^2 = (6.215)^2 + (3.029)^2 = 47.80$$

$$R = 6.91 \text{ kN}$$

and

$$\tan \theta = \frac{R_y}{R_x} = \frac{3.029}{6.215} = 0.4874 \qquad \theta = 26.0°$$

$$R = 6.91 \text{ kN} \angle 26.0° \qquad\qquad \text{Answer}$$

2.7 DIFFERENCE OF TWO FORCES—VECTOR DIFFERENCES

In this section we will define what is meant by a negative force. The definition must be consistent with what we have already learned about the addition of two forces and will agree with the usual rules of algebra. From the usual rules of algebra, we write

$$\mathbf{F} - \mathbf{F} = 0 \quad \text{or} \quad \mathbf{F} + (-\mathbf{F}) = 0$$

Therefore, the force $-\mathbf{F}$ when added to the force $\mathbf{F}$ has a resultant equal to zero. Such a force must have the same magnitude as force $\mathbf{F}$ and be directed in the opposite direction. It follows that the force $\mathbf{F}_1$ may be subtracted from force $\mathbf{F}_2$ by reversing $\mathbf{F}_1$ and adding it to force $\mathbf{F}_2$ by any of the various methods.

EXAMPLE 2.10

For the concurrent force system shown in Fig. 2.17(a), find $R = F_1 + F_2 - F_3$ by the rectangular component method.

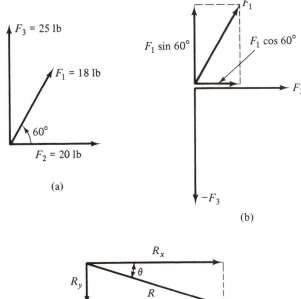

(a)

(b)

(c)

FIGURE 2.17

Solution

Replace F_3 by a $-F_3$ and proceed as in Example 2.9. The x and y components of each force are determined by trigonometry in Fig. 2.17(b). Calculations for the components are tabulated as shown. The resultants in the x and y directions are given by

$$R_x = \Sigma\, F_x = 29.0\ \text{lb} \qquad R_y = \Sigma\, F_y = -9.41\ \text{lb}$$

Force	Magnitude (lb)	Angle (deg)	x component (lb)	y component (lb)
F_1	18	60	9.0	15.59
F_2	20	0	20.0	0
$-F_3$	25	90	0	−25.0
			$\Sigma\, F_x = 29.0\ \text{lb}$ (R_x)	$\Sigma\, F_y = -9.41\ \text{lb}$ (R_y)

The magnitude and direction of the resultant is determined from the Pythagorean theorem and the definition of the tangent of an angle. From Fig. 2.17(c), we write

$$R^2 = (29.0)^2 + (-9.41)^2 = 929.5$$

$$R = 30.5 \text{ lb}$$

and

$$\tan \theta = \frac{R_y}{R_x} = \frac{9.41}{29.0} \qquad \theta = 18.0°$$

$$\mathbf{R} = 30.5 \text{ lb} \diagdown 18.0° \qquad \qquad \text{Answer}$$

PROBLEMS

2.23 through 2.26 Determine the resultant of the forces shown by the method of components.

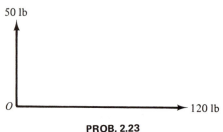

PROB. 2.23

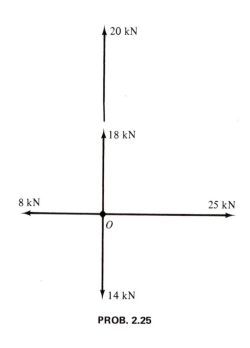

PROB. 2.25

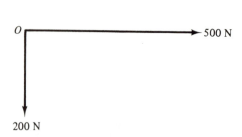

PROB. 2.24

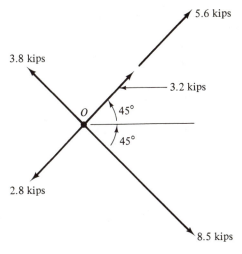

PROB. 2.26

2.27 Two forces act on the free end of a cantilever beam as shown, find the magnitude and direction of the resultant by the method of components.

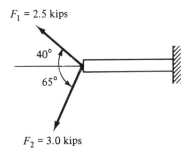

PROB. 2.27

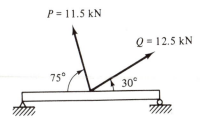

PROB. 2.28

2.28 Two forces act at the middle of a simply supported beam as shown, find the magnitude and direction of the resultant by the method of components.

2.29 Three forces act on the pile as shown. If the resultant of the three forces is equal to 300 N and is directed vertically, determine the magnitude and direction of S by the method of components.

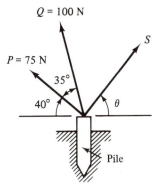

PROB. 2.29

2.30 Three forces act on the pin-connected tower as shown. If the resultant of the three forces is equal to 3200 lb and is directed horizontally, determine the magnitude and direction of V by the method of components.

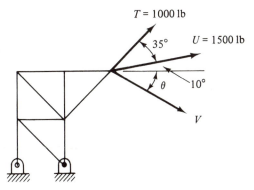

PROB. 2.30

In Probs. 2.31 through 2.50, use the method
of rectangular components.

2.31 Solve Prob. 2.1.

2.32 Solve Prob. 2.2.

2.33 Solve Prob. 2.3.

2.34 Solve Prob. 2.4.

2.35 Solve Prob. 2.5.

2.36 Solve Prob. 2.6.

2.37 Solve Prob. 2.7.

2.38 Solve Prob. 2.8.

2.39 Solve Prob. 2.9.

2.40 Solve Prob. 2.10.

2.41 Solve Prob. 2.11.

2.42 Solve Prob. 2.12.

2.43 Solve Prob. 2.13.

2.44 Solve Prob. 2.14.

2.45 Solve Prob. 2.15.

2.46 Solve Prob. 2.16.

2.47 Using the figure for Prob. 2.27, find
$F_1 - F_2$.

2.48 Using the figure for Prob. 2.28, find
$P - Q$.

2.49 Using the figure for Prob. 2.29, find
$P - Q + S$ if $S = 90$ N and $\theta = 65°$.

2.50 Using the figure for Prob. 2.30, find
$U + V - T$ if $V = 1200$ lb and $\theta = 35°$.

Equilibrium of Concurrent Forces in a Plane

3.1 CONDITIONS FOR EQUILIBRIUM

When the resultant of a force system acting on a body is zero, the body is in equilibrium. If a body is in equilibrium, the body will either remain at rest, if originally at rest, or in motion, if originally in motion. We will be concerned here with the *equilibrium of bodies at rest*.

In Chapter 2 we considered various methods for finding the resultant of concurrent force systems. We make use of those methods to solve problems involving concurrent force systems that act on bodies at rest—force systems that have a zero resultant.

3.2 ACTION AND REACTION

In the process of separating a body from its surroundings, we make use of a principle known as *Newton's third law*, which states that action equals reaction. That is, if a body A exerts a force on body B, body B exerts a force on body A equal in magnitude, opposite in direction, and having the same line of action. Newton's third law cannot be proven mathematically. It agrees with intuition, and deduction from it agrees with experiment.

Consider a block attached to a cable that is supported by the ceiling. In Fig. 3.1(a) we show the action–reaction pairs of forces. The force F_1 represents the force exerted on the cable by the block. The reaction F_2 is equal and opposite and represents the forces exerted on the block by the cable. The force F_3 represents the force exerted on the

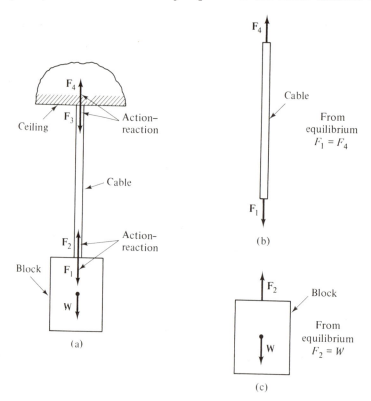

FIGURE 3.1

ceiling by the cable. Its reaction F_4 is equal and opposite and represents the force exerted on the cable by the ceiling. The forces acting on the cable and on the block are shown in Fig. 3.1(b) and (c). Forces F_1 and F_4 do not constitute an action–reaction pair since they act on the same body—the cable. The same is true for forces F_2 and W which act on the block. However, both the cable and the block are in equilibrium with $F_4 = F_1$ and $F_2 = W$. Therefore, $F_3 = W$. Thus the cable can be thought of as transmitting the weight of the block from the block to the ceiling.

3.3 SPACE DIAGRAM—FREE-BODY DIAGRAM

Problems in statics are derived from actual physical problems such as structures, machines, or other physical bodies. A sketch called a *space diagram* describes the physical problem to be solved.

Statics involves the forces or interactions of bodies on each other. The bodies must be separated from each other so that unknown forces may be determined. To this end, we select and free a body from its surroundings. A diagram showing the forces acting on the body is then drawn. Such a diagram is called a *free-body diagram*. When the free-body diagram involves a concurrent force system, the problems can be solved by the methods of this chapter.

3.4 CONSTRUCTION OF A FREE-BODY DIAGRAM

In this section we consider various physical bodies that are described by space diagrams. For each example a free body will be isolated and all the forces acting on the body will be shown.

Among the forces to be shown will be the weight. We shall see in Chapter 8 that the weight of the body acts through a point called the *center of gravity* of the body. It is directed from the center of gravity downward toward the center of the earth. For a uniform body the center of gravity is at the geometric center of the body.

EXAMPLE 3.1

The 12.5-kN block is supported by cables as shown in Fig. 3.2(a). Draw the free-body diagrams of the block and point B.

Solution

Consider the block as a free body [Fig. 3.2(b)]. The force T of the cable acts upward along the cable, away from the body, and the 12.5-kN weight of the body acts down-

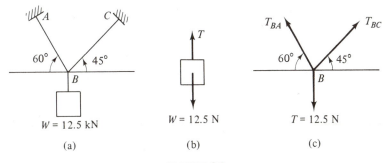

(a) (b) (c)

FIGURE 3.2

47

ward toward the center of the earth. The two forces act along the same straight line. They form a collinear force system. Since the block is at rest, the forces are in equilibrium and the resultant is equal to zero. Therefore, $T - W = 0$ or $T = W = 12.5$ kN. The vertical cable can be thought of as transmitting the weight of the block from the block to point B.

A cable can support a tensile force; therefore, in the free body of point B [Fig. 3.2(c)] each cable is in tension and acts away from point B.

EXAMPLE 3.2

The 2.55-kg block in Fig. 3.3(a) is supported by a cable that passes over a frictionless pulley. Draw the free-body diagram for the pulley.

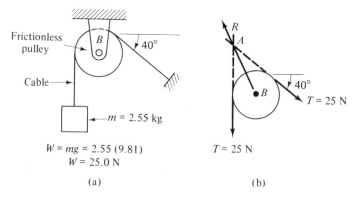

(a)   (b)

FIGURE 3.3

Solution

From Eq. (1.1) the block weighs

$$W = mg = 2.55(9.81) = 25.0 \text{ N}$$

In Example 3.1 we saw that the tension in the vertical cable was equal to the weight of the body supported by the cable. Therefore, the tension in the cable is $T = 25.0$ N.

We now draw a free-body diagram of the frictionless pulley [Fig. 3.3(b)]. The tensile forces in the cable on each side of the frictionless pulley are the same. Thus the tensile force on the left and right of the pulley is equal to 25 N. Both of the tensile forces are directed away from the pulley and their lines of action intersect at point A. For equilibrium, the reactive force R of the axle at B on the pulley must also act through point A.

Ropes, strings, and cords are analyzed in the same way as the cable in a free-body diagram.

EXAMPLE 3.3

The links AB and BC support a rope that is attached to the 5-kip load as shown in Fig. 3.4(a). Draw a free-body diagram of point B.

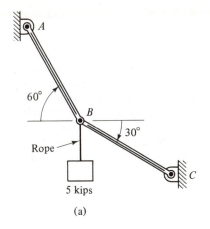

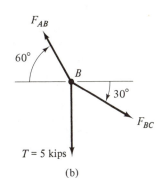

5 kips

(a)

$T = 5$ kips

(b)

FIGURE 3.4

Solution

In Example 3.1 we saw that the tension in the vertical cable was equal to the weight of the body supported by the cable. Therefore, the tension in the rope is $T = 5$ kips. The links AB and BC are weightless bodies joined to the supports at A and C and together at B by frictionless pins. Thus each link or body is subject to two forces—one at each end. As such, they are special cases of two-force bodies. The equilibrium of two-force bodies is discussed in Sec. 5.5, where we show that the force in a link acts in the direction of the link.

Consider the frictionless pin at B as a free body [Fig. 3.4(b)]. The tensile force T acts down—away from B—and the forces in the link F_{AB} and F_{BC} act in the direction of the links. The forces in the links can act either toward B or away from B. The correct directions must be found from the conditions of equilibrium.

EXAMPLE 3.4

Consider a 100-lb block that is supported by a cord and smooth plane as shown in Fig. 3.5(a). Draw the free-body diagram of the block.

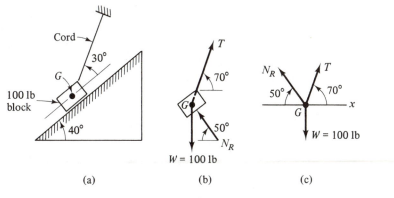

(a)

(b)

(c)

FIGURE 3.5

49

Solution

We isolate the block as the free body [Fig. 3.5(b)]. There are three forces acting on it. The tensile force T on the cord acts along the cord and away from the body. The reactive force N_R of the plane on the block acts normal to the plane at the surface of contact and toward the body. The 100-lb weight of the body acts from the center of gravity G downward toward the center of the earth. The lines of action of the three forces intersect at point G and thus represent a concurrent force system. In Fig. 3.5(c), we move force N_R along its line of action so that all the forces are directed away from point G.

PROBLEMS

Assume pulleys frictionless, struts and links weightless, and inclined planes smooth.

3.1 through 3.8 Draw a free-body diagram of point B.

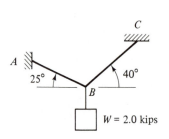

PROB. 3.1

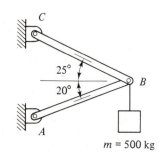

PROB. 3.3

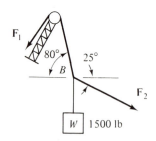

PROB. 3.2

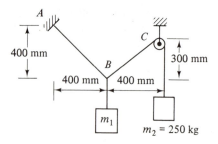

PROB. 3.4

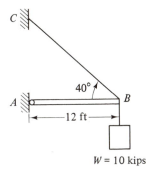

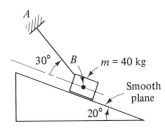

PROB. 3.5

PROB. 3.7

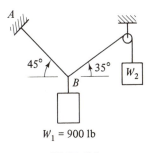

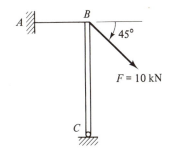

PROB. 3.6

PROB. 3.8

3.5 THREE CONCURRENT FORCES IN EQUILIBRIUM

When a body is in equilibrium under the action of three concurrent forces, the problem can be solved by drawing a force triangle. The method can best be illustrated by examples.

EXAMPLE 3.5

Three concurrent forces in equilibrium acting at point B are shown in the free-body diagram of Fig. 3.4(b). The direction of the forces in links AB and BC was assumed to act away from B. Find the magnitude of the forces F_{AB} and F_{BC}.

Solution

Draw the known force $T = 5$ kips. Through the end of T draw a line parallel to F_{BC} and through the *other* end of T draw a line parallel to F_{AB}. Two possibilities exist as shown in Fig. 3.6(a) and (b). In either case, the lines of action intersect to form a triangle. The force triangle can now be constructed around either of these triangles [Fig. 3.7(a) and (b)].

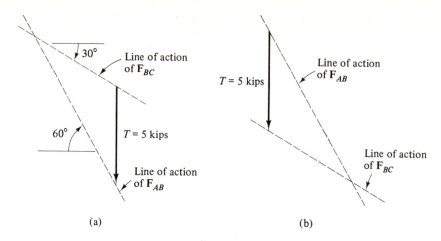

(a)   (b)

FIGURE 3.6

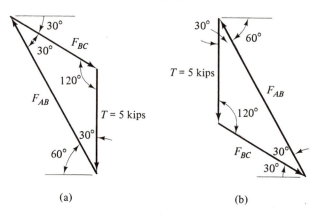

(a)   (b)

FIGURE 3.7

Since the three forces are in equilibrium and by construction form a closed triangle,

$$\mathbf{T} + \mathbf{F}_{AB} + \mathbf{F}_{BC} = 0$$

Applying the law of sines to either triangle, we have

$$\frac{F_{AB}}{\sin 120°} = \frac{F_{BC}}{\sin 30°} = \frac{5}{\sin 30°}$$

Therefore,

$$F_{AB} = \frac{\sin 120°}{\sin 30°} (5) = \frac{0.8660}{0.5} (5)$$

$$= 8.66 \text{ kips} \qquad \qquad \text{Answer}$$

and

$$F_{BC} = \frac{\sin 30°}{\sin 30°} \text{ (5)}$$

$$= 5.0 \text{ kips} \qquad\qquad \text{Answer}$$

The solution may also be obtained graphically. Select an appropriate scale and draw vector forces $T = 5$ kips, F_{BC}, and F_{AB} in a tip-to-tail fashion (Fig. 3.8). The magnitudes of the unknown forces are found to be

$$F_{BC} = 5 \text{ kips} \qquad F_{AB} = 8.7 \text{ kips} \qquad\qquad \text{Answer}$$

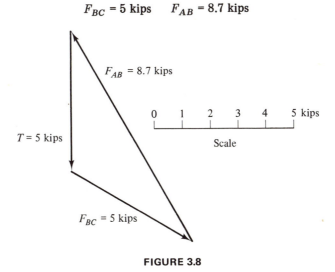

FIGURE 3.8

EXAMPLE 3.6

The three forces in Fig. 3.9 are in equilibrium. The magnitude of all three forces is known, but the directions of F_2 and F_3 are unknown. Find the direction of F_2 and F_3 graphically and mathematically.

Solution

Graphical Method

Select an appropriate scale. Draw $F_1 = 40$ N to scale from point A to point B [Fig. 3.10(a)]. With a compass set for a radius $r_2 = F_2 = 55$ N, draw an arc from the tip of F_1, point B. With a compass set for radius $r_3 = F_3 = 35$ N, draw a second arc from the tail of F_1, point A. The two arcs intersect at point C. Since the force triangle must close, F_2 acts from B to C and force F_3 acts from C to A as shown in the force triangle, Fig. 3.10(b). Measuring the direction of the forces, we have

$$\theta_2 = 39° \quad \text{and} \quad \theta_3 = 87° \qquad\qquad \text{Answer}$$

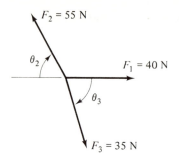

FIGURE 3.9

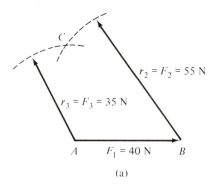

(a)

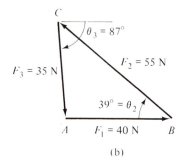

(b)

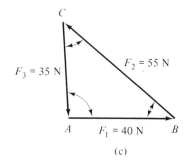

(c)

0   10   20   30   40 N
Scale

FIGURE 3.10

Mathematical Method

A force triangle is constructed in Fig. 3.10(c). Applying the law of cosines to the triangle, we have

$$F_2^2 = F_1^2 + F_3^2 - 2F_1F_2 \cos A$$

$$(55)^2 = (40)^2 + (35)^2 - 2(40)(35) \cos A$$

or

$$\cos A = -0.0714, \quad A = 94.1°$$

From the law of sines, we write

$$\frac{\sin B}{F_3} = \frac{\sin A}{F_2} \qquad \frac{\sin B}{35} = \frac{\sin 94.1°}{55}$$

$$\sin B = \frac{35}{55} \sin 94.1° = 0.6347 \qquad B = 39.4°$$

Since $A + B + C = 180°$, $C = 46.5°$. Therefore, the directions of the forces are

$$\theta_2 = B = 39.4° \qquad\qquad\qquad \text{Answer}$$
$$\theta_3 = B + C = 39.4° + 46.5° = 85.9° \qquad \text{Answer}$$

EXAMPLE 3.7

The three forces acting at A in Fig. 3.11(a) are in equilibrium. The magnitude and direction of F_1 and F_2 are known. Determine F_3 graphically and mathematically.

Solution

Graphical Method

Select the scale. Draw F_1 from A to B and F_2 from B to C. Force F_3 must be represented by a vector drawn from C to A [Fig. 3.11(b)]. Measuring the magnitude and direction of F_3, we have

$$F_3 = 8 \text{ kN} \diagup 54° \qquad\qquad \text{Answer}$$

Mathematical Method

A force triangle is constructed in Fig. 3.11(c). Angle $B = 90° + 20° = 110°$. From the law of cosines,

$$F_3^2 = F_1^2 + F_2^2 - 2F_1F_2 \cos B$$
$$= (5)^2 + (4.8)^2 - 2(5)(4.8) \cos 110° = 64.46$$
$$F_3 = 8.03 \text{ kN}$$

Applying the law of sines yields

$$\frac{\sin C}{F_1} = \frac{\sin B}{F_3} \qquad \frac{\sin C}{5} = \frac{\sin 110°}{8.03}$$

Thus

$$\sin C = \frac{5}{8.03} \sin 110° = 0.585 \qquad C = 35.8°$$

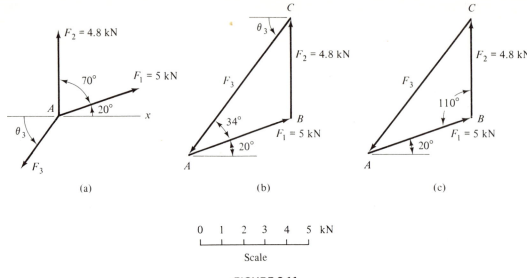

Scale

FIGURE 3.11

The angle $\theta_3 = 90° - C = 90° - 35.8° = 54.2°$.

$$\mathbf{F_3} = 8.03 \text{ kN} \nearrow 54.2° \qquad \text{Answer}$$

3.6 FOUR OR MORE FORCES IN EQUILIBRIUM

When a body is in equilibrium under the action of four or more concurrent forces, the problem may be solved *graphically* by drawing the force polygon. The force polygon method is illustrated in the following example.

EXAMPLE 3.8

The four forces acting at point B as shown in Fig. 3.12(a) are in equilibrium. The magnitude and direction of F_1 and F_2 are known. The magnitude of F_4 is known and the direction of F_3 is known. Find the direction of F_4 and the magnitude of F_3 by constructing a force polygon.

Solution

As shown in Fig. 3.12(b), we start drawing the polygon with the forces of known direction and magnitude. Draw in tip-to-tail fashion, forces F_1 and F_2 from B to C and C to D. Through point D draw a line parallel to the direction of F_3, and around point B draw an arc of radius $r = F_4 = 70$ lb. The arc and line intersect at point E. Force F_3 acts from D to E and force F_4 acts from E to B. Measuring, we obtain

$$F_3 = 52 \text{ lb} \quad \text{and} \quad \theta_4 = 70° \qquad \text{Answer}$$

For a mathematical solution, we use the method of components as described in the following section.

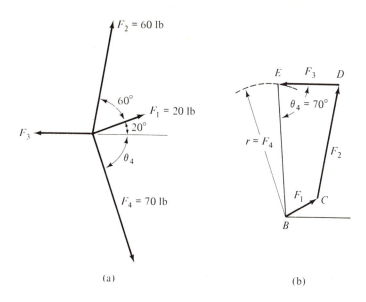

(a)   (b)

0 10 20 30 40 50 lb
Scale

FIGURE 3.12

3.7 EQUILIBRIUM BY RECTANGULAR COMPONENT METHOD

In the *rectangular component method* each force is replaced by its x and y components. The resultant in the x direction, obtained by adding the x components of the forces, must add to zero and the resultant in the y direction, obtained by adding the y components of the forces, must also add to zero. That is,

$$R_x = \Sigma\, F_x = 0 \qquad R_y = \Sigma\, F_y = 0 \qquad (3.1)$$

The following examples will illustrate the rectangular component method.

EXAMPLE 3.9

The three forces in Fig. 3.13(a) are in equilibrium. Determine F_1 and F_3 by the method of rectangular components.

Solution

The x and y components of each force are determined by trigonometry as shown in Fig. 3.13(b) and tabulated as shown. From equilibrium equations (3.1),

57

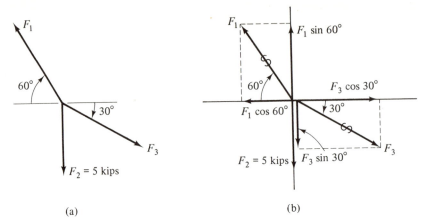

(a) (b)

FIGURE 3.13

$$\Sigma \, F_x = -0.500 F_1 + 0.866 F_3 = 0 \qquad \text{(a)}$$

and

$$\Sigma \, F_y = +0.866 F_1 - 5.0 - 0.500 F_3 = 0 \qquad \text{(b)}$$

From Eq. (a),

$$-0.500 F_1 + 0.866 F_3 = 0$$

or

$$F_3 = \frac{0.500}{0.866} F_1 = 0.577 F_1 \qquad \text{(c)}$$

Substituting the value of F_3 from Eq. (c) in terms of F_1 in Eq. (b), we have

$$0.866 F_1 - 5.000 - 0.5 (0.577) F_1 = 0$$

or

$$F_1 = 8.67 \text{ kips} \qquad \text{Answer}$$

Force	Magnitude (kips)	Angle (deg)	x component (kN)	y component (kN)
F_1	F_1	60	$-0.500 F_1$	$+0.866 F_1$
F_2	5	90	0	-5.0
F_3	F_3	30	$+0.866 F_3$	$-0.500 F_3$
			$\Sigma \, F_x = 0$ Eq. (a)	$\Sigma \, F_y = 0$ Eq. (b)

Substituting the value of F_1 into Eq. (c), we write

$$F_3 = 0.577(8.67) = 5.00 \text{ kips} \qquad \text{Answer}$$

Check:

$$\Sigma F_x = -0.500(8.67) + 0.866(5.00) = 0.005$$
$$= 0 \quad \text{OK}$$

EXAMPLE 3.10

The four forces shown in Fig. 3.14(a) are in equilibrium. Determine F_3 and F_4 by the method of rectangular components.

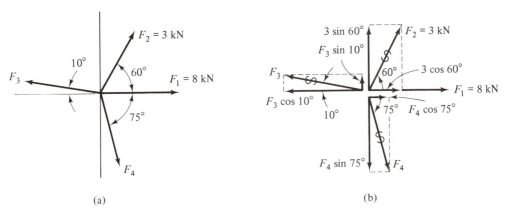

(a)   (b)

FIGURE 3.14

Solution

The x and y components of each force are determined by trigonometry in Fig. 3.14(b) and tabulated as shown. From equilibrium in the x direction, we write

$$\Sigma F_x = 8.00 + 1.500 - 0.985 F_3 + 0.259 F_4 = 0 \qquad \text{(d)}$$

Thus

$$F_4 = \frac{0.985 F_3 - 9.50}{0.259} = 3.80 F_3 - 36.68 \qquad \text{(e)}$$

and from equilibrium in the y direction,

$$\Sigma F_y = 2.598 + 0.1736 F_3 - 0.966 F_4 = 0 \qquad \text{(f)}$$

Substituting the value of F_4 from Eq. (e), in Eq. (f) yields

$$0.1736F_3 - 0.966(3.80F_3 - 36.68) = -2.598$$

or

$$F_3 = 10.87 \text{ kN} \qquad \text{Answer}$$

Force	Magnitude (kN)	Angle (deg)	x component (kN)	y component (kN)
F_1	8.0	0	8.000	0
F_2	3.0	60	1.500	2.598
F_3	F_3	10	$-0.985F_3$	$0.1736F_3$
F_4	F_4	75	$0.259F_4$	$-0.966F_4$
			$\Sigma F_x = 0$ Eq. (d)	$\Sigma F_y = 0$ Eq. (f)

Substituting the value of F_3 in Eq. (e), we obtain

$$F_4 = 3.80(10.87) - 36.68$$

or

$$F_4 = 4.63 \text{ kN} \qquad \text{Answer}$$

Check:

$$\Sigma F_x = 8.000 + 1.500 - 0.985(10.87) + 0.259(4.63) = -0.008$$
$$= 0 \quad \text{OK}$$

PROBLEMS

3.9 through 3.14 Determine the tension in the cables (a) graphically, (b) mathe-matically from the force triangle, and (c) by the method of components.

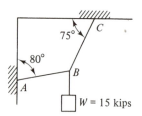

PROB. 3.9

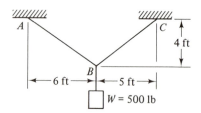

PROB. 3.10

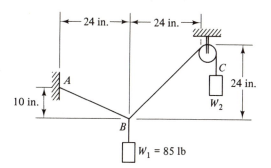

PROB. 3.11

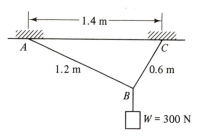

PROB. 3.14

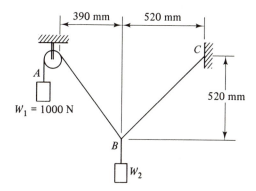

PROB. 3.12

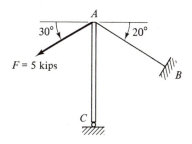

PROB. 3.15

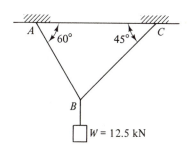

PROB. 3.13

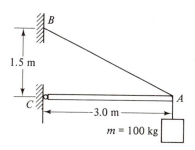

PROB. 3.16

3.15 through 3.18 Determine the tension in the cable and the tension or compression in the strut (a) graphically, (b) mathematically from the force triangle, and (c) by the method of components.

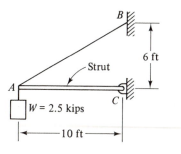

PROB. 3.17

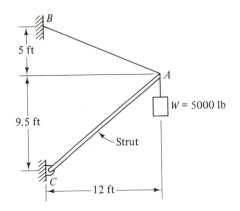

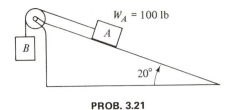

PROB. 3.18

PROB. 3.21

3.19 Determine the force F and the normal reaction of the smooth plane on block A (a) graphically, (b) mathematically from the force triangle, and (c) by the method of components.

3.22 Use the figure for Prob. 3.2. Determine the tension in the cables by the method of components.

3.23 Use the figure for Prob. 3.3. Determine the tension or compression in the struts by the method of components.

3.24 Use the figure for Prob. 3.4. By the method of components, find the tension in the cables.

3.25 Use the figure for Prob. 3.5. By the method of components, find the tension in the cable and the force in the strut.

3.26 Use the figure for Prob. 3.6. Find the tension in the cables by the method of components.

3.27 Use the figure for Prob. 3.7. Determine the tension in the cable and the normal reaction of the smooth plane on block B by the method of components.

3.28 Use the figure for Prob. 3.8. Find the tension in the cable and the force in the strut by the method of components.

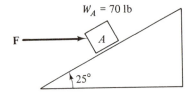

PROB. 3.19

3.20 and 3.21 Determine the tension in the cable and the normal reaction of the smooth plane on block A (a) graphically, (b) mathematically from the force triangle, and (c) by the method of components.

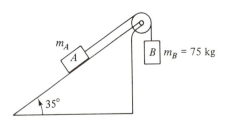

PROB. 3.20

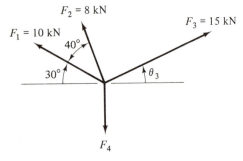

PROB. 3.29

3.29 and 3.30 The four forces are in equilibrium. Determine F_4 and θ_3 (a) graphically from the force polygon, and (b) by the method of components.

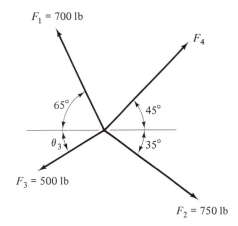

PROB. 3.30

Resultant of Nonconcurrent Forces _____

4.1 INTRODUCTION

In the last two chapters we considered concurrent force systems. In each case, the free body reduced to a single point or particle on which the forces were considered to act. When dealing with nonconcurrent force systems, the physical size of the body must be considered together with the fact that the forces act at different points on the body.

The bodies considered in statics are usually assumed to be rigid. A rigid body is one that does not deform or change shape. Of course, the actual physical body, such as a structure or machine, is never completely rigid. They deform or change shape as a result of the forces that act on them. The deformations or changes in shape are usually small. Such changes become important in the study of strength of materials, where we are concerned with the stresses and deflections or deformations produced by loads on the body. However, they can be neglected in the problems we consider in statics.

With the free body occupying more than a point, we must consider force systems that are *not concentrated* at a point but are *distributed* over a surface or throughout a volume. Examples of such systems are wind pressure, water pressure, and the weight of a body. Some distributed force systems will be discussed in Sec. 4.9.

4.2 TRANSMISSIBILITY

The *principle of transmissibility* states that a force **F** acting at a point A on a rigid body may be transmitted or moved to any other point B on its line of action without changing the effect of the force F on the rigid body. The principle is illustrated in Fig. 4.1(a) and (b).

The principle of transmissibility has limitations. Consider a bar acted on by two equal and opposite forces [Fig. 4.2(a)]. Making use of the principle of transmissibility we move the force that acts at B to A, as shown in Fig. 4.2(b). From the point of view of rigid-body statics, the systems are both in equilibrium and identical. However, in Fig. 4.2(a) the bar shown is in tension. There is an internal force between A and B equal to F, and the length of the bar will increase. In the bar shown in Fig. 4.2(b) the internal force between A and B is zero and the length of the bar will remain unchanged. Thus we see that the principle of transmissibility cannot be applied, except with care, to problems involving internal forces and deformations.

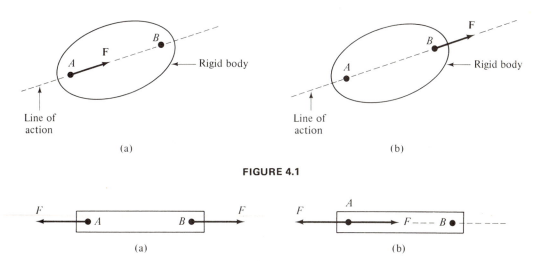

FIGURE 4.1

FIGURE 4.2

4.3 MOMENT OF A FORCE

The tendency of a force to produce rotation about a point is called the *moment* of a force about that point. Consider a wrench that has been applied to a nut as shown in Fig. 4.3(a). To obtain the maximum rotation or turning effect on the nut, we know from common experience that the force should be applied perpendicular to the handle as far away from O as possible. The following definition is consistent with common experience.

The moment of a force about a point is defined as the product of the force and the perpendicular distance from the line of action of the force to the point. The magnitude of the moment of a force will be defined

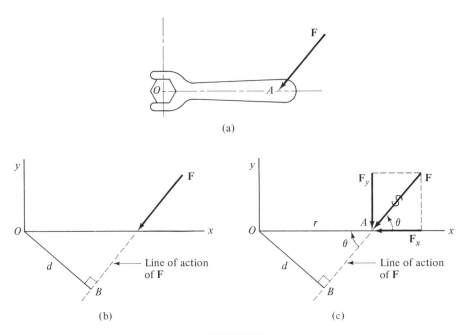

(a)

(b)   (c)

FIGURE 4.3

in equation form with the aid of Fig. 4.3(b). The perpendicular distance from the line of action of F to point O is shown as BO. Let BO be equal to d. Therefore, the moment of the force F about O is equal to the product of F and d. In equation form

$$M_O = Fd \tag{4.1}$$

The point O is the moment center and the distance d is the arm of the force. Thus a moment may be expressed in pound-feet (lb-ft), pound-inches (lb-in.), or newton·meters (N·m).

4.4 THEOREM OF MOMENTS

In Fig. 4.3(c) we resolve the force **F** into x and y components. The component F_x has no moment about O because it acts directly through O. Thus the sum of the moments about O of the components of the force are given by

$$M_O = F_y r \qquad \text{(a)}$$

From the construction we see that $d = r \sin \theta$ or $r = d/\sin \theta$ and $F_y = F \sin \theta$. Substituting into Eq. (a), we have

$$M_O = F_y r = F \sin \theta \, \frac{d}{\sin \theta} = Fd$$

The result agrees with the definition of the moment of the force about O [Eq. (4.1)]. Thus the moment of the force **F** about O is equal to the sum of the moments of the components of **F** about O. Our discussion here represents the proof of a special case of the *theorem of moments*, also known as *Varignon's theorem*.

The theorem of moments states that *the moment of a force about a point is equal to the sum of the moments of the components of that force about the same point.*

The following examples will be used to illustrate methods for finding moments of forces and application of the theorem of moments.

EXAMPLE 4.1

Find the moment of the 120-lb force at C about points A, B, C, and D for the rigid body in Fig. 4.4(a).

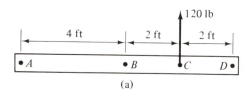

(a)

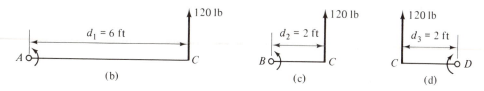

(b)   (c)   (d)

FIGURE 4.4

Solution

To help us learn the method for finding the moment of a force we will draw a sketch that shows the *force*, the *moment center*, and the *arm* of the force.

Moment about A

The force tends to produce rotation about A in the counterclockwise direction, Fig. 4.4(b). By convention, the *counterclockwise* direction will be *positive*. The moment about A is given by

$$M_A = Fd_1 = (120)(6) = 720 \text{ lb-ft}$$

$$= 720 \text{ lb-ft } \circlearrowleft \qquad\qquad\qquad \text{Answer}$$

Moment about B

As shown in Fig. 4.4(c), the force tends to produce counterclockwise rotation about B; thus the moment about B is

$$M_B = Fd_2 = (120)(2) = 240 \text{ lb-ft}$$

$$= 240 \text{ lb-ft } \circlearrowleft \qquad\qquad\qquad \text{Answer}$$

Moment about C

The moment of the force about C will be zero, since the force acts directly through C and can have no moment about C.

$$M_C = 0$$

Moment about D

The force tends to produce clockwise rotation about D [Fig. 4.4(d)]. By convention, *clockwise* rotation will be *negative*. The moment about D is given by

$$M_D = Fd_3 = -120(2) = -240 \text{ lb-ft}$$

$$= 240 \text{ lb-ft } \circlearrowright \qquad\qquad\qquad \text{Answer}$$

EXAMPLE 4.2

Determine the moment about C due to the force F of magnitude 100 N applied at point A to the bracket shown in Fig. 4.5(a).

Solution 1

We construct the diagram shown in Fig. 4.5(b) to determine the arm of the force, d, that is, the perpendicular distance from the force to the moment center at point C. From the right triangle ABE,

$$\tan 30° = \frac{AB}{BE} = \frac{0.6}{BE}$$

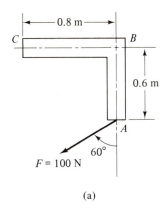

(a)

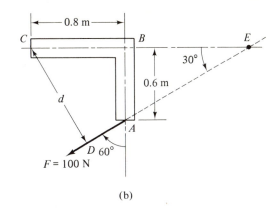

(b)

FIGURE 4.5

$$BE = \frac{0.6}{\tan 30°} = 1.039 \text{ m}$$

From the right triangle CDE,

$$\sin 30° = \frac{d}{CE} = \frac{d}{1.039 + 0.8}$$

$$d = 1.839 \sin 30° = 0.9195 \text{ m}$$

Therefore, since a clockwise rotation is negative,

$$M_C = Fd = -100(0.920) = -92.0 \text{ N} \cdot \text{m}$$

$$= 92 \text{ N} \cdot \text{m} \; \circlearrowleft \qquad\qquad \text{Answer}$$

Solution 2

We find the x and y components of the 100-N force and use the theorem of moments to find the moment about C.

The x and y components of the force in Fig. 4.6(a) are given by $F_x = F \sin \theta$ and $F_y = F \cos \theta$; therefore,

$$F_x = 100 \sin 60° = 86.6 \text{ N}$$

$$F_y = 100 \cos 60° = 50.0 \text{ N}$$

To assist us in learning the method, we draw a sketch of each component of the force together with its arm and moment center C [Fig. 4.6(b) and (c)].

The moment about C must be equal to the algebraic sum of the moments of the components about C. Both components produce clockwise or negative rotation; therefore,

$$M_C = -50(0.8) - 86.6(0.6) = -91.96 \text{ N} \cdot \text{m}$$

$$= 92 \text{ N} \cdot \text{m} \; \circlearrowleft \qquad\qquad \text{Answer}$$

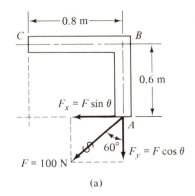

(a)

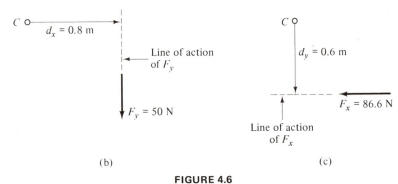

(b)   (c)

FIGURE 4.6

As expected, the answers obtained by both solutions are the same. Unless the arm of a force can be obtained by inspection, it is usually simpler to solve the problem by the theorem of moments.

EXAMPLE 4.3

Determine the moment of the three forces that act on the truss shown in Fig. 4.7(a) about point A and point B.

Solution

To find the x and y components of the force $F_1 = 10$ kips, we consider the right angle BFE [Fig. 4.7(b)]. The side BE of the right triangle may be found from the Pythagorean theorem.

$$(BE)^2 = (5)^2 + (4)^2 = 41$$

$$BE = 6.403 \text{ ft}$$

Therefore,

$$\sin \theta = \frac{BF}{BE} = \frac{5}{6.403} = 0.781$$

$$\cos \theta = \frac{FE}{BE} = \frac{4}{6.403} = 0.625$$

The slope of the force $F_1 = 10$ kips is the same as member BE and the x and y components are given by

$$F_x = F \cos \theta = 10(0.625) = 6.25 \text{ kips}$$

$$F_y = F \sin \theta = 10(0.781) = 7.81 \text{ kips}$$

We draw the truss showing F_2, F_3, and the components of F_1 in Fig. 4.7(c).

Moments about A

We see from Fig. 4.7(c) that the arm of F_x is GA, the arm of F_y and F_2 is FA, and the arm of F_3 is zero. Therefore, the algebraic sum of the moments about A (counterclockwise positive) is

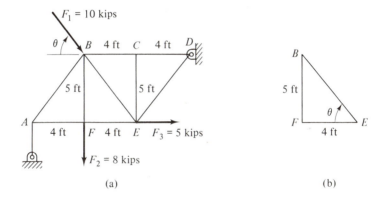

(a) (b)

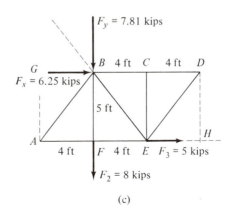

(c)

FIGURE 4.7

$$\circlearrowright \Sigma M_A = -F_x(GA) - F_y(FA) - F_2(FA) + F_3(0)$$

$$= -6.25(5) - 7.81(4) - 8(4)$$

$$= -31.25 - 31.24 - 32.00 = -94.49 \text{ kip-ft}$$

$$= 94.5 \text{ kip-ft } \circlearrowleft \qquad\qquad\qquad \textbf{Answer}$$

Moment about D

The arm of F_x is zero, the arm of F_y and F_2 is BD, and the arm of F_3 is HD [Fig. 4.7(c)]. Therefore, the algebraic sum of the moments about D (counterclockwise positive) is

$$\circlearrowright \Sigma M_D = +F_x(0) + F_y(BD) + F_2(BD) + F_3(HD)$$

$$= 7.81(8) + 8(8) + 5(5)$$

$$= 62.48 + 64 + 25 = +151.48 \text{ kip-ft}$$

$$= 151.5 \text{ kip-ft } \circlearrowright \qquad\qquad\qquad \textbf{Answer}$$

PROBLEMS

4.1 Determine the moment of the forces about points A, B, and C for the beam shown.

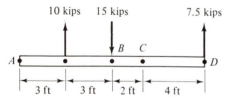

PROB. 4.1

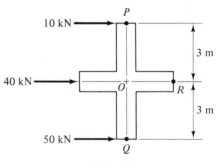

PROB. 4.2

4.2 For the parallel-force system shown, determine the moment of the forces about points $O, P, Q,$ and R.

4.3 Determine the moment of the forces about points A, B, and C for the cantilever beam shown.

4.4 Determine the moment of the forces shown about points $D, E, F,$ and G.

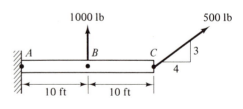

PROB. 4.3

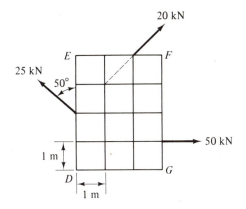

PROB. 4.4

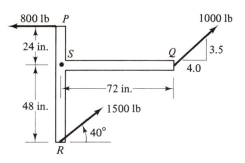

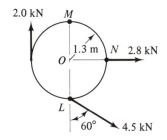

PROB. 4.5

4.5 Determine the moment of the forces shown about points P, Q, R, and S.

4.6 For the circular disk shown, determine the moment of the forces about points L, M, N, and O.

PROB. 4.6

4.5 RESULTANT OF PARALLEL FORCES

The resultant of a parallel system of forces must have the same or equivalent effect as the parallel system. That is, the resultant force must be equal to the sum of the parallel forces. Also, the sum of the moments of the forces about any point must be equal to the moment of the resultant about that same point.

Consider the system of parallel forces $\mathbf{F}_1$, $\mathbf{F}_2$, and $\mathbf{F}_3$ with moment arms of r_1, r_2, and r_3 measured from O in Fig. 4.8(a). The resultant of this force system must produce the same effect as the original force system. To produce the same effect, the sum of the forces must be equal to the resultant force and the sum of the moments of the forces about a point must be equal to the moment of the resultant about that same point. Thus the force system in Fig. 4.8(b) is the resultant of the force system in Fig. 4.8(a) if

$$\Sigma \, F = F_1 + F_2 + F_3 = R \tag{a}$$

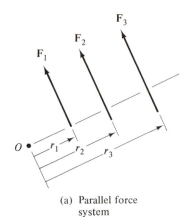

(a) Parallel force
system

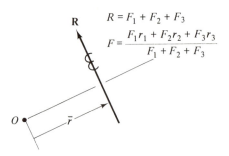

$$R = F_1 + F_2 + F_3$$

$$F = \frac{F_1 r_1 + F_2 r_2 + F_3 r_3}{F_1 + F_2 + F_3}$$

(b) Resultant or
equivalent force

FIGURE 4.8

and

$$\Sigma \, M_O = F_1 r_1 + F_2 r_2 + F_3 r_3 = R\bar{r}$$

or

$$\bar{r} = \frac{F_1 r_1 + F_2 r_2 + F_3 r_3}{F_1 + F_2 + F_3} \tag{b}$$

The magnitude of the resultant R is given by Eq. (a), and the location of the resultant from point O is given by Eq. (b).

Not all parallel-force systems have a single force as the resultant. A system of two equal parallel forces acting in opposite directions is called a *couple*. The couple (Sec. 4.7) does not have a single force as a resultant.

EXAMPLE 4.4

For the parallel-force system acting on the rod in Fig. 4.9(a), determine the resultant or equivalent force.

Solution

The resultant or equivalent force acting on the rod is shown in Fig. 4.9(b). To be equivalent,

$$R = \uparrow \Sigma \, F = -F_1 + F_2 - F_3 = -10 + 12 - 25 = -23 \text{ kN}$$

or

$$R = 23 \text{ kN} \downarrow$$

and

$$\supset \Sigma \, M_A = F_2 \, (\overline{AB}) - F_3 \, (\overline{AC}) = - \, R\overline{r}$$

$$12 \, (4) - 25 \, (10) = -23 \, \overline{r}$$

or

$$\overline{r} = \frac{-202}{-23} = 8.78 \text{ m}$$

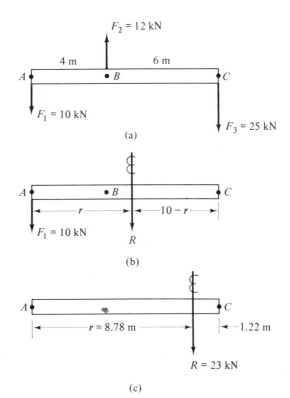

(a)

(b)

(c)

FIGURE 4.9

We check the results by calculating moments with respect to C.

$$\circlearrowright \Sigma M_C = R(10 - r)$$

$$10(10) - 12(6) = 23(10 - 8.78)$$

$$28 = 28.06 \quad \text{OK}$$

The magnitude and location of the resultant is shown in Fig. 4.9(c).

PROBLEMS

4.7 For the wheel loads shown, determine the resultant. Locate the resultant with respect to A and B.

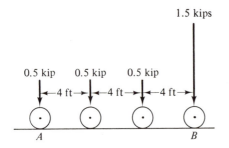

PROB. 4.7

4.8 Determine the resultant of the parallel-force system shown. Locate the resultant with respect to A and B.

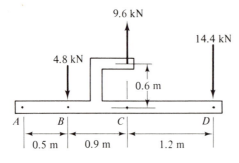

PROB. 4.8

4.9 Use the figure for Prob. 4.1. Find the resultant of the parallel-force system shown. Locate the resultant with respect to A and D.

4.10 The pin-connected truss is acted on by forces as shown. Determine the resultant and locate with respect to points A and B.

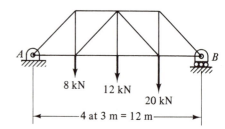

PROB. 4.10

4.11 Use the figure for Prob. 4.2. Find the resultant for the forces shown. Locate the resultant with respect to points P and Q.

4.12 Determine the resultant of the parallel-force system shown. Locate the resultant with respect to points A and D.

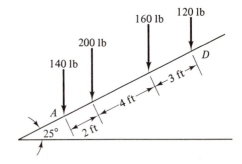

PROB. 4.12

4.6 RESULTANT OF NONPARALLEL FORCES

The resultant of a nonparallel system of forces is found in essentially the same way as the resultant of a parallel-force system. By resolving each force into rectangular components we have two sets of parallel forces at right angles to each other. We illustrate the procedure in the following example.

EXAMPLE 4.5

Determine the resultant of the force system shown in Fig. 4.10(a). Locate the resultant with respect to point A.

Solution

The x and y components of the forces are calculated as follows:

$$F_{x_1} = -F_1 \cos 60° = -12(0.500) = -6.0 \text{ kips}$$
$$F_{x_2} = +F_2 \cos 45° = +20(0.707) = 14.14 \text{ kips}$$
$$F_{x_3} = +F_3 \cos 15° = +25(0.966) = 24.15 \text{ kips}$$

and

$$F_{y_1} = +F_1 \sin 60° = +12(0.866) = 10.39 \text{ kips}$$
$$F_{y_2} = +F_2 \sin 45° = +20(0.707) = 14.14 \text{ kips}$$
$$F_{y_3} = -F_3 \sin 15° = -25(0.259) = -6.47 \text{ kips}$$

In the free-body diagram [Fig. 4.10(b)] each force has been replaced by its x and y components. Adding forces in the x and y directions we have

$$\rightarrow \Sigma \, F_x = R_x = -6.0 + 14.14 + 24.15 = 32.29 \text{ kips}$$
$$\uparrow \Sigma \, F_y = R_y = 10.39 + 14.14 - 6.47 = 18.06 \text{ kips}$$

The resultant is given by

$$R^2 = R_x^2 + R_y^2 = (32.29)^2 + (18.06)^2 = 1369$$
$$R = 37.0 \text{ kips}$$
$$\tan \theta = \frac{18.06}{32.29} = 0.559 \qquad \theta = 29.2°$$
$$\mathbf{R} = 37.0 \text{ kips} \angle 29.2°$$

Therefore, the resultant is directed upward and to the right. To find the position of the resultant with respect to A, we assume that the line of action of R is as shown in

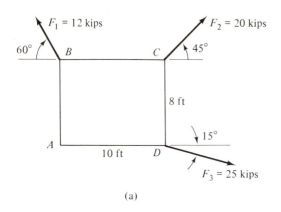

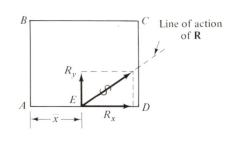

(a)

(c) Resultant or equivalent force

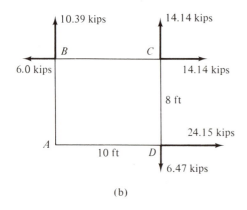

(b)

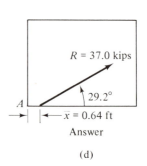

(d)

FIGURE 4.10

Fig. 4.10(c). From the principle of transmissibility, the resultant **R** can be moved to any point on its line of action. We select point E, for convenience, since the x component of the resultant will have no moment about A.

To conclude our solution, we sum the moments about A of the forces shown in Fig. 4.10(b) and equate them to the moments about A of the components of the resultant as shown in Fig. 4.10(c). Thus

$$\circlearrowright \Sigma M_A = R_y \bar{x} + R_x(0)$$

$$\circlearrowright \Sigma M_A = 6(8) + 14.14(10) - 14.14(8) - 6.47(10)$$

$$= 11.58 \text{ kip-ft}$$

$$R_y \bar{x} = 18.06 \bar{x}$$

$$18.06 \bar{x} = 11.58$$

$$x = \frac{11.58}{18.06} = 0.641 \text{ ft}$$

$$R = 37.0 \text{ kips} \measuredangle 29.2°$$
0.64 ft to the right of A **Answer**

The answer is shown in Fig. 4.10(d).

4.13 A cantilever truss is acted on by forces as shown. Find the resultant and locate with respect to A and C.

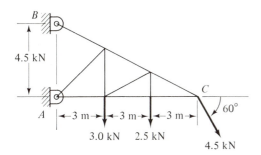

PROB. 4.13

4.14 Use the figure for Prob. 4.3. Determine the resultant of the force system shown. Locate the resultant with respect to A and C.

4.15 Determine the resultant of the forces acting on the hook-shaped member shown. Locate the resultant with respect to B and C.

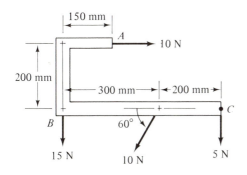

PROB. 4.15

4.16 For the force system acting on the T-shaped member shown, find the resultant and locate with respect to points A and B.

4.17 Use the figure for Prob. 4.4. Determine the resultant of the force system shown. Locate the resultant with respect to points D and E.

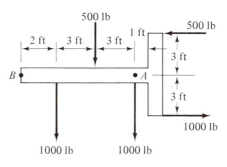

PROB. 4.16

4.18 Forces act on the L-shaped member shown. Determine the resultant and locate with respect to points A and B.

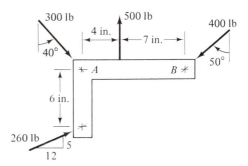

PROB. 4.18

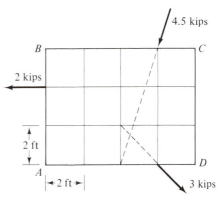

PROB. 4.19

79

4.19 Determine the resultant of the force system shown on page 79. Locate the resultant with respect to points A and D.

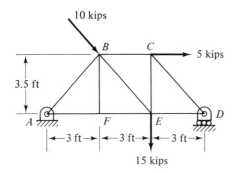

PROB. 4.21

4.20 Use the figure for Prob. 4.5. Determine the resultant of the force system shown. Locate the resultant with respect to S and Q.

4.21 A pin-connected truss is acted on by forces as shown. Find the resultant and locate with respect to points A and D.

4.22 Determine the resultant of the forces acting on the pin-connected tower shown. Locate the resultant with respect to the right side of the tower—FJ.

4.23 For the force system acting on the member shown, find the resultant and locate with respect to M and N.

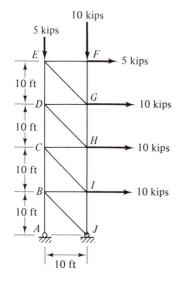

PROB. 4.22

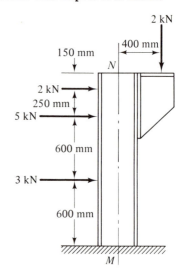

PROB. 4.23

4.24 Use the figure for Prob. 4.6. Determine the resultant of the forces acting on the disk shown. Locate the resultant with respect to points L and M.

4.7 MOMENT OF A COUPLE

We consider a special parallel force system consisting of two parallel forces of equal magnitude acting in opposite directions as shown in Fig. 4.11. Such a system is called a couple. The plane in which the forces act is called the plane of the couple and the perpendicular distance d, between the line of action of the forces, is called the *arm of the couple*.

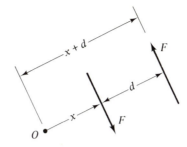

FIGURE 4.11

Adding the forces of the couple to find the resultant force, we have $R = F - F = 0$. Thus the resultant force is equal to zero. Next, we sum the moments of the forces about a point O to find the moment of the couple about O.

$$\circlearrowright \Sigma\, M_O = -Fx + F(x + d) = Fd \circlearrowright \tag{4.2}$$

Therefore, we see that the *moment of a couple depends only* on the *forces* in the couple and the *arm of the couple*. That is, the moment of a couple is the same about any point in the plane of the couple. A couple produces a pure moment—a turning effect only.

In Fig. 4.12 we show several equivalent couples. The couples are equivalent because their moments are equal. From Fig. 4.12(a)-(c),

$$M = Fd = 15(6) = 12(7.5) = 10(9) = 90 \text{ lb-ft}$$
$$= 90 \text{ lb-ft}$$

To add two or more couples, we simply calculate the algebraic sum of their moments.

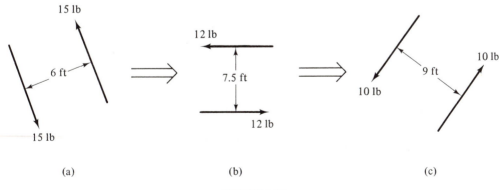

FIGURE 4.12

EXAMPLE 4.6

Find the resultant moment of the couple that acts on the plate shown in Fig. 4.13(a).

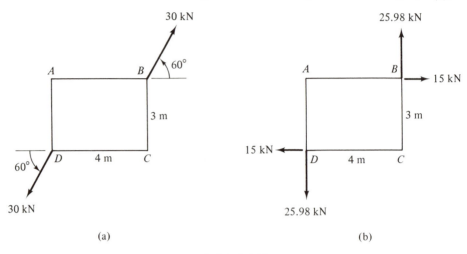

FIGURE 4.13

Solution

The x and y components of the 30-kN forces are calculated and displayed on the free-body diagram of the plate shown in Fig. 4.13(b). The 15-kN forces have an arm of 3 m and form a clockwise couple. The 25.98-kN forces have an arm of 4 m and form a counterclockwise couple. Therefore,

$$M = 25.98(4) - 15(3) = 58.92 \text{ kN} \cdot \text{m}$$

$$= 58.9 \text{ kN} \cdot \text{m} \qquad\qquad \text{Answer}$$

4.8 RESOLUTION OF A FORCE INTO A FORCE AND COUPLE

Consider a force **P** acting on a body at point A which is located a distance d away from point O as shown in Fig. 4.14(a).

In certain physical applications it is convenient to replace the force **P** at A by a force **P** at O and a couple. The replacement or resolution may be described as follows. Introduce two forces **P** and $-$**P** at O [Fig. 4.14(b)]. The force system is unchanged because the two forces at O add to zero. However, the force directed up at O and the force directed down at A form a couple with a clockwise moment $M = Pd$. Thus we are left with a force downward at O and a clockwise couple with moment equal to Pd as shown in Fig. 4.14(c). The resolution is complete. We have re-placed the force at A by a force at O and a couple.

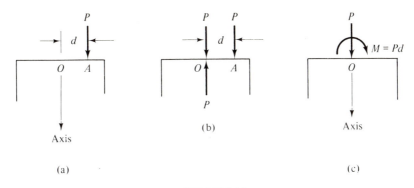

FIGURE 4.14

EXAMPLE 4.7

Replace the force at B by a force at C and a couple [Fig. 4.15(a)].

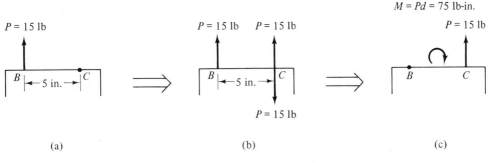

FIGURE 4.15

Solution

We introduce two 15-lb forces at C, one up and the other down, as shown in Fig. 4.15(b). The force up at B and down at C form a clockwise couple with a moment $M = Pd = 15(5) = 75$ lb-in. The resolution is shown in Fig. 4.15(c), where $M = 75$ lb-in. $\circlearrowright$ and $P_{at\,C} = 15$ lb $\uparrow$.

PROBLEMS

4.25 A couple acts at A as shown. Replace the couple by (a) two vertical forces at B and C, and (b) two horizontal forces at D and E.

4.26 The three couples shown on page 84 are equivalent. Determine (a) the distance d for the second couple, and (b) the force F for the third couple.

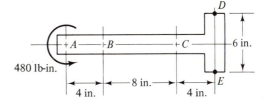

PROB. 4.25

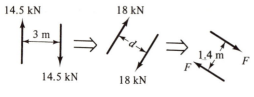

PROB. 4.26

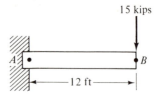

PROB. 4.29

4.27 Find the resultant moment for the couples that act on the plate shown.

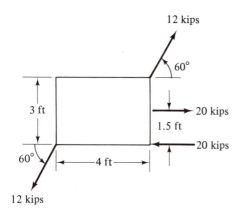

PROB. 4.27

4.30 Replace the force at C by (a) a force at A and a couple, and (b) a force at B and a couple.

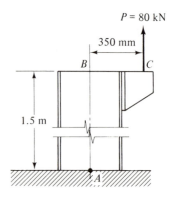

PROB. 4.30

4.28 Couples act at A and E as shown. Replace the couples by (a) two horizontal forces at B and C, and (b) two vertical forces at D and E.

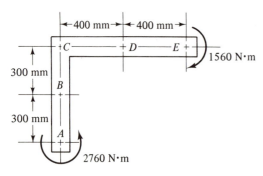

PROB. 4.28

4.29 Replace the force at B by a force at A and a couple.

4.31 Replace the force at A by (a) a force at B and a couple, and (b) a force at C and a couple.

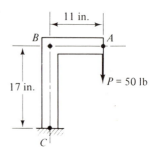

PROB. 4.31

4.32 The bracket is attached to a plate by two bolts as shown. (a) Replace the force at D by a force at C and a couple. (b) Replace the couple from part (a) by two vertical forces at A and B.

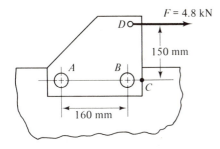

PROB. 4.32

a force at M and a couple. Add the forces and couples to find a single force at M and a single couple.)

4.33 The disk is acted on by forces as shown. Replace the forces by a force at O and a couple. (*Hint:* Replace each force by a force at O and a couple. Add the forces at O and the couples to find a single force at O and a single couple.)

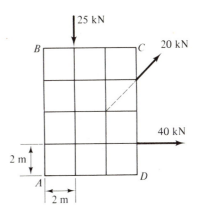

PROB. 4.34

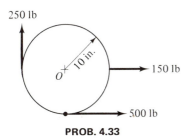

PROB. 4.33

4.34 The grid is acted on by forces as shown. Replace the forces by a force at A and a couple. (*Hint:* Replace each force by a force at A and a couple. Add the forces at A and the couples to find a single force at A and a single couple.)

4.35 A section cut from an offset link has dimensions and a load as shown. Replace the force at A by a force at B and a couple. (*Hint:* Find the components of the load. Replace each component by a force at B and a couple. Add the forces and couples to find a single force at B and a single couple.)

4.36 For the beam column shown replace the forces by a resultant force at M and a couple. (*Hint:* Replace each force by

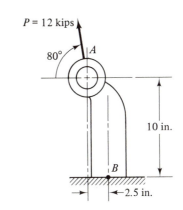

PROB. 4.35

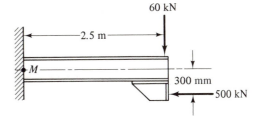

PROB. 4.36

4.9 RESULTANT OF DISTRIBUTED LOADING

In many practical problems the surface or volume of a body may be subjected to distributed loading. The loading may be the result of wind, water, the weight of material supported by the body, or the weight of the body itself.

We shall consider two kinds of distributed loads: the *uniform load* and the *triangular load*. Each unit length of a homogeneous beam with a uniform cross section has the same weight. Such a beam is a good example of the uniformly distributed load. The space diagram and the idealized space diagram for such a beam is shown in Fig. 4.16(a) and (b). The beam weight will be expressed in lb/ft or N/m. The force of water on a seawall is an example of a triangular load [Fig. 4.17(a) and (b)]. The maximum force of the water at the base of the wall for one unit length of wall will be expressed in lb/ft or N/m.

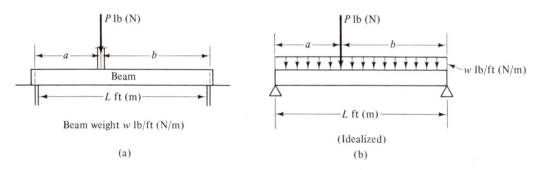

FIGURE 4.16

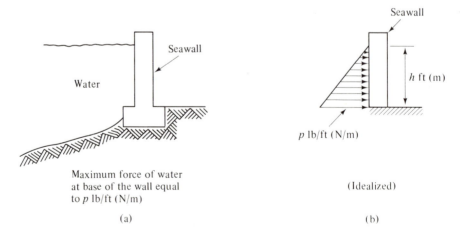

FIGURE 4.17

Resultant of a Uniform Load

A uniform load may be represented by a rectangle. The height of the rectangle represents the load per unit of length. It may be expressed in lb/ft or N/m. The length of the rectangle is the distance over which the load acts and may be expressed in feet or meters. The resultant of a uniform load is equal in magnitude to the area of the rectangle, that is, $R = wL$, and it acts at the middle of the rectangle a distance from either end of $L/2$. See Fig. 4.18(a).

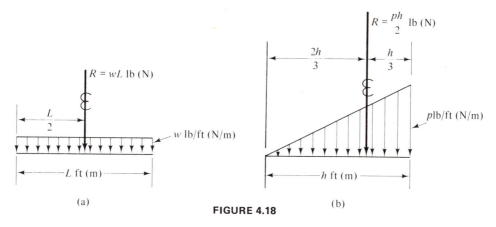

(a)   (b)

FIGURE 4.18

Resultant of a Triangular Load

A triangular load varies uniformly from zero on one end to a maximum on the other end. The height represents the load per unit of length and the length is the distance over which the load acts. The resultant of a triangular load is equal in magnitude to the area of the triangle, $R = hp/2$ and it acts a distance $h/3$ from one end of the triangle and a distance $2h/3$ from the other end as shown in Fig. 4.18(b).

EXAMPLE 4.8

Find the resultant of the uniform load and the resultant of the triangular load acting on the beam in Fig. 4.19(a).

Solution

The resultant of the uniform load acting on the beam from A to C is equal to the area of the rectangle, $R = wL = 5(1.5) = 7.5$ kN. It acts at a distance $L/2 = 1.5/2 = 0.75$ m from either A or C. The resultant of the triangular load acting on the beam from C to B is equal to the area of the triangle, $R = hp/2 = 10(4.5)/2 = 22.5$ kN. It acts at a distance $L/3 = 1.5$ m from C or $2L/3 = 2(4.5)/3 = 3.0$ m from B. The resultant forces are shown on the beam in Fig. 4.19(b).

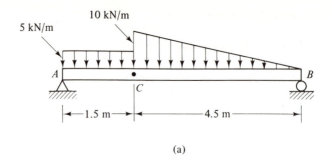

(a)

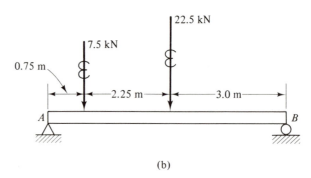

(b)

FIGURE 4.19

PROBLEMS

4.37 through 4.42 (a) Determine the resultant of the force system shown. Locate the resultant with respect to point A. (b) Replace the resultant force by a force at A and a couple.

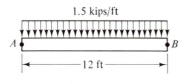

1.5 kips/ft

PROB. 4.37

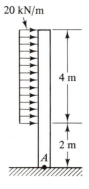

20 kN/m

PROB. 4.38

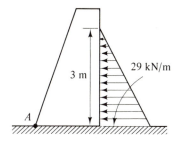

PROB. 4.39

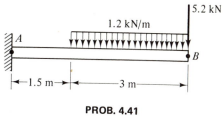

PROB. 4.41

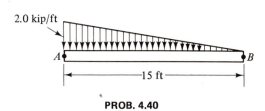

PROB. 4.40

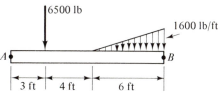

PROB. 4.42

5

Equilibrium of a Rigid Body ————————

5.1 INTRODUCTION

A particle is in *equilibrium* if the resultant force acting on the particle is equal to zero. In the case of a rigid body, equilibrium requires that both the resultant force and the resultant moment acting on the body be equal to zero.

The rigid body, such as a structure or machine, may consist of a single member AB [Fig. 5.1(a)] or several members such as CD and DE joined together to form a single system [Fig. 5.1(b)].

Forces and moments may be classified as *external* or *internal*. Forces that hold a member or members together are internal and may occur either at a hinge such as D or between any two parts of a single member. External forces can be either *applied* or *reactive*. Applied forces act directly on the member, such as F_1, F_2, and F_3, and the weight of the

90

members, W_1, W_2, and W_3. Reactive forces are those produced by the members' supports, R_A, R_B, R_C, and R_E. Applied forces or loads are usually known quantities, whereas reactive forces or reactions are unknown quantities.

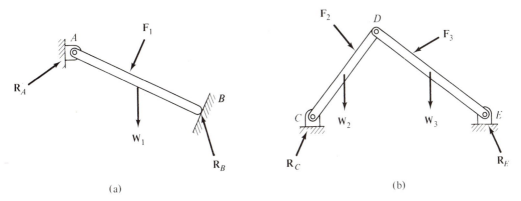

(a)

(b)

FIGURE 5.1

5.2 SUPPORT CONDITIONS

The supports develop reactions in response to the weight of the body and to loads (external forces or moments) that are applied to the body. They prevent the body from moving. That is, the body is in equilibrium under the action of the loads and reactions. There are several types of supports for bodies loaded by forces *acting in a plane*. They may be classified by the kind of resistance they offer to the forces, as follows.

One Reactive Force

These supports can prevent translational motion along a specific line of action. They include smooth surfaces, rollers, cables, and links. The smooth surface and cable provide resistance in one direction only, whereas the roller and link provide resistance in either direction along the same line of action.

Two Reactive Forces

These supports prevent translational motion in a plane, that is, motion in two perpendicular directions. They include smooth pins or hinges and rough surfaces. The smooth pin provides resistance in any direction in a plane, whereas the rough surface provides resistance normal to the surface and tangent to the surface.

Type of support	Reaction force	Number and kind of unknown reaction
 Smooth surface	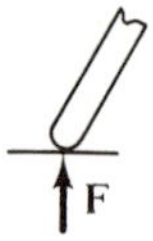 $\uparrow F$	One unknown — magnitude of force **F**. Force normal to surface and toward body
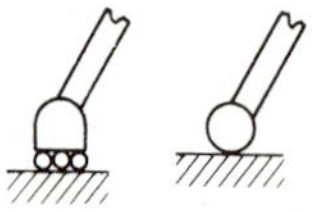 Rollers	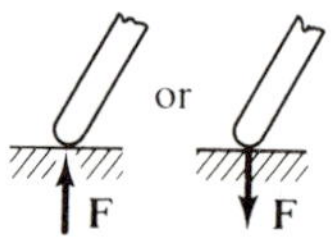 $\uparrow F$ or $\downarrow F$	One unknown — magnitude of force **F**. Force normal to surface and directed toward body or away from body.
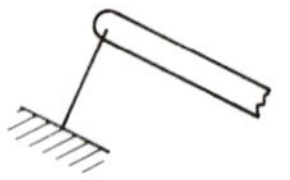 Cables	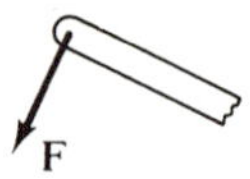 F	One unknown — magnitude of force **F**. Force acts in the direction of the cable away from the body.
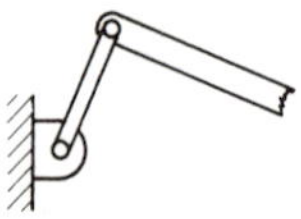 Links	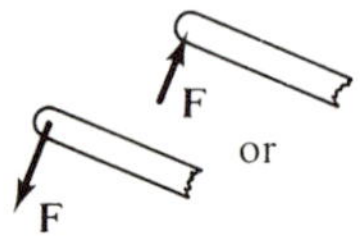 F or F	One unknown — magnitude of force **F**. Force acts in the direction of the link directed either toward or away from body.
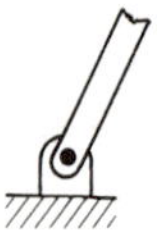 Smooth pin or hinge	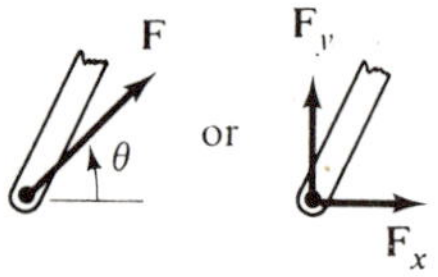 F, θ or F_y, F_x	Two unknowns — magnitude of the force **F** and direction θ or components of the force F_x and F_y.
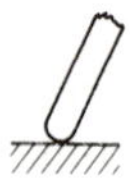 Rough surface	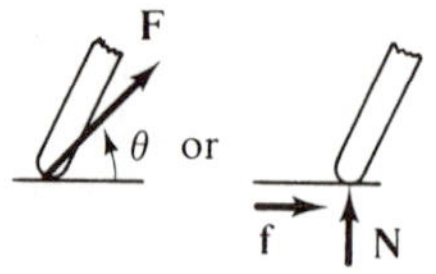 F, θ or f N	Two unknowns — magnitude and direction of resultant force or the force **N** normal to the surface toward the body and the friction force f tangent to surface.
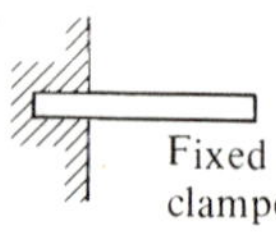 Fixed or clamped	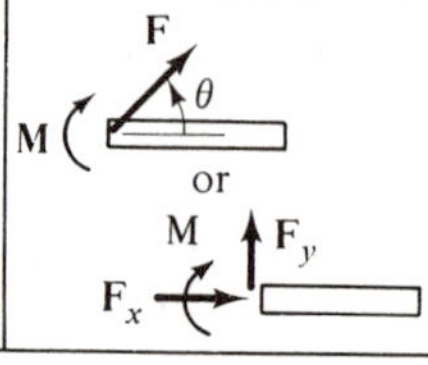 M F θ or M F_y F_x	Three unknowns — magnitude of the couple **M** and the direction and magnitude of the resultant force **F** or the magnitude of the couple and the components of the force F_x and F_y.

FIGURE 5.2

Three Reactive Forces

These supports prevent translational motion in two directions as well as rotational motion. Such a support is called fixed, built-in, or clamped. The actual support is produced by building the body into a wall, casting the body as part of a larger body or welding or mechanically attaching the body to a larger body.

The various supports described are shown in Fig. 5.2 together with the number and kind of unknown reactions.

5.3 CONSTRUCTION OF FREE-BODY DIAGRAMS

We consider here various physical problems as described by space diagrams. In each problem one or more of the supports shown in Fig. 5.2 will be used. A free body will be isolated, and all the loads and reactions acting on the body will be shown.

EXAMPLE 5.1

Loads P_1 and P_2 are applied to a bent which may be supported in four different ways, as shown in Fig. 5.3. Draw the free-body diagram and determine the total number of unknown forces and moments. Neglect the weight of the bent.

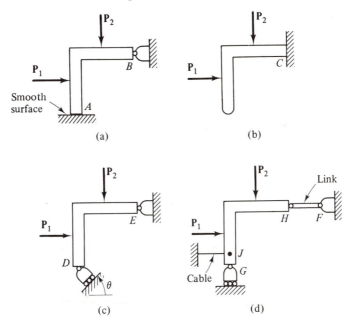

FIGURE 5.3

Solution

The free-body diagrams for the various parts of Fig. 5.3 correspond with the same part of Fig. 5.4. Since the support at A is a *smooth surface*, the reaction is perpendicular to the surface and directed toward the bent. The *rollers* at D and G provide reactions that are perpendicular to the surface on which they roll and are directed toward the bent or away from the bent as required by equilibrium. The reaction of the *link* at H is directed along the link and can be directed toward the bent or away from the bent as required by equilibrium. The *cable* at J is in tension and the reaction is in the direction of the cable and away from the bent. For the *pins* at B and E a reaction of unknown

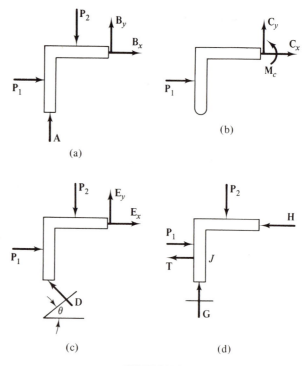

FIGURE 5.4

magnitude and direction occurs. For convenience, the x and y components of the reaction are shown. The direction of the components was assumed. Their actual direction will depend on the requirements of equilibrium. The support at C is *fixed*; therefore, three reactions occur. For convenience, the components of the reaction and the moment of the couple were assumed in a positive direction. Their actual direction will be found from equilibrium. In each bent there are three unknown reactions, consisting of either three forces or two forces and a moment couple.

EXAMPLE 5.2

Draw the free-body diagram for the *uniform* beam shown in Fig. 5.5(a). The total weight of the beam is 100 lb.

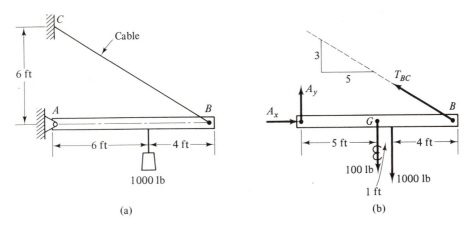

FIGURE 5.5

(a)   (b)

Solution

In Fig. 5.5(b) we draw the free-body diagram of the beam. The 100-lb force represents the weight of the beam. The beam is uniform; therefore, the geometric center and center of gravity G are at the same point. The 1000-lb force represents the tension in the cable supporting the 1000-lb weight. The unknown force T_{BC} represents the tension in the cable BC and the two unknown forces A_x and A_y represent the horizontal and vertical components of the pin reaction at A. There are three unknown reactions, T_{BC}, A_x, and A_y.

EXAMPLE 5.3

The beam shown in Fig. 5.6(a) carries a distributed load as shown. Neglect the weight of the beam. Draw the free-body diagram for the beam.

Solution

The free-body diagram of the beam is shown in Fig. 5.6(b). The resultant of the uniformly distributed load is equal to

$$F_1 = 300 \, \frac{\text{lb}}{\text{ft}} \times 8 \text{ ft} = 2400 \text{ lb}$$

and the resultant of the triangular load is equal to

$$F_2 = \tfrac{1}{2}(300) \, \frac{\text{lb}}{\text{ft}} \times 12 \text{ ft} = 1800 \text{ lb}$$

We have replaced the uniformly distributed load by the resultant at its center and the triangular load by the resultant at the third point. The unknown force F_C at the roller C is normal to the surface on which the rollers move and assumed to be directed toward the beam. The two unknown forces A_x and A_y represent the horizontal and vertical components of the pin reaction at A. There are three unknown reactions, F_C, A_x, and A_y.

95

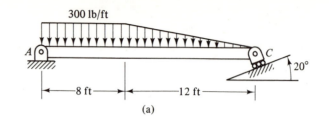

(a)

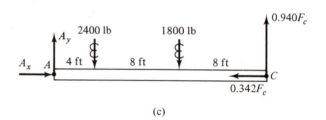

(b)

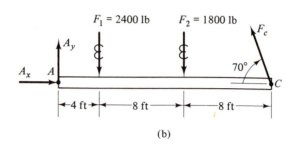

(c)

FIGURE 5.6

5.4 EQUATIONS FOR EQUILIBRIUM OF A RIGID BODY

Equilibrium of a rigid body requires that the resultant force and resultant moment acting on the body vanish. After resolving forces into their x and y components, the conditions for equilibrium in a plane can be expressed in terms of three equations

$$\Sigma F_x = 0, \quad \Sigma F_y = 0, \quad \text{and} \quad \Sigma M_A = 0 \tag{5.1}$$

where A is any point in the plane of the body. These three equations are independent of each other. That is, no two of the equations can be combined to form the third equation. In fact, for equilibrium in two directions, we can write *only three independent equations.* However, an alternative set of equilibrium equations may be useful. One of the force equations has been replaced by a moment equation.

$$\Sigma F_x = 0, \quad \Sigma M_A = 0, \quad \text{and} \quad \Sigma M_B = 0 \tag{5.2}$$

where A and B are any two points in the plane of the body not on a line perpendicular to the x axis.

In the third possible set of equations, both force equations have been replaced by a moment equation. That is,

$$\Sigma\, M_A = 0, \quad \Sigma M_B = 0, \quad \text{and} \quad \Sigma\, M_C = 0 \qquad (5.3)$$

where A, B, and C are any three points in the plane of the body not on a straight line.

Any one of the three sets of equations may be used to solve an equilibrium problem. The problem can be simplified, however, if we select equations of equilibrium that result in only one unknown in each equation. Moment equations can frequently be used to eliminate unknowns by summing moments about the point of intersection of the lines of action of two unknown forces. In the problem shown in Fig. 5.6, unknowns A_x and A_y can be eliminated if we sum moments with respect to A. Unknowns A_x and F_C can be eliminated if we sum moments with respect to C. In fact, we can write any number of force and moment equations. However, *only three of the equations can be independent.*

EXAMPLE 5.4

A force of 6500 N is applied to the beam as shown in Fig. 5.7(a). The uniform beam weighs 600 N. What are the reactions at A and B?

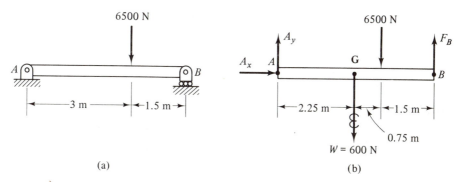

FIGURE 5.7

Solution

The solution will be outlined step by step.

1. Draw a free-body diagram. The beam is uniform; therefore, the weight of 600 N acts at the geometric center as shown. The unknown horizontal and vertical reaction of the pin at A is shown as A_x and A_y. The unknown force F_B is perpendicular to the surface on which the roller moves.

2. Summing moments about A, we have

$$\Sigma M_A = -600(2.25) - 6500(3) + F_B(4.5) = 0$$
$$= -1350 - 19\ 500 + 4.5\,F_B = 0$$

or

$$4.5\,F_B = 20\ 850 \qquad F_B = 4630\ \text{N} \qquad\qquad \text{Answer}$$

3. Summing moments about B, we have

$$\Sigma M_B = -A_y(4.5) + 600(2.25) + 6500(1.5) = 0$$
$$= -4.5\,A_y + 1350 + 9750 = 0$$

or

$$4.5\,A_y = 11\ 100 \qquad A_y = 2470\ \text{N} \qquad\qquad \text{Answer}$$

4. Summing forces in the x direction, we have

$$\to \Sigma F_x = A_x = 0 \qquad A_x = 0 \qquad\qquad \text{Answer}$$

5. We *check* our answers by using one additional equilibrium equation. Summing forces in the y direction, we have

$$\uparrow \Sigma F_y = A_y - 600 - 6500 + F_B = 0$$

or

$$2470 - 600 - 6500 + 4630 = 0 \qquad \text{OK}$$

EXAMPLE 5.5

For the beam in Example 5.3, find the reactions at A and B.

Solution

1. Draw the free-body diagram. [The free-body diagram was drawn in Fig. 5.6(b).]

2. We replace the unknown reaction F_C by its x and y components [Fig. 5.6(c)].

$$F_C \cos 70° = 0.342\,F_C \quad \text{and} \quad F_C \sin 70° = 0.940\,F_C$$

3. Summing moments about A, we have

$$\Sigma M_A = -2400(4) - 1800(12) + 0.940\,F_C(20) = 0$$
$$= -9600 - 21{,}600 + 18.80\,F_C = 0$$

or

$$18.80 F_C = 31,200 \qquad F_C = 1660 \text{ lb} \qquad\qquad \text{Answer}$$

4. Summing moments about C, we have

$$\circlearrowright \Sigma\, M_C = -A_y(20) + 2400(16) + 1800(8) = 0$$

$$= -20 A_y + 38,400 + 14,400 = 0$$

$$20 A_y = 52,800$$

$$A_y = 2640 \text{ lb} \qquad\qquad \text{Answer}$$

5. Summing forces in the x direction, we have

$$\rightarrow \Sigma\, F_x = A_x - 0.342 F_C = 0$$

Substituting for F_C yields

$$A_x = 0.342 F_C = 0.342(1660)$$

$$= 568 \text{ lb} \qquad\qquad \text{Answer}$$

6. We *check* our answer by using one additional equilibrium equation. Summing forces in the y direction, we have

$$\uparrow \Sigma\, F_y = A_y - 2400 - 1800 + 0.940 F_C$$

or

$$2640 - 2400 - 1800 + 1560 = 0 \qquad \text{OK}$$

EXAMPLE 5.6

Determine the reaction at the fixed support A for the loaded bent [Fig. 5.8(a)].

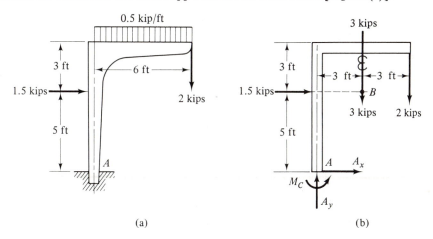

(a)   (b)

FIGURE 5.8

Solution

1. The free-body diagram is shown in Fig. 5.8(b). The uniformly distributed load was replaced by its resultant. At the fixed support A we have three unknowns, the two forces A_x and A_y and the moment couple M_C.

2. Summing moments about A, we have

$$\Sigma\, M_A = M_C - 1.5\,(5) - 3\,(3) - 2\,(6) = 0$$

$$= M_C - 7.5 - 9 - 12 = 0$$

$$M_C = 28.5 \text{ kip-ft} \qquad\qquad\qquad\qquad\qquad \text{Answer}$$

3. Summing the forces in the x direction,

$$\rightarrow \Sigma\, F_x = 1.5 + A_x = 0$$

$$A_x = -1.5 \text{ kips}$$

$$= 1.5 \text{ kips} \leftarrow \qquad\qquad\qquad\qquad\qquad \text{Answer}$$

4. Summing forces in the y direction,

$$\uparrow \Sigma\, F_y = A_y - 3 - 2 = 0$$

$$A_y = 5 \text{ kips}$$

$$= 5 \text{ kips} \uparrow \qquad\qquad\qquad\qquad\qquad \text{Answer}$$

5. We *check* our answer by using one additional equilibrium equation. Recall that the moment of a couple has the same value about any point in the plane of the couple. Summing moments about B,

$$\Sigma\, M_B = A_x\,(5) + M_C - 2\,(3) - A_y\,(3) = 0$$

$$= -1.5\,(5) + 28.5 - 6 - 5\,(3)$$

$$= -7.5 + 28.5 - 6 - 15 = 0 \qquad \text{OK}$$

PROBLEMS

In Probs. 5.1 through 5.30, draw the appropriate free-body diagram for each problem. Neglect the weight of the member unless a weight is given.

5.1 Determine the reactions at A and B for the simply supported beam shown.

5.2 The cantilever beam supports two loads as shown. Determine the reactions at A.

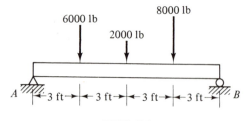

PROB. 5.1

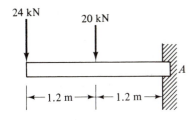

PROB. 5.2

5.3 The total weight for the uniform beam *ABC* is 200 lb. Determine the reactions at *A* and the tension in the cable *BD*.

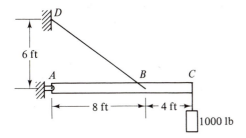

PROB. 5.3

5.4 The simply supported beam is loaded as shown. Determine the reactions at *A* and *B*.

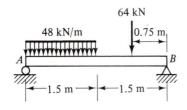

PROB. 5.4

5.5 Determine the reactions at *A* and the tension or compression in member *IJ* for the pin-connected tower shown.

5.6 The T-shaped member is supported as shown. Determine the reactions at *A* and the tension or compression in link *BC*.

5.7 Determine the reactions at *A* and *B* for the beam with the overhang as shown.

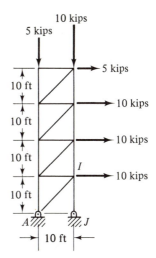

PROB. 5.5

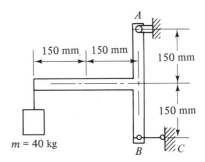

PROB. 5.6

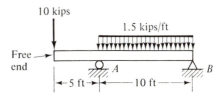

PROB. 5.7

5.8 The cantilever beam supports two concentrated loads and a couple as shown on page 102. Determine the reactions at *A*.

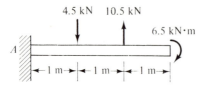

4.5 kN   10.5 kN

6.5 kN·m

A

|← 1 m →|← 1 m →|← 1 m →|

PROB. 5.8

5.9 The simply supported beam supports loads as shown. Determine the reactions at *A* and *B*.

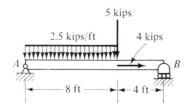

5 kips

2.5 kips/ft   4 kips

A   B

|← 8 ft →|← 4 ft →|

PROB. 5.9

5.10 The cantilever beam supports a uniform load as shown. Find the reactions at *B*.

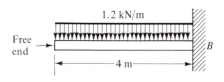

1.2 kN/m

Free
end

B

|← 4 m →|

PROB. 5.10

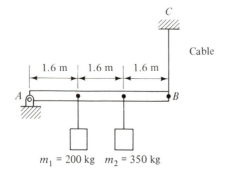

C

Cable

1.6 m | 1.6 m | 1.6 m

A   B

$m_1 = 200$ kg $m_2 = 350$ kg

PROB. 5.11

5.11 The beam is supported by a pin and cable as shown. Find the reactions at *A* and the tension in the cable *BC*.

5.12 The pin-connected tower supports a horizontal load as shown. Determine the reactions at *A* and the tension or compression in member *HJ*.

5.13 The cantilever truss supports loads as shown. Determine the reactions at *A* and the tension or compression in member *FG*.

5.14 The mechanism is acted on by two torques (couples) as shown. Determine the forces exerted on the mechanism by the bolts at *A* and *B*. (*Hint:* Assume a pin support at *A* and a roller at *B*.)

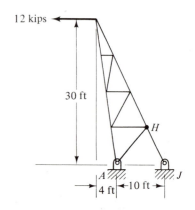

12 kips

30 ft

H

A   J
|4 ft|← 10 ft →|

PROB. 5.12

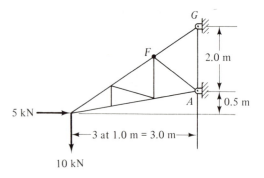

G

F   2.0 m

A

0.5 m

5 kN

|← 3 at 1.0 m = 3.0 m →|

10 kN

PROB. 5.13

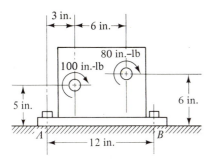

PROB. 5.14

5.15 The boom is supported by a pin and cable as shown. Find the reactions at A and the tension in the cable BC.

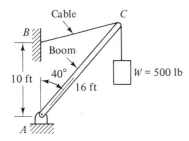

PROB. 5.15

5.16 The frame is acted on by two forces and a couple as shown. Determine the reactions at A and B.

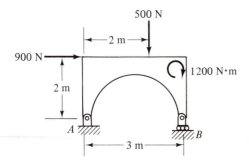

PROB. 5.16

5.17 The loaded bracket shown is supported by a pin at A and a frictionless roller at B. Find the reactions at A and B.

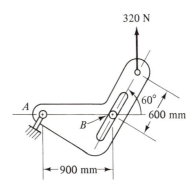

PROB. 5.17

5.18 The triangular truss supports loads as shown. Determine the reactions at E and G.

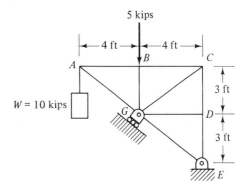

PROB. 5.18

5.19 The four-panel truss supports loads as shown. Find the reactions at A and E.

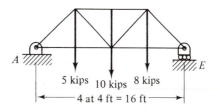

PROB. 5.19

5.20 The boom *OA* supports a load of 10 kN as shown. The uniform boom weighs 2 kN. Find the reactions at *O* and the tension *T* in the lifting cable.

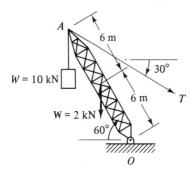

PROB. 5.20

5.21 Member *ABC* has a uniform cross section and weighs 250 lb. Determine the reactions at *C* and *B*.

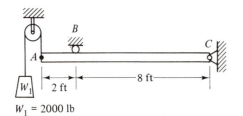

$W_1 = 2000$ lb

PROB. 5.21

5.22 Beam *ABC* has a uniform cross section and weighs 80 N. Determine (a) the reactions at *A* and the tension in the cable and (b) the normal reaction between the block and the beam. (*Hint:* Draw a free-body diagram of the beam and the block together. Solve for the tension in the cable and the reactions at *A*. With the tension known, draw a free-body diagram of the block and solve for the normal reaction.)

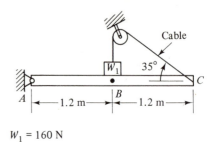

$W_1 = 160$ N

PROB. 5.22

5.23 The hook-shaped bar is acted on by a force as shown. Find the reactions at *B* and *C*.

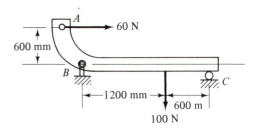

PROB. 5.23

5.24 The lever shown is supported by a pin at *A* and a frictionless roller at *B*. Determine the reactions at *A* and *B*.

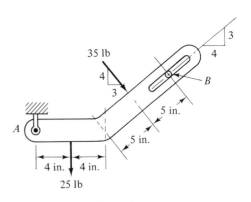

PROB. 5.24

5.25 The 400-lb block A rests on a bar as shown. The bar is uniform and weighs 300 lb. Determine (a) the reactions at B and the tension in the cable and (b) the normal reaction between the block and the bar. (*Hint:* Draw the free-body diagram of the bar and block together. Solve for the tension in the cable and the reactions at B. With the tension known, draw the free-body diagram of the block and solve for the normal reaction.)

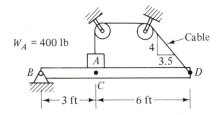

PROB. 5.25

5.26 The boom is supported by a cable and pin as shown. Neglect the size of the pulley at D. Find the reactions at C and the tension in the cable.

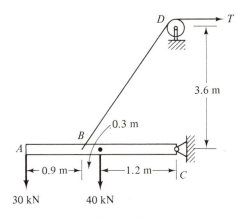

PROB. 5.26

5.27 The truss supports a sign board as shown. Calculate the reactions at A and the force in member FG produced by the horizontal wind load of 60 lb per vertical foot of sign.

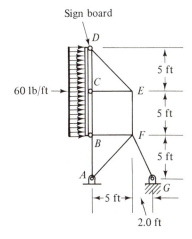

PROB. 5.27

5.28 A loaded bracket is supported by a pin at A and a frictionless roller at B. The bracket is in equilibrium. Determine the reactions at A and B.

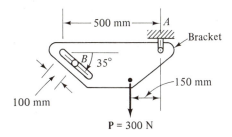

PROB. 5.28

5.29 The truss supports loads as shown on page 106. Determine the reactions at E and the force in link AF.

5.30 A uniform member ABC weighing 500 lb is being lifted by a cable as shown on page 106. When $\theta = 55°$, find the tension in the lifting cable and the tension in the anchor cable. Neglect the size of the pulley at D.

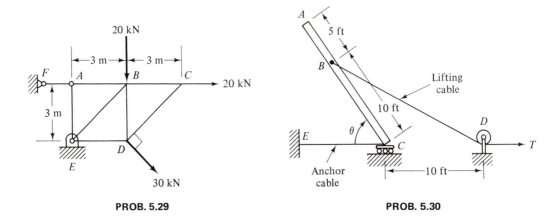

PROB. 5.29   PROB. 5.30

5.5 EQUILIBRIUM OF A TWO-FORCE BODY

A body acted on by two forces is called a two-force body. For equilibrium of a two-force body the force acting at one point must be *equal in magnitude*, *opposite in direction*, and *have the same line of action* as the force acting at the other point. For proof, consider the two-force body acted on by forces **P** and **Q** which are directed at angles θ_A and θ_B as shown in Fig. 5.9(a). We replace **P** and **Q** by components parallel and perpendicular to the line joining A and B [Fig. 5.9(b)]. Since the body is in equilibrium, moments about any point and the resultant in any direction must vanish. Summing moments about A and then about B, $Q_y = 0$ and $P_y = 0$. Sum-

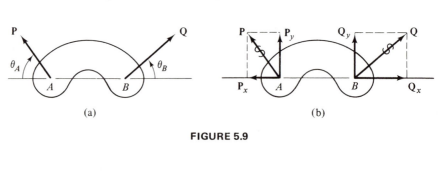

FIGURE 5.9

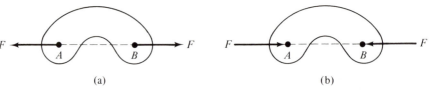

FIGURE 5.10

ming forces in the direction of the line AB, $P_x = Q_x$. Therefore, only two possibilities exist for a two-force body, as shown in Fig. 5.10(a) and (b). The forces must be equal, opposite, and have the same line of action.

The *link* is a special case of a two-force body.

5.6 EQUILIBRIUM OF A THREE-FORCE BODY

A body acted on by three forces is called a *three-force body*. For equilibrium, the line of action of the three forces must be either *concurrent* or *parallel*.

Consider a body in equilibrium acted on by three nonparallel forces P, Q, and S [Fig. 5.11(a)]. The line of action of two of the forces must intersect at some point B. Since the three forces are in equilibrium, the sum of their moments about B must be equal to zero. Therefore, the line of action of P must also act through B as in Fig. 15.11(b). Thus the line of action of all three forces P, Q, and S act through B; they are therefore concurrent.

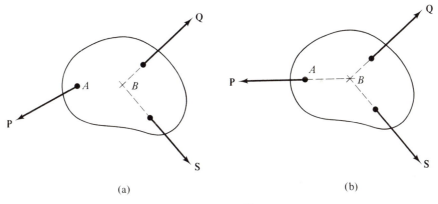

(a)   (b)

FIGURE 5.11

The three-force body problem can be solved by the triangular method, which is based on the fact that the three forces are concurrent (Sec. 3.4). The problem can also be solved by the equilibrium equations (Sec. 5.4).

EXAMPLE 5.7

A truss supports a sign board as shown in Fig. 5.12(a). Calculate the reactions at A and the force in member FG produced by the horizontal wind load of 0.8 kN per vertical meter of sign.

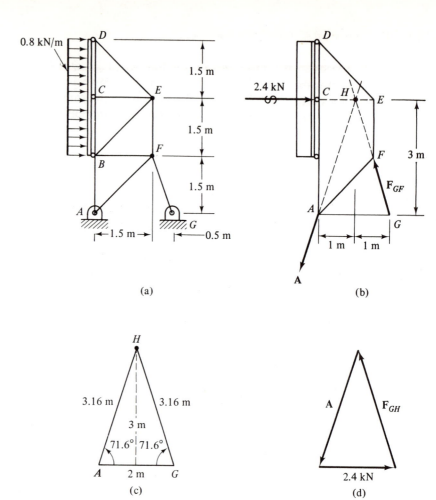

(a)

(b)

(c)

(d)

FIGURE 5.12

Solution

The truss $ADEF$ is a three-force body acted on by the resultant of the wind forces, the force along the two-force member FG and the pin reaction at A. The resultant of the wind force $R = (0.8 \text{ kN/m})(3 \text{ m}) = 2.4 \text{ kN}$ acts at the center C of the sign board. The forces are concurrent and intersect at H as shown [Fig. 5.12(b)]. Dimensions of the triangle AHG are shown in Fig. 5.12(c). The force triangle is constructed around AHG in Fig. 5.12(d). The triangles are similar; therefore, $A = F_{GH}$ and

$$\frac{A}{2.4} = \frac{AH}{AG} = \frac{3.16}{2} \quad \text{or} \quad A = 3.79 \text{ kN}$$

Thus

$$F_{GH} = 3.79 \text{ kN} \qquad A = 3.79 \text{ kN} \ \angle 71.6° \qquad \text{Answer}$$

PROBLEMS

In Probs. 5.31 through 5.40, use the method of Sec. 5.6.

5.31 Solve Prob. 5.3.

(*Hint:* First replace the 200 lb weight of beam *ABC* and the 1000 lb weight at *C* by their resultant.)

5.32 Solve Prob. 5.6.

5.33 Solve Prob. 5.12.

5.34 Solve Prob. 5.13.

5.35 Solve Prob. 5.15.

5.36 Solve Prob. 5.17.

5.37 Solve Prob. 5.22.

5.38 Solve Prob. 5.23.

5.39 Solve Prob. 5.27.

5.40 Solve Prob. 5.28.

5.7 STATICAL DETERMINACY AND CONSTRAINT OF A RIGID BODY

In this section we examine various bodies to determine if the supports provide complete constraints against motion and if the reactions at the supports can be determined from the equations of equilibrium.

The analysis is simplified if we replace the supports by a set of links or hinged bars which have actions that are *equivalent* to those of the supports. The replacement is made as follows.

A *single-reaction support* such as a smooth surface, roller, or cable is replaced by a single link. The link must have the same direction as the reaction [Fig. 5.13(a)].

A *two-reaction support*, the smooth pin or hinge, is replaced by two links at right angles to each other as in Fig. 5.13(b).

A *three-reaction support*, the fixed or clamped end, is replaced by three links as shown in Fig. 5.13(c). The horizontal force F_x and

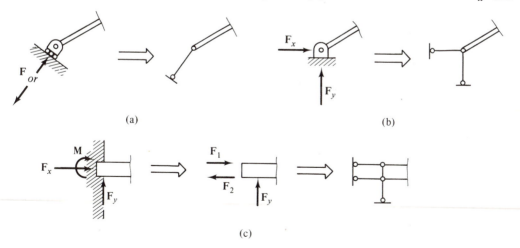

(a)

(b)

(c)

FIGURE 5.13

couple M are first replaced by two horizontal forces, F_1 and F_2. The three forces are then replaced by two horizontal links and a vertical link as shown in the figure.

We emphasize again that *the links have actions which are equivalent to those of the supports they replace*.

With the aid of linkage diagrams we consider the bodies shown in Figs. 5.14 and 5.15. We wish to determine if the supports provide complete constraints against motion and if the reactions at the supports can be found from the equations of equilibrium.

With two links at point A [Fig. 5.14(a)] the only possibility of motion is rotation of the body about A. Any arrangement of the link BF that prevents motion perpendicular to line AB will constrain the body. With the link as shown, the body is *constrained*. The three equations of equilibrium can be used to solve for the three unknown reactions. Therefore, the body is statically *determinate*. If link BF is rotated so that it lies along the line through A and B, the body can rotate and would only be *partially constrained*. The three reactions are concurrent and we have only two equations of equilibrium. Thus the body is statically *indeterminate*.

In Fig. 5.14(b) we add one additional link GC. Support of the

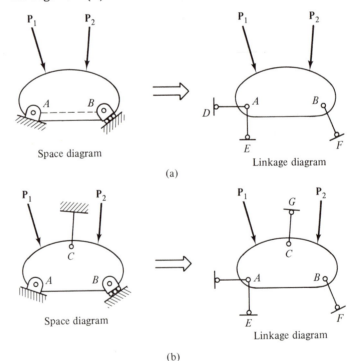

Space diagram

Linkage diagram

(a)

Space diagram

Linkage diagram

(b)

FIGURE 5.14

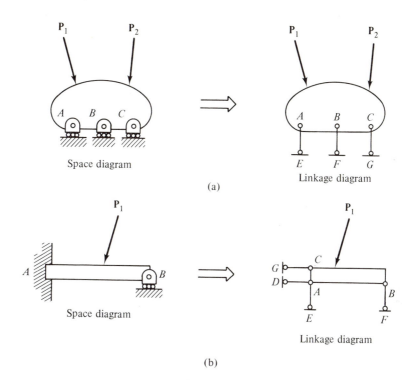

Space diagram

Linkage diagram

(a)

Space diagram

Linkage diagram

(b)

FIGURE 5.15

body exceeds that required for complete constraint and one of the supports is *redundant*. With three equilibrium equations and four unknown reactions, the body is statically *indeterminate*.

With the three parallel links as shown in Fig. 5.15(a), the body is free to move to the right. Therefore, the body is *partially constrained*. Although there are only three reactions, the body is statically *indeterminate*. This is true because one of the equilibrium equations, $\Sigma F_x = 0$, does not contain any of the unknown reactions. We are left with two equilibrium equations and three unknown reactions. If link BF is removed, the body is supported by two parallel links and the body is still *partially constrained*. However, the two reactions can be determined from the three equations of equilibrium. Thus the body is statically *determinate*.

In Fig. 5.15(b) the body is supported by four links. One of the links is *redundant* and the body is statically *indeterminate*. By removing any one of the links the body remains constrained and becomes statically *determinate*.

It follows from the preceding examples that complete constraint is provided if a body is supported by three links whose axes are neither parallel nor intersect at a common point. In such cases the three reactions can be found by the equilibrium equations and the body is statically determinate.

Force Analysis of Structures and Machines

6.1 INTRODUCTION

In Chapter 5 we were concerned mainly with a single structural member or machine element. Here we will study problems involving the force analysis of structures and machines.

A *structure* consists of a series of connected structural members or rigid bodies that are designed to *support* loads or forces. The loads may be either stationary or moving, but the structure is usually at rest. If the structure is moved, the motion is executed slowly. *Machines* are made up of several machine elements connected together. They are designed to *transmit* and *transform* loads or forces rather than support them. The machine generally has moving parts. With structures and machines we must determine not only the external forces acting on them but also the forces that hold the parts together.

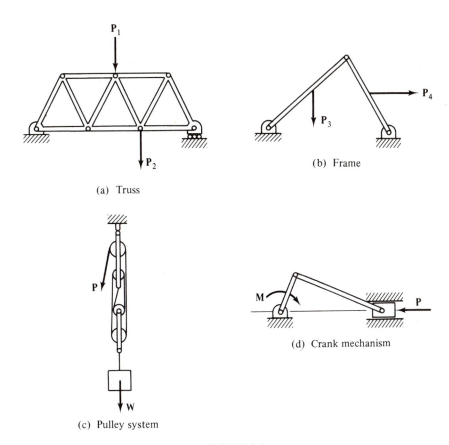

(a) Truss

(b) Frame

(c) Pulley system

(d) Crank mechanism

FIGURE 6.1

In this chapter we consider simple trusses, pin-connected frames, and machines or mechanisms that can be analyzed in a plane. Examples of each are shown in Fig. 6.1.

6.2 SIMPLE PLANE TRUSSES

A *truss* is a structure usually consisting of straight uniform bars or members arranged in a series of adjoining triangles and fastened together at their ends to form *joints*. The joints may be pinned, welded, riveted, bolted, or nailed, depending on the type of truss. They may be used for such things as bridges, buildings, and towers and support both stationary and moving loads.

To simplify our calculations, we make three assumptions about the *simple plane truss*.

1. The bars are connected at their ends to form joints that behave like frictionless pins.

2. All forces acting on the truss are applied at the joints.
3. The members, joints, and loads all lie in a plane.

The first assumption represents a substantial departure from the real truss. In fact, the ends of the members are attached to each other so

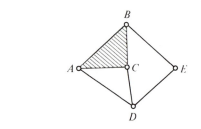

FIGURE 6.2

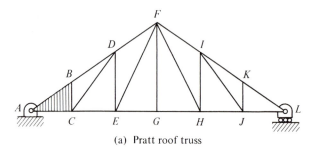

(a) Pratt roof truss

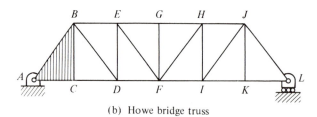

(b) Howe bridge truss

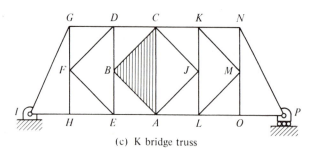

(c) K bridge truss

FIGURE 6.3

that bending of the member must occur. However, the forces in the members calculated on the basis of pin construction do not differ substantially from their true value. The second assumption is justified in most practical cases. Except for the weight of a member that acts at its center of gravity, the loads are usually transmitted by other structures or members to the joints of the truss. The weight of a member is small in comparison with other loads on the truss and can be either neglected or divided equally between the joints at its end. With careful design and fabrication of the truss, the third assumption is also justified.

The first and second assumption ensures that all bars are two-force bodies or members. Thus no bending can occur in the members. Each member must be either in tension or compression. The third assumption ensures that we have a problem in plane statics.

A *simple truss* can be constructed by the following method. As shown in Fig. 6.2, we start with the simplest rigid structure, consisting of three bars and three joints, triangle *ABC*. By adding pairs of bars *AD* and *CD*, the structure remains rigid. To the rigid structure *ABCD* we add the

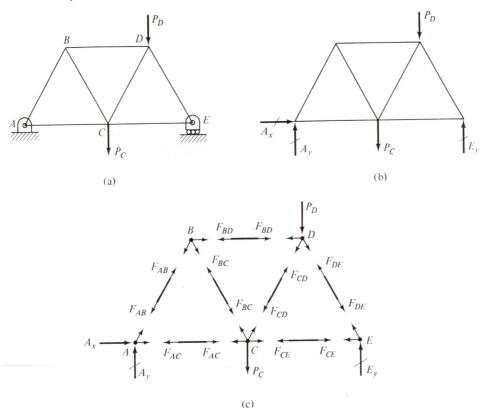

FIGURE 6.4

pair of bars *BE* and *DE*, and the structure still remains rigid. In fact, the process may be continued until the truss is complete. The term *rigid* here means that the truss functions as a whole like a rigid body.

The trusses shown in Fig. 6.3 were all constructed in this manner. The shaded triangle *ABC* has been taken as the starting point for each truss. Adding pairs of members *CD* and *BD*, we establish joint *D*. By addition of each new pair of members, we establish the remaining joints in alphabetical order. The trusses shown are all simple trusses.

Consider the Warren truss shown in Fig. 6.4(a). The pin joints are labeled *A*, *B*, *C*, *D*, and *E*. The member joining pins *A* and *B* is referred to as *AB*, and the force in the member is F_{AB}. Using these symbols, the free-body diagram for the entire truss is shown in Fig. 6.4(b) and the free-body diagram for each joint and member is shown in Fig. 6.4(c). All members of the truss are shown in tension. If, in fact, the member is in tension, the equilibrium equations will give a positive value for the force in the member. If the member is in compression, the equilibrium equations will give a negative value for the forces in the member. Notice that between each joint and member the forces are in opposite directions. For example, the force of member *AB* on joint *A* is equal in magnitude and opposite in direction to the force of joint *A* on member *AB*. This is in agreement with Newton's third law, which states that action equals reaction.

6.3 MEMBERS UNDER SPECIAL LOADING

Consider three members of a truss joined together at a joint with two of the members lying along a straight line [Fig. 6.5(a)]. Summing forces perpendicular to the straight line, we see that a component of F_{AB} is equal to zero. Therefore, F_{AC} must be zero, provided that there are no external forces applied to the joint in question. Since $F_{AC} = 0$, $F_{AB} = F_{AD}$. Member *AC* is a zero-force member.

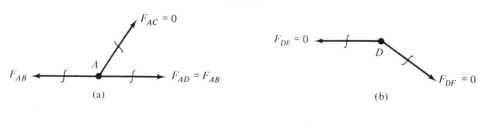

FIGURE 6.5

When two members are joined together as shown in Fig. 6.5(b), the force in each member is zero. If the members are collinear as shown in Fig. 6.5(c), the forces are equal.

6.4 METHOD OF JOINTS

The method of joints requires that a free-body diagram of each joint be drawn. Since the forces at each joint are concurrent, equilibrium requires that $\Sigma F_x = 0$ and $\Sigma F_y = 0$. Thus only two unknowns can be determined at each joint. Therefore, the method requires that we begin our analysis by drawing a free-body diagram of a joint with only two unknown forces; then proceeding from joint to joint, so that each new free-body diagram contains only two unknown forces. The method is self-checking, since the last joint will involve only known forces and they must be in equilibrium.

The method of joints will be illustrated by several examples.

EXAMPLE 6.1

Shown in Fig. 6.6(a) is a simple plane truss. Using the method of joints, find the force in each member of the truss.

Solution

The free-body diagram of the entire truss is shown in Fig. 6.6(b). The x and y components of the reaction at pin A and the reaction at roller D are shown. For equilibrium of the entire truss:

$$\circlearrowleft \Sigma M_A = D_y 9 - 5(4) - 15(6) = 0$$
$$D_y = 12.22 \text{ kN} \uparrow \qquad\qquad\qquad\qquad \text{Answer}$$

$$\circlearrowleft \Sigma M_D = 15(3) - A_y 9 - 5(4) = 0$$
$$A_y = 2.78 \text{ kN} \uparrow \qquad\qquad\qquad\qquad \text{Answer}$$

$$\rightarrow \Sigma F_x = A_x + 5 = 0$$
$$A_x = -5 \text{ kN} = 5 \text{ kN} \leftarrow \qquad\qquad\qquad\qquad \text{Answer}$$

Check:

$$\uparrow \Sigma F_y = A_y + D_y - 15 = 2.78 + 12.22 - 15 = 0 \qquad \text{OK}$$

In Fig. 6.6(c)-(h) the free-body diagrams for the joints are drawn in sequence so that each new free-body diagram contains only two unknown forces. The internal forces have all been assumed to be tensile. Starting with joint A, we write the equilibrium equation for each joint.

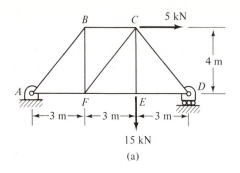

(a)

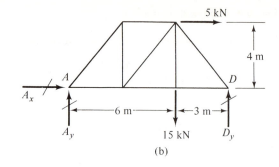

(b)

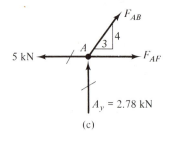

(c)

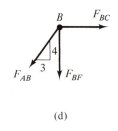

(d)

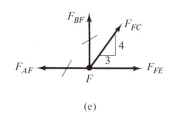

(e)

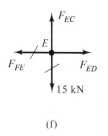

(f)

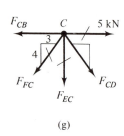

(g)

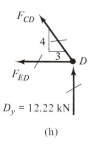

(h)

FIGURE 6.6

Joint A

$$\uparrow \Sigma\, F_y = 2.78 + 0.8F_{AB} = 0$$

$$F_{AB} = -3.48 \text{ kN} \quad \text{(compression)} \qquad \qquad \text{Answer}$$

$$\rightarrow \Sigma\, F_x = -5 + F_{AF} + 0.6F_{AB} = -5 + F_{AF} + 0.6(-3.48) = 0$$

$$F_{AF} = 7.08 \text{ kN} \quad \text{(tension)} \qquad \qquad \text{Answer}$$

Joint B

$$\rightarrow \Sigma\, F_x = F_{BC} - 0.6F_{AB} = F_{BC} - 0.6(-3.48) = 0$$

$$F_{BC} = -2.08 \text{ kN} \quad \text{(compression)} \qquad \qquad \text{Answer}$$

$$\uparrow \Sigma\ F_y = -F_{BF} - 0.8 F_{AB} = -F_{BF} - 0.8(-3.48) = 0$$

$$F_{BF} = 2.78 \text{ kN} \quad \text{(tension)} \qquad \text{Answer}$$

Joint F

$$\uparrow \Sigma\ F_y = F_{BF} + 0.8 F_{FC} = 2.78 + 0.8 F_{FC} = 0$$

$$F_{FC} = -3.48 \text{ kN} \quad \text{(compression)} \qquad \text{Answer}$$

$$\rightarrow \Sigma\ F_x = F_{FE} + 0.6 F_{FC} - F_{AF} = F_{FE} + 0.6(-3.48) + 7.08 = 0$$

$$F_{FE} = 9.17 \text{ kN} \quad \text{(tension)} \qquad \text{Answer}$$

Joint E

$$\rightarrow \Sigma\ F_x = F_{ED} - F_{FE} = F_{ED} - 9.17 = 0$$

$$F_{ED} = 9.17 \text{ kN} \quad \text{(tension)} \qquad \text{Answer}$$

$$\uparrow \Sigma\ F_y = F_{EC} - 15 = 0$$

$$F_{EC} = 15 \text{ kN} \quad \text{(tension)} \qquad \text{Answer}$$

Joint C

$$\rightarrow \Sigma\ F_x = 5 + 0.6 F_{CD} - 0.6 F_{FC} - F_{BC} = 5 + 0.6 F_{CD} - 0.6(-3.48) - (-2.08) = 0$$

$$F_{CD} = -15.28 \text{ kN} \quad \text{(compression)} \qquad \text{Answer}$$

First check:

$$\uparrow \Sigma\ F_y = -0.8 F_{FC} - F_{EC} - 0.8 F_{CD} = -0.8(-3.48) - F_{EC} - 0.8(-15.28) = 0$$

$$F_{EC} = 15.00 \text{ kN} \quad \text{(tension)} \quad \text{OK}$$

This agrees with the value calculated from joint E.

Joint D

Second check:

$$\rightarrow \Sigma\ F_x = -F_{ED} - 0.6 F_{CD} = -9.17 - 0.6(-15.28) = 0 \quad \text{OK}$$

This checks forces F_{ED} and F_{CD}.

Third check:

$$\uparrow \Sigma\ F_y = 0.8 F_{CD} + 12.22 = 0.8(-15.28) + 12.22 = 0 \quad \text{OK}$$

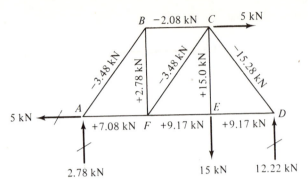

FIGURE 6.7

This provides a further check of force F_{CD}. The answers are shown on the truss in Fig. 6.7. Positive values indicate tension and negative values indicate compression.

EXAMPLE 6.2

Using the method of joints, find the force in each member of the truss shown in Fig. 6.8(a).

Solution

The member DF is a two-force member; therefore, the reaction at D is directed along the member. Members EG and EC lie along a straight line; therefore, the force in member ED is zero and the forces in EG and EC are equal. The free-body diagram of the entire truss is shown in Fig. 6.8(b). For equilibrium of the *entire truss*:

$$\circlearrowright \Sigma M_G = D_x(6.93) - 5(8) - 10(12) = 0$$
$$D_x = 23.1 \text{ kips} \rightarrow \qquad \text{Answer}$$

$$\rightarrow \Sigma F_x = -G_x + D_x = 0$$
$$G_x = 23.1 \text{ kips} \leftarrow \qquad \text{Answer}$$

$$\uparrow \Sigma F_y = G_y - 5 - 10 = 0$$
$$G_y = 15 \text{ kips} \uparrow \qquad \text{Answer}$$

Check:

$$\circlearrowright \Sigma M_D = G_x(6.93) - G_y(4) - 5(4) - 10(8) = 160 - 60 - 20 - 80 = 0 \qquad \text{OK}$$

In Fig. 6.8(c)–(g) the free-body diagrams of the joints are drawn in sequence so that each new free-body diagram contains only two unknown forces. Starting with joint A, we write the equilibrium equations for each joint.

Joint A

$$\uparrow \Sigma F_y = F_{AC} \sin 60° - 10 = 0$$
$$F_{AC} = 11.55 \text{ kips} \quad \text{(tension)} \qquad \text{Answer}$$

120

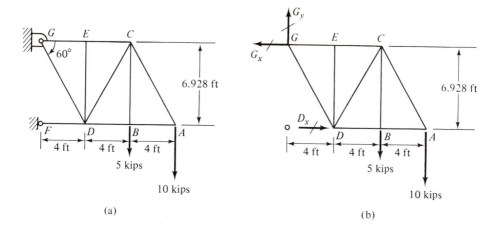

(a)   (b)

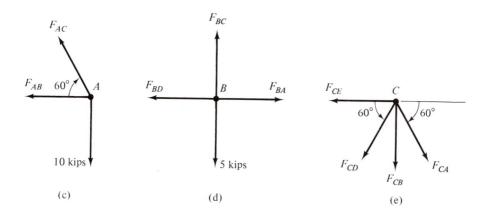

(c)   (d)   (e)

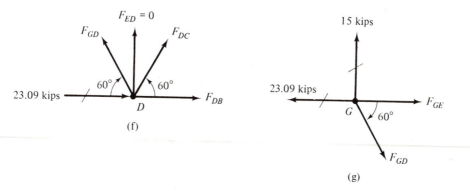

(f)   (g)

FIGURE 6.8

121

$$\rightarrow \Sigma \ F_x = -F_{AB} - F_{AC} \cos 60° = -F_{AB} - 11.55 \cos 60° = 0$$

$$F_{AB} = -5.78 \text{ kips} \quad \text{(compression)} \qquad \text{Answer}$$

Joint B

$$\rightarrow \Sigma \ F_x = F_{BA} - F_{BD} = -5.78 - F_{BD} = 0$$

$$F_{BD} = -5.78 \text{ kips} \quad \text{(compression)} \qquad \text{Answer}$$

$$\uparrow \Sigma \ F_y = F_{BC} - 5 = 0$$

$$F_{BC} = 5 \text{ kips} \quad \text{(tension)} \qquad \text{Answer}$$

Joint C

$$\uparrow \Sigma \ F_y = -F_{CD} \sin 60° - F_{CB} - F_{CA} \sin 60°$$

$$= -11.55 \sin 60° - 5 - F_{CD} \sin 60° = 0$$

$$F_{CD} = -17.32 \text{ kips} \quad \text{(compression)} \qquad \text{Answer}$$

$$\rightarrow \Sigma \ F_x = F_{CA} \cos 60° - F_{CD} \cos 60° - F_{CE}$$

$$= 11.55 \cos 60° - (-17.32) \cos 60° - F_{CE} = 0$$

$$F_{CE} = 14.44 \text{ kips} \quad \text{(tension)} \qquad \text{Answer}$$

Joint E

Member *ED* is a zero-force member and members *CE* and *EG* are collinear, therefore

$$F_{ED} = 0 \quad \text{and} \quad F_{CE} = F_{EG} = 14.44 \text{ kips} \quad \text{(tension)} \qquad \text{Answer}$$

Joint D

$$\rightarrow \Sigma \ F_x = F_{DB} + F_{DC} \cos 60° + 23.09 - F_{GD} \cos 60°$$

$$= -(5.78) + (-17.32) \cos 60° + 23.09 - F_{GD} \cos 60° = 0$$

$$F_{GD} = 17.32 \text{ kips} \quad \text{(tension)} \qquad \text{Answer}$$

First check:

$$\uparrow \Sigma \ F_y = F_{DC} \sin 60° + F_{GD} \sin 60° = 0$$

$$F_{DC} = -F_{GD} = -17.32 \text{ kips} \qquad \text{OK}$$

Joint G

Second check:

$$\uparrow \Sigma \ F_y = 15 - F_{GD} \sin 60° = 15 - 17.32 \sin 60° = 0 \qquad \text{OK}$$

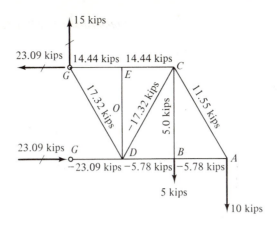

15 kips

23.09 kips 14.44 kips 14.44 kips

G   *E*   *C*

17.32 kips

O   −17.32 kips

5.0 kips

11.55 kips

23.09 kips *G*   *D*   *B*   *A*

−23.09 kips −5.78 kips −5.78 kips

5 kips

10 kips

FIGURE 6.9

Third check:

$$\rightarrow \Sigma\ F_x = F_{GE} + F_{GD} \cos 60° - 23.09 = 14.44 + 17.32 \cos 60° - 23.09 = 0 \quad \text{OK}$$

Answers are shown on the truss in Fig. 6.9.

PROBLEMS

6.1 through 6.14 Using the method of joints, determine the force in each member of the truss shown. Indicate whether the member is in tension or compression.

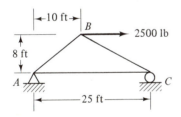

PROB. 6.1

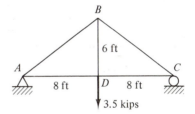

PROB. 6.3

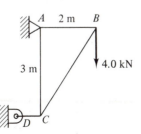

PROB. 6.2

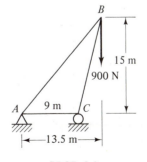

PROB. 6.4

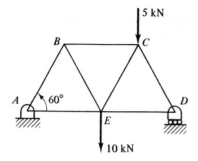

PROB. 6.5

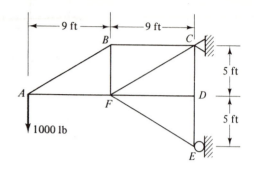

PROB. 6.8

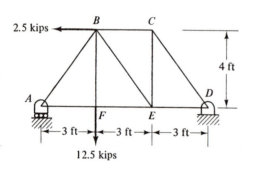

PROB. 6.6

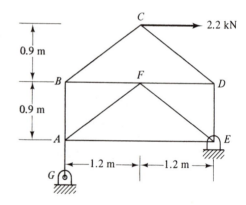

PROB. 6.9

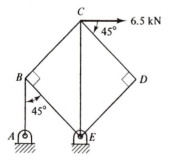

PROB. 6.7

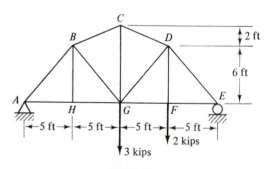

PROB. 6.10

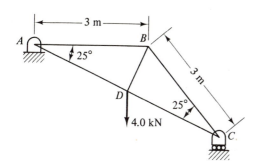

PROB. 6.11

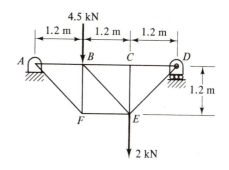

PROB. 6.13

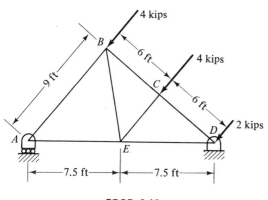

PROB. 6.12

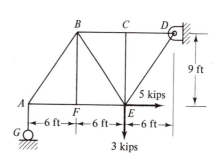

PROB. 6.14

6.5 METHOD OF SECTIONS

A second method for finding the forces in the members of a truss consist of cutting a section through the truss and drawing a free-body diagram of

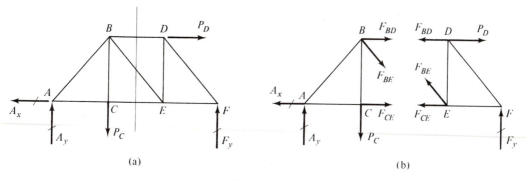

(a)   (b)

FIGURE 6.10

either part of the truss. The forces in the members cut become external forces. The forces are nonconcurrent. Therefore, the three independent equations of equilibrium can be used to solve for no more than three unknown member forces. In Fig. 6.10(a) we show a section cutting through the truss and in Fig. 6.10(b) we show free-body diagrams of both parts of the truss. Forces in the members are shown in tension.

The method will be used to solve several examples.

EXAMPLE 6.3

Use the method of sections to find the forces in members BC, BG, HG, and DG for the truss shown in Fig. 6.11(a).

Solution

The slope of the diagonal members of the truss is calculated in Fig. 6.11(c). The free-body diagram of the entire truss is shown in Fig. 6.11(b). For equilibrium of the *entire truss*:

$$\text{)} \Sigma M_A = E_y 8 - 24(4) - 8(3) = 0$$

$$E_y = 15.0 \text{ kN} \uparrow \qquad\qquad \text{Answer}$$

$$\text{)} \Sigma M_E = -A_y 8 - 8(3) + 24(4) = 0$$

$$A_y = 9.0 \text{ kN} \uparrow \qquad\qquad \text{Answer}$$

$$\rightarrow \Sigma F_x = -A_x + 8 = 0$$

$$A_x = 8.0 \text{ kN} \leftarrow \qquad\qquad \text{Answer}$$

Check:

$$\uparrow \Sigma F_y = A_y + E_y - 24 = 9 + 15 - 24 = 0$$

The free-body diagram for part of the truss to the left of section 1–1 is shown in Fig. 6.11(d). The x and y components of the force F_{BG} are shown on the diagram. For equilibrium of the part of the truss to the left of section 1–1:

$$\text{)} \Sigma M_B = F_{HG} 3 - 8(3) - 9(2) = 0$$

$$F_{HG} = 14.0 \text{ kN} \quad \text{(tension)} \qquad\qquad \text{Answer}$$

$$\text{)} \Sigma M_G = -F_{BC} 3 - 9(4) = 0$$

$$F_{BC} = -12.0 \text{ kN} \quad \text{(compression)} \qquad\qquad \text{Answer}$$

$$\uparrow \Sigma F_y = 9.0 - 0.832 F_{BG} = 0$$

$$F_{BG} = 10.82 \text{ kN} \quad \text{(tension)} \qquad\qquad \text{Answer}$$

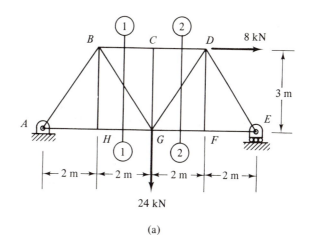

(a)

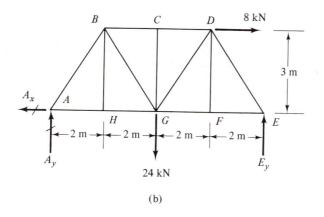

(b)

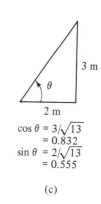

$\cos \theta = 3/\sqrt{13}$
$\quad = 0.832$
$\sin \theta = 2/\sqrt{13}$
$\quad = 0.555$

(c)

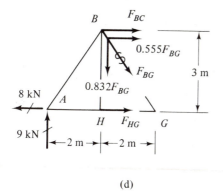

(d)

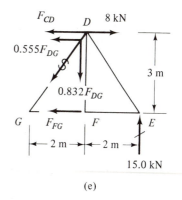

(e)

FIGURE 6.11

Check:

$$\rightarrow \Sigma\ F_x = F_{HG} + 0.555 F_{BG} + F_{BC} - 8$$
$$= 14.0 + 0.555\,(10.82) + (-12) - 8 = 0 \quad \text{OK}$$

The free-body diagram for part of the truss to the right of section 2–2 is shown in Fig. 6.11(e). The x and y components of the force F_{DG} are shown on the diagram. For equilibrium of the part of the truss to the right of section 2–2:

$$\uparrow \Sigma\ F_y = 15 - 0.832\ F_{DG} = 0$$
$$F_{DG} = 18.03 \text{ kN} \quad \text{(tension)} \qquad\qquad \text{Answer}$$

EXAMPLE 6.4

Use the method of sections to find the force in members BC, BG, and HG of the truss shown in Fig. 6.12(a).

Solution

The free-body diagram of the entire truss is shown in Fig. 6.12(b). The slope of the diagonal members is calculated in Fig. 6.12(c). For equilibrium of the *entire truss:*

$$\Sigma\ M_A = E_y\,16 - 2\,(4) - 4\,(8) - 4\,(12) = 0$$
$$E_y = 5.5 \text{ kips} \uparrow \qquad\qquad \text{Answer}$$

$$\Sigma\ M_E = -A_y\,16 + 2\,(12) + 4\,(8) + 4\,(4) = 0$$
$$A_y = 4.5 \text{ kips} \uparrow \qquad\qquad \text{Answer}$$

Check:

$$\uparrow \Sigma\ F_y = A_y + E_y - 2 - 4 - 4 = 5.5 + 4.5 - 2 - 4 - 4 = 0 \quad \text{OK}$$

The truss to the left of section 1–1 is shown in Fig. 6.12(d). For equilibrium of the part of the truss to the left of section 1–1:

$$\Sigma\ M_A = -0.658 F_{BG}\,(4.0) - 0.753 F_{BG}\,(3.5) - 2\,(4) = 0$$
$$F_{BG} = -1.52 \text{ kips} \quad \text{(compression)} \qquad\qquad \text{Answer}$$

$$\Sigma\ M_B = -4.5\,(4) + F_{HG}\,3.5 = 0$$
$$F_{HG} = 5.14 \text{ kips} \quad \text{(tension)} \qquad\qquad \text{Answer}$$

$$\Sigma\ M_G = -4.5\,(8) - 0.658 F_{BC}\,(4) - 0.753 F_{BC}\,(3.5) + 2\,(4) = 0$$
$$F_{BC} = -5.32 \text{ kips} \quad \text{(compression)} \qquad\qquad \text{Answer}$$

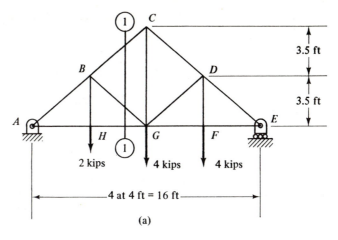

(a)

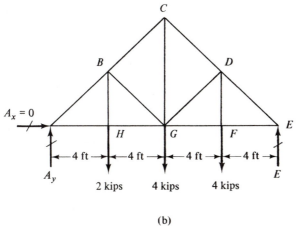

(b)

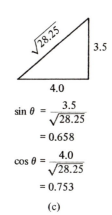

$$\sin \theta = \frac{3.5}{\sqrt{28.25}}$$
$$= 0.658$$
$$\cos \theta = \frac{4.0}{\sqrt{28.25}}$$
$$= 0.753$$

(c)

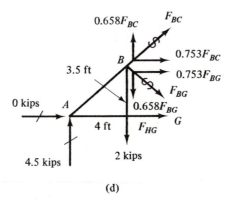

(d)

FIGURE 6.12

Check:

$$\rightarrow \Sigma \, F_x = F_{HG} + 0.753 F_{BC} + 0.753 F_{BG}$$

$$= 5.14 + 0.753(-5.32) + 0.753(-1.52) = 0 \quad \text{OK}$$

PROBLEMS

6.15 For the truss shown, determine by the method of sections the force in members (a) *CD*, *CF*, and *GF*, and (b) *BC*, *BG*, and *AG*.

6.17 Use the figure for Prob. 6.6. Determine the force in members *BC*, *BE*, and *FE* of the truss shown. Use the method of sections.

6.18 Using the method of sections, determine the force in members (a) *AB*, *AF*, and *GF*, and (b) *BC*, *BE*, and *FE* of the truss shown.

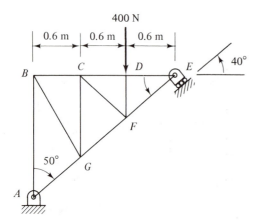

PROB. 6.15

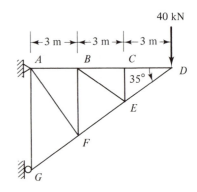

PROB. 6.18

6.16 For the truss shown, determine by the method of sections the force in members *BC*, *BF*, and *AF*.

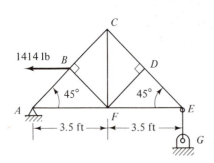

PROB. 6.16

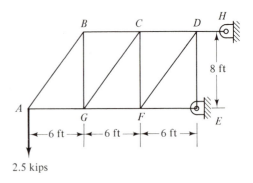

PROB. 6.20

6.19 Use the figure for Prob. 6.10. Determine the force in members BC, BG, and HG of the truss shown. Use the method of sections.

6.20 For the truss shown, determine by the method of sections the force in members (a) BC, GC, and GF, and (b) CD, FD, and FE.

6.21 Use the figure for Prob. 6.13. Determine the force in members BC, BE, and EF of the truss shown. Use the method of sections.

6.22 Use the figure for Prob. 6.14. Determine the force in members BC, BE, and EF of the truss shown. Use the method of sections.

6.23 through 6.26 Determine the zero-force members in the trusses loaded as shown.

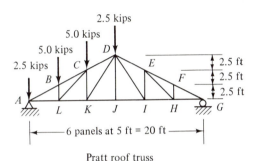

Pratt roof truss

PROB. 6.23

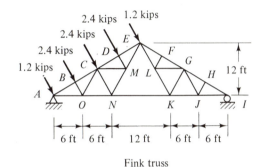

Fink truss

PROB. 6.25

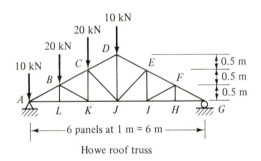

Howe roof truss

PROB. 6.24

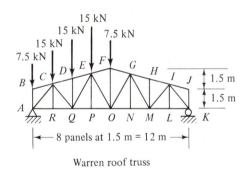

Warren roof truss

PROB. 6.26

6.6 FRAMES AND MACHINES

A *pin-connected frame* differs from a truss in that the members are no longer connected at their ends and the loads may be applied at any point to the structure. At least one of the members must be acted on by three or more forces. Thus the forces at the joints that hold the frame together must have x and y components.

A machine differs from a pin-connected frame in that it is designed to transmit and transform *input forces* into *output forces* rather than support loads. Frames are rigid structures while machines are non-rigid structures. In our study of machines, we will be concerned with the relationship between input and output forces necessary for equilibrium. The same methods will be used for the solution of frames and machines.

Consider the frame illustrated in Fig. 6.13(a), which supports a load W. The frame consists of three members joined together by frictionless pins. It is supported by a pin at A and a roller at D.

A free-body diagram of the entire frame is shown in Fig. 6.13(b). The external loads and reactions appear on the frame. They consist of the weight W at F, A_x and A_y, the horizontal and vertical components of the

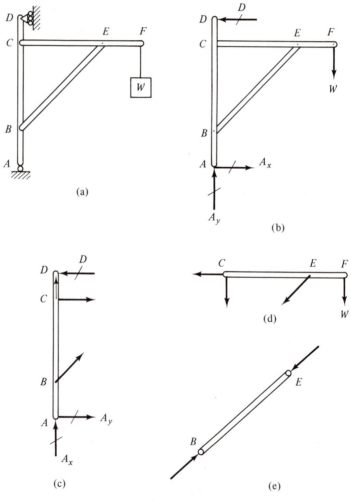

FIGURE 6.13

pin reaction at A, and the reaction of the roller at D. The internal reactions holding the frame together are not shown on the free-body diagram. To determine the forces that hold the frame together, we separate the members of the frame and draw free-body diagrams for each member. The forces at the hinges are now external and must be shown on the free-body diagrams.

The free-body diagrams for the three parts of the frame are shown in Fig. 6.13(c)-(e). In accordance with Newton's third law, the forces at C on member AD are in opposite directions from the forces on member CF. Similarly, the force at B on member AD is opposite the force on member BE, and the force at E on member CF is opposite the force on member BE. Member BE is a short link—a special case of a two-force member—and the force in the member acts along the member. If the three free-body diagrams for the parts of the frame are combined, the forces at B, C, and E cancel and we have a free-body diagram for the entire frame.

The following example problems illustrate the step-by-step procedure required to solve frame problems.

EXAMPLE 6.5

The frame shown in Fig. 6.14(a) is supported by a pin at D and a link AB. Neglect the weight of the member. Find the components of all the forces acting on each member of the frame.

Solution

Entire Structure

A free-body diagram of the entire structure is drawn in Fig. 6.14(b). The x and y components of the 4-kN force are shown on the diagram. Member AB is a two-force member. From equilibrium:

$$\circlearrowright \Sigma M_D = 2(4) + 3.46(2) - B_x 6 = 0$$

$$B_x = 2.49 \text{ kN} \qquad \text{Answer}$$

$$\circlearrowright \Sigma M_B = -2(2) + 3.46(2) + D_x 6 = 0$$

$$D_x = -0.488 \text{ kN} = -0.49 \text{ kN} \qquad \text{Answer}$$

$$\uparrow \Sigma F_y = D_y - 3.46 = 0$$

$$D_y = 3.46 \text{ kN} \qquad \text{Answer}$$

Check:

$$\rightarrow \Sigma F_x = D_x + B_x - 2$$

$$= 2.49 + (-0.49) - 2 = 0 \qquad \text{OK}$$

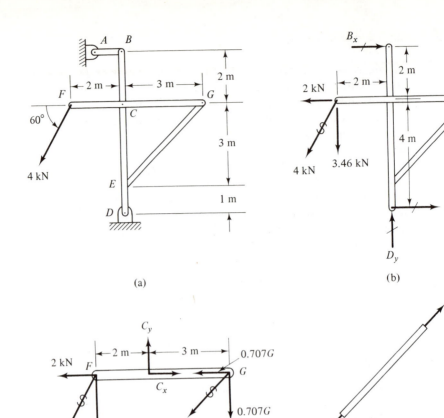

(a)

(b)

(c)

(d)

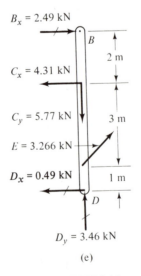

(e)

FIGURE 6.14

Member EG

Since member *EG* is a two-force member, we draw a free-body diagram and show the member in tension [Fig. 6.14(d)]. Therefore,

$$G = E$$

Member FG

The free-body diagram of member *FG* [Fig. 6.14(c)] shows the x and y components of the reaction at C and the reaction G is directed at $45°$. Components of G are shown on the diagram. From equilibrium:

$$\circlearrowright \Sigma \; M_C = 3.46(2) - 0.707 G(3) = 0$$
$$G = 3.26 \text{ kN} \qquad\qquad \text{Answer}$$

$$\circlearrowright \Sigma \; M_G = 3.46(5) - C_y 3 = 0$$
$$C_y = 5.77 \text{ kN} \qquad\qquad \text{Answer}$$

$$\rightarrow \Sigma \; F_x = 2 - C_x + 0.707 G = 0$$
$$C_x = 2 + 0.707(3.26) = 4.31 \text{ kN} \qquad\qquad \text{Answer}$$

Check:

$$\uparrow \Sigma \; F_y = -3.46 + C_y - 0.707 G$$
$$= -3.46 + 5.77 - 0.707(3.26) = 0 \quad \text{OK}$$

Member BD

The free-body diagram is shown in Fig. 6.14(e).

Check:

$$\rightarrow \Sigma \; F_x = B_x - C_x + 0.707 E + D_x$$
$$= 2.49 - 4.31 + 0.707(3.26) + (-0.49) = 0 \quad \text{OK}$$

$$\uparrow \Sigma \; F_y = -C_y + 0.707 E + D_y$$
$$= -5.77 + 0.707(3.26) + 3.46 = 0 \quad \text{OK}$$

The pin reactions on member *BD* are:

$$B_x = 2.49 \text{ kN} \rightarrow$$
$$C_x = 4.31 \text{ kN} \leftarrow$$
$$C_y = 5.77 \text{ kN} \downarrow$$
$$D_x = 0.49 \text{ kN} \leftarrow$$
$$D_y = 3.46 \text{ kN} \uparrow$$

and

$$E = 3.27 \text{ kN} \angle 45°$$

EXAMPLE 6.6

The frame shown in Fig. 6.15(a) is supported by a pin at A and rollers at E. Find the components of all the forces acting on each member of the frame.

Solution

We draw free-body diagrams for the entire structure and for each member of the structure. They are shown in Fig. 6.15(b)–(e).

Entire Structure

For equilibrium:

$$\circlearrowright \Sigma M_A = -1000(9) - 600(3) + E_y 9 = 0$$
$$E_y = 1200 \text{ lb} \hspace{3cm} \text{Answer}$$

$$\rightarrow \Sigma F_x = 1000 - A_x = 0$$
$$A_x = 1000 \text{ lb} \hspace{3cm} \text{Answer}$$

$$\uparrow \Sigma F_y = A_y - 600 + E_y = A_y - 600 + 1200 = 0$$
$$A_y = -600 \text{ lb} \hspace{3cm} \text{Answer}$$

Member AC

$$\circlearrowright \Sigma M_B = -1000(4) - C_x(8) - A_x(5) = 0$$

Substituting the value of $A_x = 1000$ lb yields

$$C_x = -1125 \text{ lb} \hspace{3cm} \text{Answer}$$

$$\circlearrowright \Sigma M_C = 1000(4) + B_x(8) - A_x(13) = 0$$
$$B_x = 1125 \text{ lb} \hspace{3cm} \text{Answer}$$

$$\uparrow \Sigma F_y = A_y - B_y - C_y = -600 - B_y - C_y = 0$$
$$B_y + C_y = -600 \hspace{3cm} \text{(a)}$$

Check:

$$\rightarrow \Sigma F_x = C_x + 1000 + B_x - A_x$$
$$= -1125 + 1000 + 1125 - 1000 = 0 \hspace{1cm} \text{OK}$$

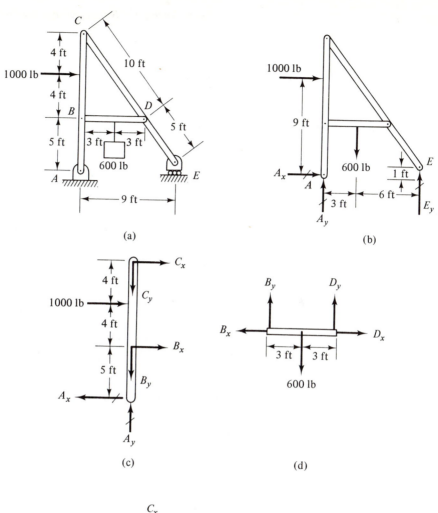

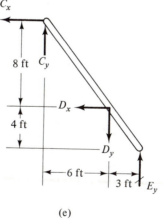

(e)

FIGURE 6.15

Member BD

$$\text{)} \, \Sigma \, M_D = -B_y(6) + 600(3) = 0$$

$$B_y = 300 \text{ lb} \qquad\qquad\qquad \text{Answer}$$

Substituting into Eq. (a), we obtain

$$C_y = -900 \text{ lb} \qquad\qquad\qquad \text{Answer}$$

$$\uparrow \Sigma \, F_y = B_y + D_y - 600 = 300 + D_y - 600 = 0$$

$$D_y = 300 \text{ lb} \qquad\qquad\qquad \text{Answer}$$

$$\rightarrow \Sigma \, F_x = -B_x + D_x = -1125 + D_x = 0$$

$$D_x = 1125 \text{ lb} \qquad\qquad\qquad \text{Answer}$$

Member CE

Check:

$$\rightarrow \Sigma \, F_x = -C_x - D_x = -(-1125) - 1125 = 0 \quad \text{OK}$$

and

$$\uparrow \Sigma \, F_y = E_y - D_y + C_y = 1200 - 300 + (-900) = 0 \quad \text{OK}$$

The pin reactions on member *AC* are:

$$A_x = 1000 \text{ lb} \leftarrow \qquad B_y = 300 \text{ lb} \downarrow$$
$$A_y = 600 \text{ lb} \downarrow \qquad C_x = 1125 \text{ lb} \rightarrow$$
$$B_x = 1125 \text{ lb} \rightarrow \qquad C_y = 900 \text{ lb} \uparrow$$

The pin reactions on member *BD* are:

$$B_x = 1125 \text{ lb} \leftarrow \qquad D_x = 1125 \text{ lb} \rightarrow$$
$$B_y = 300 \text{ lb} \uparrow \qquad D_y = 300 \text{ lb} \uparrow$$

The pin reactions on member *CE* are:

$$C_x = 1125 \text{ lb} \leftarrow \qquad D_y = 300 \text{ lb} \downarrow$$
$$C_y = 900 \text{ lb} \downarrow \qquad E_y = 1200 \text{ lb} \uparrow$$
$$D_x = 1125 \text{ lb} \leftarrow$$

EXAMPLE 6.7

The three-hinged structure in Fig. 6.16(a) is loaded as shown. Find the components of all the forces acting on each member of the frame.

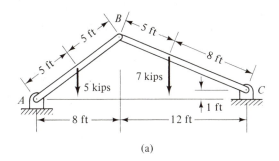

(a)

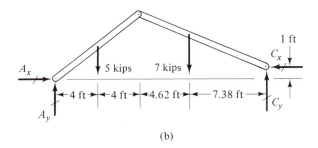

(b)

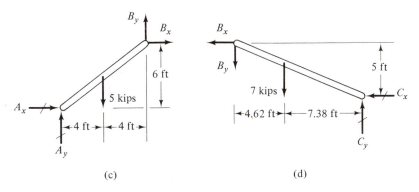

(c)   (d)

FIGURE 6.16

Solution

The free-body diagram of the entire structure and members AB and BC are shown in Fig. 6.16(b)–(d)

Entire Structure

For equilibrium:

$$\circlearrowright \Sigma \, M_A = -5(4) - 7(12.615) + C_x + C_y 20 = 0$$

$$C_x + 20C_y = 108.3 \tag{a}$$

$$\circlearrowright \Sigma \, M_C = -A_y 20 + A_x(1) + 5(16) + 7(7.38) = 0$$

$$A_x - 20A_y = -131.7 \tag{b}$$

$$\rightarrow \Sigma \, F_x = A_x - C_x = 0$$

$$A_x = C_x \tag{c}$$

Member AB

$$\circlearrowright \Sigma \, M_B = A_x 6 + 5(4) - A_y 8 = 0$$

$$6A_x - 8A_y = -20 \tag{d}$$

From Eq. (b),

$$A_x = 20A_y - 131.7 \tag{e}$$

Substituting for A_x in Eq. (d), we have

$$6(20A_y - 131.7) - 8A_y = -20$$

$$120A_y - 790 - 8A_y = -20$$

$$A_y = 6.88 \text{ kips} \qquad \text{Answer}$$

From Eq. (e),

$$A_x = 20(+6.88) - 131.7$$

$$= +5.84 \text{ kips} \qquad \text{Answer}$$

From Eq. (c),

$$C_x = +5.84 \text{ kips} \qquad \text{Answer}$$

From Eq. (a),

$$20C_y = 108.3 - C_x$$

$$C_y = 5.12 \text{ kips} \qquad \text{Answer}$$

$$\rightarrow \Sigma \, F_x = A_x + B_x = 5.84 + B_x = 0$$

$$B_x = -5.84 \text{ kips} \qquad \text{Answer}$$

$$\uparrow \Sigma \, F_y = A_y + B_y - 5 = 6.88 + B_y - 5 = 0$$

$$B_y = -1.88 \text{ kips} \qquad\qquad\qquad \text{Answer}$$

Member BC

Check:

$$\rightarrow \Sigma \, F_x = -B_x - C_x = -(-5.84) - (5.84) = 0 \quad \text{OK}$$

$$\uparrow \Sigma \, F_y = -B_y - 7 + C_y = -(-1.88) - 7 + 5.12 = 0 \quad \text{OK}$$

$$\Sigma \, M_B = -7(4.62) - C_x 5 + C_y 12 = -7(4.62) - 5.84(5) + 5.12(12) = 0 \quad \text{OK}$$

The pin reactions on member *AB* are:

$$A_x = 5.84 \text{ kips} \rightarrow \qquad B_x = 5.84 \text{ kips} \leftarrow$$

$$A_y = 6.88 \text{ kips} \uparrow \qquad B_y = 1.88 \text{ kips} \downarrow$$

The pin reactions on member *BC* are:

$$B_x = 5.84 \text{ kips} \rightarrow \qquad C_x = 5.84 \text{ kips} \leftarrow$$

$$B_y = 1.88 \text{ kips} \uparrow \qquad C_y = 5.12 \text{ kips} \uparrow$$

EXAMPLE 6.8

Determine the force required to maintain equilibrium of the frictionless pulley system shown in Fig. 6.17(a) and the reaction at the ceiling.

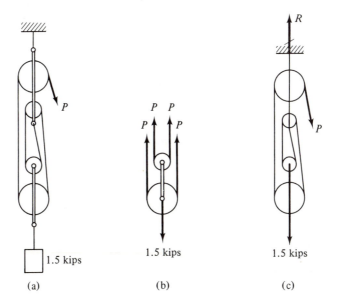

(a)   (b)   (c)

FIGURE 6.17

Solution

The cable is continuous and the pulleys are frictionless; therefore, the cable has a constant tension. The free-body diagram for the lower two pulleys is shown in Fig. 6.17(b). Equilibrium requires that

$$\uparrow \Sigma \, F_y = 4P - 1.5 = 0$$

$$P = 0.375 \text{ kips} \qquad\qquad \text{Answer}$$

The free-body diagram for the entire system is shown in Fig. 6.17(c). Equilibrium requires that

$$\uparrow \Sigma \, F_y = R - P - 1.5 = R - 0.375 - 1.5 = 0$$

$$R = 1.875 \text{ kips} \qquad\qquad \text{Answer}$$

EXAMPLE 6.9

A force is applied to the handle of the pliers shown in Fig. 6.18(a). Find the force applied to the bolt and the horizontal and vertical reactions at the hinge A.

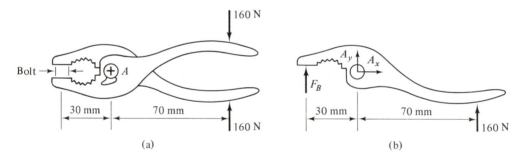

(a)   (b)

FIGURE 6.18

Solution

A free-body diagram for one member of the pliers is shown in Fig. 6.18(b). Equilibrium requires that

$$\circlearrowright \Sigma \, M_A = 160\,(70) - F_B\,30 = 0$$

$$F_B = 373 \text{ N} \qquad\qquad \text{Answer}$$

$$\rightarrow \Sigma \, F_x = A_x = 0 \qquad\qquad \text{Answer}$$

$$\uparrow \Sigma \, F_y = F_B + A_y + 160 = 373 + A_y + 160 = 0$$

$$A_y = -533 \text{ N} \qquad\qquad \text{Answer}$$

PROBLEMS

6.27 through 6.36 Determine the components of the forces acting on each member of the pin-connected frame shown.

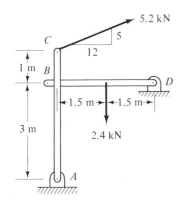

PROB. 6.27

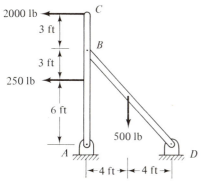

PROB. 6.28

PROB. 6.29

PROB. 6.30

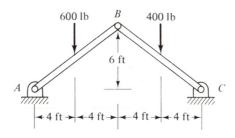

PROB. 6.31

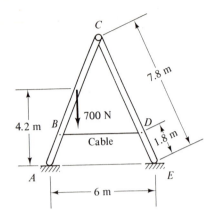

PROB. 6.32

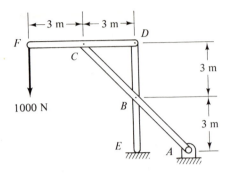

PROB. 6.34

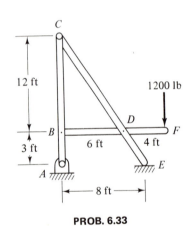

PROB. 6.33

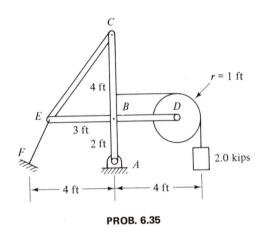

PROB. 6.35

PROB. 6.36

7

Friction _____

7.1 INTRODUCTION

In Chapter 5 we considered various support conditions. Among these are the smooth and rough surfaces. For the *smooth surface* the reactive force is normal to the surface at the point of contact. The body and surface on which the body rests are free to move with respect to each other. With the *rough surface* the reactive force is oblique to the surface at the point of contact. For convenience the reactive force can be resolved into components normal and tangential to the surface. The tangential component acts to prevent the free motion of the body and surface with respect to each other. The direction of the friction force acting on the body is opposed to the motion of the body over the surface.

Friction can be both a liability and an asset. With such things as brakes, belt drives, and walking, friction is indispensable. On the other hand, the loss of power and wear of surfaces in contact for gears, bearings, and other machine elements, due to friction, is undesirable.

In this chapter we consider dry friction, that is, friction which occurs between a body and surface that are in contact along dry surfaces. Dry friction should not be confused with friction between surfaces that are completely or partially separated by a fluid.

7.2 DRY OR COULOMB FRICTION

The nature of *dry friction* is complex and not well understood. However, it does result from the interactions between the surface layers in contact. The interactions are made up of a number of processes, including molecular attraction between the surfaces. Although dry friction is complex, empirical laws are available which can be used to determine values of the dry friction force. The laws are usually attributed to Coulomb, who published the results of numerous friction experiments in 1781.

A simple example involving the sliding of a block along a rough surface will serve to introduce Coulomb's laws of dry friction. In Fig. 7.1(a) we show a block of weight W resting on a rough surface. A gradually increasing force P is applied to the block and the resulting friction force is measured. Free-body diagrams for the block and rough surface are shown in Fig. 7.1(b) and (c). As we increase P, the frictional force increases to maintain equilibrium, $F = P$. When the friction reaches a maximum value F_m, the block is about to move. Motion of the block is impending at that point $P = F_m$. For values of $P > F_m$, the block moves and the friction force drops rapidly to a kinetic value F_k. After motion occurs, the friction force remains essentially constant for increasing values of P. The relationship between the applied force and friction force as described is shown in Fig. 7.2.

It has been found from experiment that for a given pair of surfaces, the maximum value of friction F_m is proportional to the normal

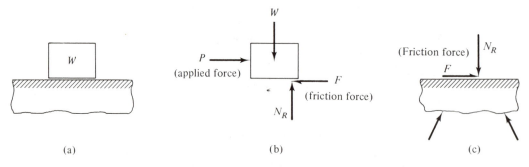

(a)   (b)   (c)

FIGURE 7.1

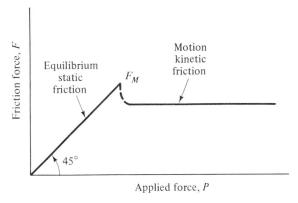

FIGURE 7.2

reaction N_R. That is,

$$F_m = f_s N_R \tag{7.1}$$

where f_s is the *coefficient of static friction*. Also, from experiment, the kinetic value of friction F_k is proportional to the normal reaction N_R. Thus

$$F_k = f_k N_R \tag{7.2}$$

where f_k is the *coefficient of kinetic friction*.

Both coefficients of friction are almost independent of the area of contact between the block and surface. Thus if the block tipped and only an edge was in contact with the surface, we would have nearly the same coefficients of static friction for the same dry surfaces. Some typical values of the coefficient of static friction are listed in Table 7.1. The values shown are representative only. Experimentation would be required to determine values for an actual engineering system.

TABLE 7.1

Coefficient of Static Friction

Substance	f_s
Metal on metal	0.15–0.75
Metal on wood	0.25–0.65
Metal on stone	0.25–0.70
Metal on ice	0.02–0.03
Wood on wood	0.40–0.70
Rubber on concrete	0.60–0.90

Note: Values of the coefficient of kinetic friction are approximately 75 percent of those for the corresponding coefficient of static friction.

7.3 ANGLE OF FRICTION

In Fig. 7.3 we show that the friction force F and the normal reaction N_R are components of the total reaction R of the surface on the block. The angle between R and N_R represented by the Greek letter φ (phi) increases with increasing values of F. When F reaches the maximum value F_m, motion impends. The angle $\varphi = \varphi_s$ between F_m and N_R is given by

$$\tan \varphi_s = \frac{F_m}{N_R} \tag{a}$$

Substituting for F_m from Eq. (7.1) in Eq. (a), we have

$$\tan \varphi_s = \frac{f_s N_R}{N_R}$$

$$\tan \varphi_s = f_s \tag{7.3}$$

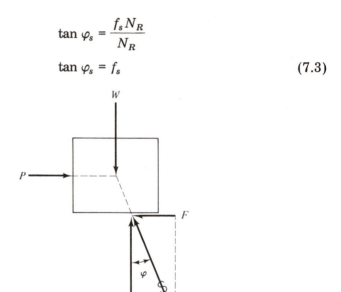

FIGURE 7.3

The angle φ_s is called the *angle of friction*. The tangent of the angle of friction is equal to the coefficient of friction. If the force P changes direction but remains horizontal, the reaction R sweeps out a cone which is called the *cone of friction*.

EXAMPLE 7.1

A block rests on a rough inclined plane. Determine the angle of inclination of the plane if the block is on the verge of sliding down the plane.

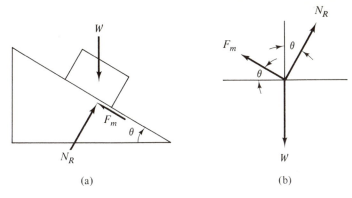

(a)   (b)

FIGURE 7.4

Solution

The free-body diagram for the block is shown in Fig. 7.4(a). The block is in equilibrium under the action of forces N_R, W, and the maximum static friction F_m. In Fig. 7.4(b) the forces have been moved along their line of action until they are directed away from a common point. From equilibrium:

$$\rightarrow \Sigma\, F_x = 0 \qquad -F_m \cos\theta + N_R \sin\theta = 0$$

$$\frac{\sin\theta}{\cos\theta} = \frac{F_m}{N_R}$$

$$\tan\theta = \frac{F_m}{N_R}$$

Substituting for F_m from Eq. (7.1), we write

$$\tan\theta = \frac{f_s N_R}{N_R}$$

$$\tan\theta = f_s$$

Comparing this result with Eq. (7.3), we see that

$$\tan\theta = \tan\varphi_s$$

$$\theta = \varphi_s \qquad\qquad\qquad \text{Answer}$$

Therefore, if the plane is inclined at the angle of friction φ_s, the body will be on the verge of motion. For this reason, the angle of friction is also called the *angle of repose*.

EXAMPLE 7.2

A ladder with a mass of 8.66 kg and a length of 2.5 m rests in a vertical plane with its lower end on a horizontal floor ($f_s = 0.3$) and its upper end against a vertical wall ($f_s = 0.2$). Find the smallest angle θ that the ladder can make with the floor before slipping begins. The center of gravity of the ladder is two-fifths of the length of the ladder from the bottom.

Solution

The free-body diagram for the ladder is shown in Fig. 7.5. The mass of the ladder is 8.66 kg. From Eq. (1.1) the weight of the ladder

$$W = mg = 8.66\,(9.81) = 85.0 \text{ N}$$

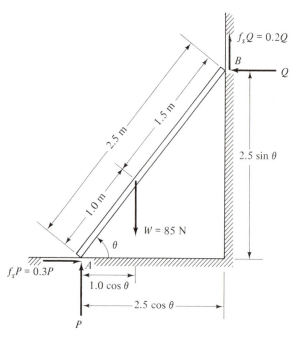

FIGURE 7.5

Since the ladder is about to move, the friction forces will have reached their maximum values and will be directed so as to oppose impending motion. Apply the equations of equilibrium:

$$\rightarrow \Sigma F_x = 0 \qquad 0.3P - Q = 0 \qquad \text{(a)}$$

$$\uparrow \Sigma F_y = 0 \qquad P + 0.2Q - 85 = 0 \qquad \text{(b)}$$

$$\circlearrowright \Sigma M_A = 0$$

$$Q(2.5 \sin \theta) + 0.2Q(2.5 \cos \theta) - 85\,(1.0 \cos \theta) = 0 \qquad \text{(c)}$$

Eliminate P between Eqs. (a) and (b):

$$Q = 24.1 \text{ N}$$

Substitute Q in Eq. (c):

$$\frac{\sin \theta}{\cos \theta} = 1.214$$

$$\tan \theta = 1.214$$

$$\theta = 50.5° \qquad\qquad \text{Answer}$$

EXAMPLE 7.3

A packing crate weighing 150 lb and resting on a floor is pushed by a horizontal force P as shown in Fig. 7.6(a). The coefficient of friction between the crate and floor is 0.35. How high above the floor can the force P act so that the crate moves without tipping?

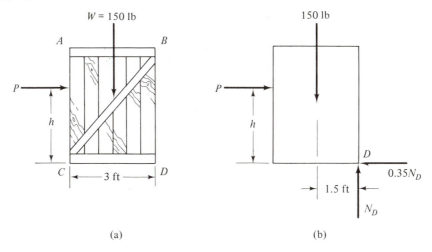

FIGURE 7.6

Solution

The free-body diagram for conditions that give the limiting position of P are shown in Fig. 7.6(b). Impending tipping requires that the friction and normal reaction be concentrated at D—the right edge of the crate. Impending slipping requires that the friction force $F_m = f_s N_R = 0.35 N_D$. From the equations of equilibrium, we write

$$\uparrow \Sigma F_y = 0 \qquad N_D - 150 = 0$$

$$N_D = 150 \text{ lb}$$

$$\rightarrow \Sigma F_x = 0 \qquad P - 0.35 N_D = 0$$

$$P = 0.35(150)$$

$$= 52.5 \text{ lb}$$

$$\Sigma\ M_D = 0 \qquad -Ph + 150(1.5) = 0$$

$$h = \frac{150(1.5)}{52.5}$$

$$= 4.29\ \text{ft}$$

The height must be less than 4.29 ft. Answer

PROBLEMS

7.1 A 250-lb block rests on a horizontal surface as shown. Determine the friction force if the applied load P is equal to (a) 30 lb, (b) 45 lb, and (c) 90 lb. $(f_s = 0.3$ and $f_k = 0.2.)$

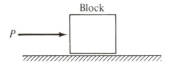

PROB. 7.1 and PROB. 7.2

7.2 The block resting on a horizontal surface as shown has a mass of 100 kg. Determine the friction force if the applied load P is equal to (a) 150 N, (b) 390 N, and (c) 600 N. $(f_s = 0.4$ and $f_k = 0.3.)$

7.3 A 100-lb block rests on the inclined plane shown. Determine the maximum and minimum value of P for which the block is in equilibrium. $(f_s = 0.3.)$

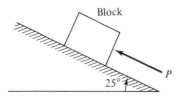

PROB. 7.3 and PROB. 7.4

7.4 The block resting on the inclined plane shown has a mass of 40 kg. Determine the maximum and minimum value of P for which the block is in equilibrium. $(f_s = 0.35.)$

7.5 The block resting on the inclined plane

shown has a mass of 60 kg. Determine the maximum and minimum value of P for which the block is in equilibrium. $(f_s = 0.35.)$

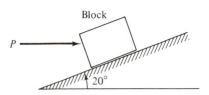

PROB. 7.5 and PROB. 7.6

7.6 A 200-lb block rests on the inclined plane shown. Determine the maximum and minimum value of P for which the block is in equilibrium. $(f_s = 0.25.)$

7.7 The 500-kg trunk rests on the floor as shown. If $b = 1$ m, $H = 2.5$ m, and $P = 1400$ N, determine the maximum height h the force can be moved above the floor before the trunk starts to tip over. $(f_s = 0.3.)$

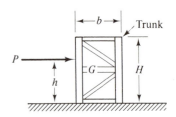

PROB. 7.7 and PROB. 7.8

7.8 The 1000-lb trunk rests on the floor as shown. If $b = 36$ in., $H = 75$ in., and $P = 280$ lb, determine the maximum height h the force can be moved above the floor before the trunk starts to tip over. $(f_s = 0.3.)$

7.9 A 650-lb packing crate rests on a ramp as shown. If $b = 36$ in., $H = 60$ in., and $h = 48$ in., determine the maximum and minimum value of P for which the crate will be in equilibrium. ($f_s = 0.25$.)

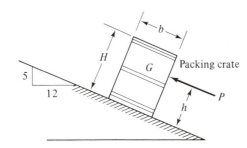

PROB. 7.9 and PROB. 7.10

7.10 A 300-kg packing crate rests on a ramp as shown. If $b = 0.9$ m, $H = 1.5$ m, and $h = 1.2$ m, determine the maximum and minimum value of P for which the crate will be in equilibrium. ($f_s = 0.25$.)

7.11 A 10-kg uniform ladder 2.5 m long rests against a smooth wall as shown. An 80-kg man climbs up the ladder to a point A, a distance $L = 2.0$ m from the bottom of the ladder, before the ladder slips. What is the coefficient of friction between the ladder and the ground?

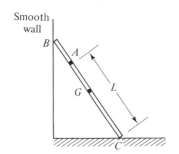

PROB. 7.11 and PROB. 7.12

7.12 A 25-lb uniform ladder 8 ft long rests against a smooth wall as shown. A 175-lb man climbs up the ladder to a point A, a distance L from the bottom of the ladder, before the ladder slips. If the coefficient of friction between the ladder and the ground is $f_s = 0.4$, what is the distance L?

7.13 A movable bracket can be raised or lowered on the vertical 1-in.-diameter rod shown. If $P = 80$ lb, $H = 3$ in., and $L = 10$ in., what minimum coefficient of friction is required to prevent the bracket from slipping downward? Assume that contact is made between the bracket and rod at points M and N.

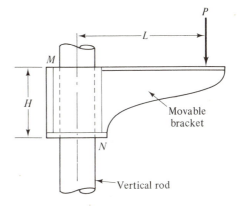

PROB. 7.13 and PROB. 7.14

7.14 A movable bracket can be raised or lowered on the 50-mm-diameter rod shown. If $H = 75$ mm, $L = 275$ mm, $P = 350$ lb, and the coefficient of friction between the bracket and rod is 0.4, will the bracket slip downward? Assume that contact is made between the bracket and the rod at points M and N.

7.4 WEDGES

The *wedge* is a simple machine. It may be used to transform horizontal input forces into vertical output forces. The angle of friction may be used to simplify the solution, as will be seen in the following example.

EXAMPLE 7.4

A wedge B of negligible weight is to be used to move block A upward [Fig. 7.7(a)]. The coefficient of friction between all surfaces of contact is 0.3. Determine the input force P required to produce an output force $Q = 2150$ lb. (Q includes the weight of block A.)

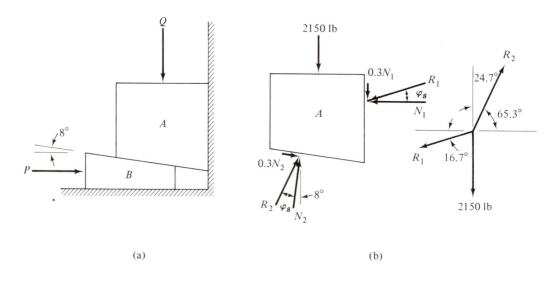

(a)   (b)

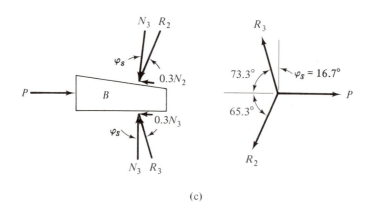

(c)

FIGURE 7.7

Solution

The free-body diagrams together with their force diagrams for impending motion of block A and wedge B have been shown in Fig. 7.7(b) and (c). The maximum static friction forces and normal reactions have been replaced by their resultants acting at the angle of friction $\varphi_s = \arctan 0.3 = 16.7°$ with the normal. For equilibrium of body A:

$$\rightarrow \Sigma \, F_x = 0 \quad R_2 \cos 65.3° - R_1 \cos 16.7° = 0$$

$$R_1 = 0.4363 R_2 \qquad \text{(a)}$$

$$\uparrow \Sigma \, F_y = 0 \quad R_2 \sin 65.3° - R_1 \sin 16.7° = 2150 \qquad \text{(b)}$$

Substituting Eq. (a) for R_1 in Eq. (b), we have

$$R_2 = 2745 \text{ lb}$$

For equilibrium of body B:

$$\uparrow \Sigma \, F_y = 0 \quad R_3 \sin 73.3° - R_2 \sin 65.3° = 0$$

Substituting for R_2, we have

$$R_3 = 2604 \text{ lb}$$

$$\rightarrow \Sigma \, F_x = 0 \quad P - R_3 \cos 73.3° - R_2 \cos 65.3° = 0$$

Substituting for R_2 and R_3 yields

$$P = 1895 \text{ lb} \qquad\qquad \text{Answer}$$

7.5 SQUARE-THREADED SCREWS—SCREW JACKS

The *screw jack* has a threaded stem that turns in a fixed base [Fig. 7.8(a)]. As a force P is applied to the handle, the stem turns and moves up (or down) lifting (or lowering) the load W. The screw can be analyzed by considering the screw thread as an inclined plane rolled on a cyclinder [Fig. 7.8(b)]. The inclination of the plane is given by

$$\tan \theta = \frac{p}{\pi d} \qquad\qquad (7.4)$$

where p is the pitch of the thread and d the mean diameter of the threads.
We isolate a small element of the screw and draw a free-body diagram for raising the load as shown in Fig. 7.8(b). The corresponding force diagram is shown in Fig. 7.8(c). The friction force F_m and normal force N_R have been replaced by their resultant R which acts at an angle φ_s (angle of static friction) from the normal N_R. From the equations of equilibrium

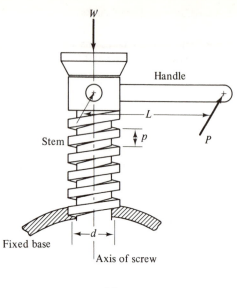

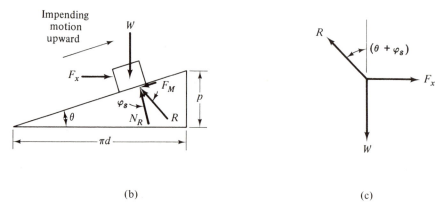

FIGURE 7.8

$$\rightarrow \Sigma \, F_x = 0 \qquad F_x - R \sin(\theta + \varphi_s) = 0 \qquad \text{(a)}$$

$$\uparrow \Sigma \, F_y = 0 \qquad R \cos(\theta + \varphi_s) - W = 0 \qquad \text{(b)}$$

Eliminating R between Eqs. (a) and (b), we obtain

$$F_x = \frac{W \sin(\theta + \varphi_s)}{\cos(\theta + \varphi_s)}$$

$$F_x = W \tan(\theta + \varphi_s) \qquad \text{(c)}$$

To produce motion the moment of the force P on the handle of the screw jack must be equal to the moment of the force F_x at the mean radius $d/2$ of the screw—both about the axis of the screw. Thus

$$M = PL = F_x \frac{d}{2}$$

Substituting for F_x from Eq. (c), we obtain

$$M = PL = W \frac{d}{2} \tan (\theta + \varphi_s) \qquad (7.5)$$

where P is the force applied to the handle, L the length of the handle, W the load to be *lifted* by the jack screw, θ the inclination of the plane of the screw given by Eq. (7.4), and φ_s the angle of friction.

To lower the load, the friction force is reversed and Eq. (7.5) must be modified as follows:

$$M = PL = W \frac{d}{2} \tan (\theta - \varphi_s) \qquad (7.6)$$

EXAMPLE 7.5

A screw jack having two threads per 10 mm and a mean diameter of 25 mm is used to lift a weight. If force of 2.5 kN is applied at the end of a handle 300 mm long, what weight can be lifted by the screw jacks? ($f_s = 0.35$.)

Solution

Since the screw moves through two turns and travels 10 mm, the pitch $p = 10/2 = 5$ mm. From Eq. (7.4),

$$\tan \theta = \frac{p}{\pi d} = \frac{5}{\pi (25)}$$

$$\theta = 3.64°$$

The angle of friction

$$\varphi_s = \arctan f_s = \arctan 0.35$$

$$= 19.29°$$

From Eq. (7.5),

$$W = \frac{2PL}{d \tan (\theta + \varphi_s)} = \frac{2(2.5)(300)}{25 \tan (3.64° + 19.29°)}$$

$$= 141.8 \text{ kN} \qquad \qquad \text{Answer}$$

PROBLEMS

In Probs. 7.15 through 7.25, assume that all screw threads are square.

7.15 Determine the force P required to move the 500-lb block shown if the coefficient of friction between all contact surfaces is 0.3.

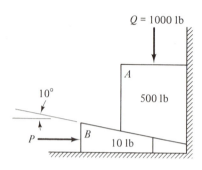

PROB. 7.15

7.16 The coefficient of friction between all contact surfaces is 0.4 except as noted. Determine the force P required to move the 2000-N block.

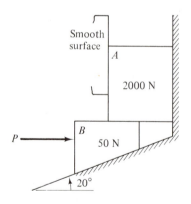

PROB. 7.16

7.17 A 5° wedge is used to raise the end of a beam as shown. If the coefficient of friction between all surfaces is $f_s = 0.25$, determine the force P required to move the wedge. Assume that the beam is horizontal and does not move.

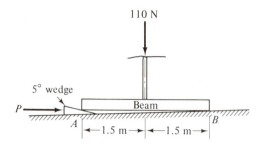

PROB. 7.17

7.18 The base of a torsion machine is to be lifted by a 7° wedge as shown. If the coefficient of friction between all surfaces is $f_s = 0.3$, determine the force P required to move the wedge. Assume that the machine base is horizontal and does not move.

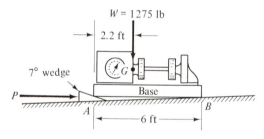

PROB. 7.18

7.19 and 7.20 Determine the vertical force Q required to move block B as shown in the figure. ($f_s = 0.2$ for all surfaces.)

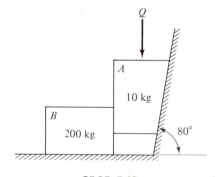

PROB. 7.19

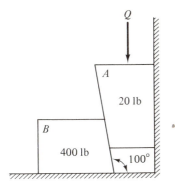

PROB. 7.20

7.21 A screw jack having 4 threads per inch and a mean diameter of 1.2 in. is used to raise or lower a weight of 5000 lb as shown. What force must be applied at the end of a 12-in.-long handle to (a) raise the weight and (b) lower the weight? ($f_s = 0.4$.)

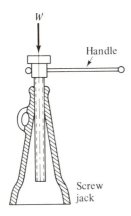

PROB. 7.21 and PROB. 7.22

7.22 A screw jack with two threads per 15 mm, a mean radius of 30 mm, and a handle length of 0.3 m is used to raise or lower a weight as shown. (a) If a force of 0.3 kN is applied to the handle to raise a weight of 20 kN, what is the coefficient of friction? (b) If a force of 0.25 kN is applied to the handle to lower a weight of 20 kN, what is the coefficient of friction?

7.23 The C-clamp is used to clamp two blocks of plastic together as shown with a force of 600 lb. The clamp has 10 threads per inch and a mean diameter of 0.5 in. If the coefficient of friction is 0.25, what moment is required?

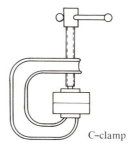

C–clamp

PROB. 7.23 and PROB. 7.24

7.24 The C-clamp shown is used to hold two blocks together. The clamp has a pitch of 0.3 mm and a mean diameter of 20 mm. The coefficient of friction is 0.2. What force will be applied to the blocks if a moment of 60 N·m is applied to the handle?

7.25 A turnbuckle is used to apply a tension T of 1200 lb to a rod as shown. The coefficient of friction is 0.4. The rod has 6 threads per inch and a mean radius of 0.75 in. (a) What maximum moment must be applied to the turnbuckle to increase the tension in the rod? (b) What moment must be applied to decrease the tension in the rod? The rod has right-hand threads at A and left-hand threads at B.

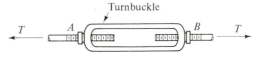

PROB. 7.25 and PROB. 7.26

7.26 A turnbuckle is used to apply a tension to a rod as shown. The coefficient of friction is 0.3. The rod has a pitch of 1 mm and a mean diameter of 8 mm. (a) If a moment of the 10 N·m is required to tighten the turnbuckle, what was the tension in the rod? (b) If a moment of 3 N·m is required to loosen the turnbuckle, what was the tension in the rod? The rod has right-hand threads at A and left-hand threads at B.

7.6 AXLE FRICTION—JOURNAL BEARINGS

Journal bearings are commonly used to provide lateral support to rotating machines such as shafts and axles. If the bearing is partially lubricated or not lubricated at all, the methods of this chapter may be applied.

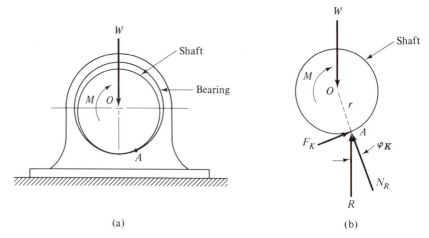

(a)   (b)

FIGURE 7.9

The journal bearing shown in Fig. 7.9(a) supports a shaft that is rotating at a constant speed. To maintain the rotation, a torque or moment couple M must be applied. The load W and moment couple M cause the shaft to touch the bearing at A. This can be explained by the fact that the shaft climbs up on the bearing until the forces and couple are in equilibrium. The shaft touching the bearing at A gives rise to reaction N_R and friction force F_k as shown in the free-body diagram of Fig. 7.9(b). For equilibrium:

$$\uparrow \Sigma \ F_y = 0 \quad W = R$$

$$\circlearrowright \Sigma \ M_o = 0 \quad M = Rr \sin \varphi_k \qquad (7.7)$$

For small angles of friction, $\sin \varphi_k \approx \tan \varphi_k = f_k$. Thus

$$M \approx Rrf_k \qquad (7.8)$$

EXAMPLE 7.6

A drum of 300-mm radius on a shaft with a diameter of 75 mm and supported by two bearings as shown in Fig. 7.10(a) is used to hoist a 250-kg load. The drum and shaft have a mass of 50 kg and the coefficient of friction for the bearing is 0.25. Determine the moment that must be applied to the drum shaft in order to raise the load at uniform speed.

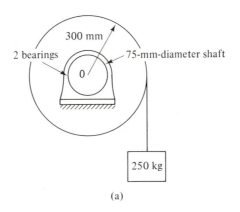

(a)

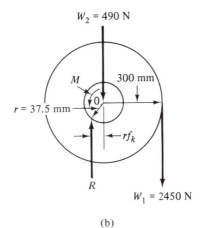

(b)

FIGURE 7.10

Solution

The masses are changed to weights by Eq. (1.1):

$$W_1 = m_1 g = 250(9.81) = 2450 \text{ N}$$

$$W_2 = m_2 g = 50(9.81) = 490 = \text{N}$$

The free-body diagram of the drum and shaft is shown in Fig. 7.10(b). From equilibrium:

$$\uparrow \Sigma F_y = 0 \quad R - W_1 - W_2 = 0$$

$$R - 2450 - 490 = 0$$

$$R = 2940 \text{ N}$$

$$\circlearrowright \Sigma M_o = 0 \quad M - Rrf_k - W_1 L = 0$$

$$M - (2940)(37.5)0.25 - 2450(300) = 0$$

$$M = 763 \text{ N} \cdot \text{m} \qquad \text{Answer}$$

7.7 SPECIAL APPLICATIONS

We will now consider the application of dry friction laws to several special devices. The derivations of the governing equations are beyond the scope of this book since they depend on calculus.

Disk Friction; Thrust Bearings

Consider a hollow rotating shaft whose end is in contact with a fixed surface as shown in Fig. 7.11. We assume that the force between the rotating shaft and fixed surface is distributed uniformly. The external torque M required to cause slipping to occur is given by

$$M = f_k P \, \frac{2(R_o^3 - R_i^3)}{3(R_o^2 - R_i^2)} \tag{7.9}$$

where P is the thrust force, R_o the outside radius, and R_i the inside radius of the hollow shaft.

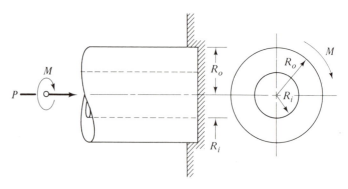

FIGURE 7.11

For a solid shaft of radius R,

$$M = f_k P \tfrac{2}{3} R \qquad (7.10)$$

The largest torque, without slipping, transmitted by a disk clutch can be found from Eqs. (7.9) and (7.10), where f_k has been replaced by f_s.

Belt Friction

Consider a flat belt passing over a fixed rough cylinder as shown in Fig. 7.12. The belt has impending motion as shown. The relationship between

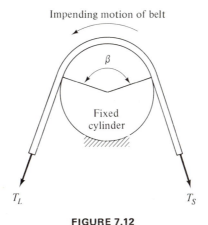

Impending motion of belt

β

Fixed cylinder

T_L T_S

FIGURE 7.12

the tensions T_L (larger tension) and T_S (smaller tension) at the ends of the belt is given by

$$\frac{T_L}{T_S} = e^{\pi f \beta / 180^{\circ}} \qquad (7.11)$$

where e is the base of the natural logarithm ($e = 2.718$) and the Greek letter β (beta) is the angle of contact between the belt and cylinder in degrees.

EXAMPLE 7.7

Determine the torque capacity of the clutch shown in Fig. 7.13 if the coefficient of friction $f_k = 0.25$ and the axial force is 1580 lb. Assume uniform force distribution on the contact surfaces.

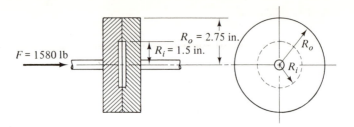

FIGURE 7.13

Solution

From Eq. (7.9),

$$M = f_k P \frac{2(R_o^3 - R_L^3)}{3(R_o^2 - R_L^2)} = 0.25\,(1580)\,\frac{2[(2.75)^3 - (1.5)^3]}{3[(2.75)^2 - (1.5)^2]}$$

$$= 864 \text{ lb-in.} \qquad\qquad \text{Answer}$$

EXAMPLE 7.8

Three rough round pins are attached to a 150-kg block B which can move vertically between smooth guides. A rope is placed over the pins as shown in Fig. 7.14(a). Determine the mass of the block A if block B is on the point of impending motion. ($f_s = 0.35$.)

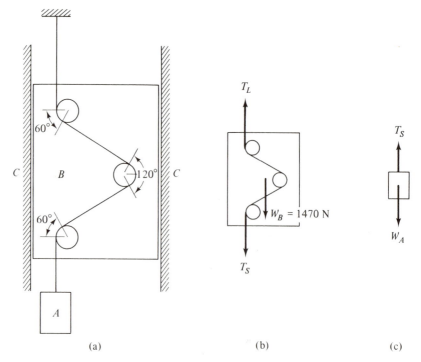

(a) (b) (c)

FIGURE 7.14

Solution

Free-body diagrams of blocks B and A are shown in Fig. 7.14(b) and (c). The weight of block B is found from Eq. (1.1).

$$W_B = m_B g = 150(9.81) = 1470 \text{ N}$$

The total angle in contact between the rope and the three pins,

$$\beta = 60° + 120° + 60° = 240°$$

For equilibrium of block B:

$$\uparrow \Sigma \, F_y = 0 \qquad T_L - 1470 - T_S = 0 \qquad\qquad \text{(a)}$$

From Eq. (7.11),

$$\frac{T_L}{T_S} = e^{\pi f \beta / 180°} = e^{\pi (0.35) \, 240° / 180°}$$

$$= e^{1.466} = 4.332 \qquad\qquad \text{(b)}$$

Eliminating T_L between Eqs. (a) and (b), we obtain

$$4.332 T_S - T_S = 1470$$

$$T_S = 441 \text{ N}$$

For equilibrium of block A, $T_S = W$. Therefore, $W_A = 441$ N. From Eq. (1.1),

$$W_A = m_A g \quad \text{or} \quad m_A = \frac{W_A}{g} = \frac{441}{9.81} = 45 \text{ kg} \qquad\qquad \text{Answer}$$

7.8 ROLLING RESISTANCE

If a wheel is moved without slipping over a horizontal surface while supporting a load, a force is required to maintain uniform motion. Thus some kind of rolling resistance must be present. The usual method for describing rolling resistance is based on the deformation of the surface as shown in Fig. 7.15. Let the resultant reaction between the surface and rollers act at point A. From equilibrium:

$$\circlearrowright \Sigma \, M_A = 0 \qquad Wa - Ph = 0$$

Since h is nearly equal to r, we write

$$Wa - Pr = 0$$

$$P = \frac{Wa}{r} \qquad\qquad \text{(7.12)}$$

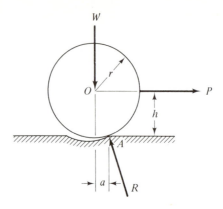

FIGURE 7.15

The distance a is called the *coefficient* of *rolling resistance*. The values of a are given in millimeters or inches. They vary substantially for the same material with different values of W and r. Therefore, caution must be used with this method for finding rolling resistance.

EXAMPLE 7.9

What horizontal force must be applied to the center of a wheel with a mass of 250 kg and a diameter of 0.7 m to just cause motion of the wheel? The coefficient of rolling resistance is 0.25 mm.

Solution

From Eq. (1.1), the weight of the wheel

$$W = mg = 250(9.81) = 2450 \text{ N}$$

From Eq. (7.12),

$$P = \frac{Wa}{r} = \frac{2450(0.25)}{700}$$

$$= 0.875 \text{ N} \qquad \text{Answer}$$

PROBLEMS

7.27 A 20-lb rotor is attached to a 2-in.-diameter shaft which is supported by two bearings as shown. If the moment required to start the rotor turning is 1.5 lb-ft, determine the coefficient of friction between the shaft and bearings.

7.28 If the coefficient of friction for Prob. 7.27 is 0.3, what is the required starting moment?

7.29 A drum of 300-mm radius on a shaft with a diameter of 80 mm is supported by two bearings as shown. It is used to hoist a 500-kg load. The shaft and drum have a mass of 75 kg. The coefficient of friction for the bearings is 0.3. Determine the moment that must be applied to the drum shaft to raise the load at uniform speed.

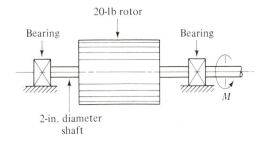

PROB. 7.27

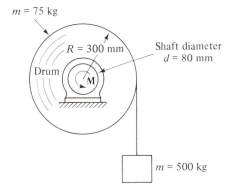

PROB. 7.29

7.30 What moment must be applied to the drum shaft in Prob. 7.29 to lower the load at uniform speed?

7.31 A thrust of 1000 lb is supported by a collar bearing as shown. Determine the moment required to maintain a constant speed of rotation if the coefficient of friction is 0.1.

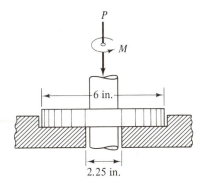

PROB. 7.31

7.32 If the moment for Prob. 7.31 is 300 lb-in., what is the coefficient of friction?

7.33 A floor polisher with a mass of 25 kg is operated on a surface with a coefficient of friction of 0.125. (a) Determine the moment required to keep the polisher from turning. (b) If the handles are separated by 0.45 m, what force on each handle is required to keep the handles from turning?

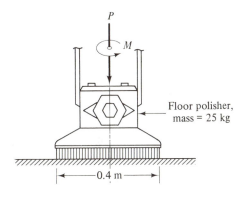

PROB. 7.33

7.34 A rope is wrapped around a capstan to secure a ship. The force on the free end of the rope is 800 lb and on the ship end 20 kips. Determine (a) the coefficient of friction between the rope and capstan if the rope makes three full turns and (b) the number of full turns required if the coefficient of friction is $f_s = 0.3$.

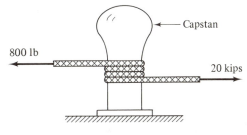

PROB. 7.34

7.35 (a) If the coefficient of friction is 0.15, what clockwise moment is required to start rotating the pulley shown?

(b) If a clockwise moment of 225 N·m is required to start the pulley rotating, what is the coefficient of friction?

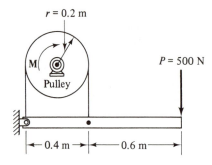

PROB. 7.35

7.36 Four rough round pins are attached to a 100-lb block B which can move vertically between smooth guides. A rope is placed over the pins as shown. (a) If the coefficient of friction is $f_s = 0.25$, determine the weight of block A if block B is about to move. (b) If block A weighs 50 lb and block B is about to move, what is the coefficient of friction?

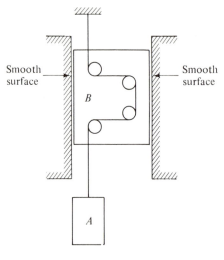

PROB. 7.36

Center of Gravity, Centroids, and Moments of Inertia of Areas _____

8.1 INTRODUCTION

In Chapter 3 we defined the center of gravity of a body to be the point at which the weight of the body could be considered to act. We shall consider the center of gravity in more detail in this chapter. In addition, we consider centroids and moments of inertia of areas. The centroid and moments of inertia are concepts that are useful in the study of strength of materials.

8.2 MOMENT OF A FORCE ABOUT AN AXIS

For problems involving forces in a plane, we determine the moment of the force with respect to a point. When considering the general case of forces in space, we must determine the moment of a force with respect to an axis.

Consider the force $\mathbf{F}_z$ parallel to the z axis, acting at a point with coordinates x, y, and z (Fig. 8.1). The moment of a force about an axis is defined as the product of the force and the perpendicular distance from the line of action of the force to the axis. Therefore, the moment of the force $\mathbf{F}_z$ about the x axis is

$$M_x = F_z\,y$$

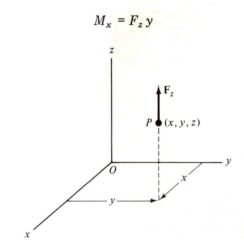

FIGURE 8.1

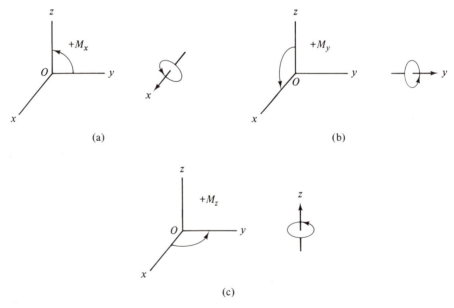

(a)

(b)

(c)

FIGURE 8.2

and about the y axis it is

$$M_y = -F_z x$$

The moment of the force about the z axis is zero, since the force is parallel to the z axis. The algebraic sign of the moment indicates the direction of rotation.

      A moment is defined to be positive if it tends to produce a clockwise rotation when viewed from the origin in the direction of the axis and negative if the rotation is reversed. The direction of rotation for positive moments about the x, y, and z axes are shown in Fig. 8.2.

EXAMPLE 8.1

Find the moments of the three forces $F_x = 10$ kN, $F_y = 5$ kN, and $F_z = 15$ kN located at $x = 2$ m, $y = 3$ m, and $z = 5$ m about the x, y, and z axes (Fig. 8.3).

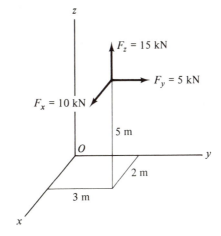

FIGURE 8.3

Solution

The sum of the moments about the x, y, and z axes are as follows:

$$M_x = F_z y - F_y z = 15\,(3) - 5\,(5) = 20 \text{ kN·m}$$
$$M_y = F_x z - F_z x = 10\,(5) - 15\,(2) = 20 \text{ kN·m}$$
$$M_z = F_y x - F_x y = 5\,(2) - 3\,(10) = -20 \text{ kN·m} \quad \text{or} \quad 20 \text{ kN·m}$$

8.3 CENTER OF GRAVITY OF A SYSTEM OF PARTICLES

Consider three particles distributed along the x axis [Fig. 8.4(a)]. The weights of the particles are W_1, W_2, and W_3 and their algebraic distances

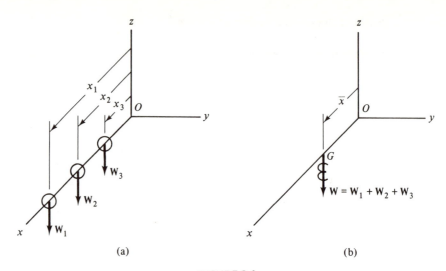

(a)   (b)

FIGURE 8.4

from the origin are x_1, x_2, and x_3. Thus we have a system of parallel forces. The parallel-force system was discussed in Sec. 4.5.

The center of gravity of the system of particles can be determined by finding the resultant of this parallel system of forces. The resultant forces or total weight is shown in Fig. 8.4(b). The magnitude of the total weight is found from the equation

$$\uparrow \Sigma\, F = -W_1 - W_2 - W_3 = -W$$

or

$$W = W_1 + W_2 + W_3 \tag{8.1}$$

and the location of the total weight from the origin is found from the theorem of moments

$$\curvearrowright \Sigma\, M_O = W_1 x_1 + W_2 x_2 + W_3 x_3 = W\bar{x}$$

or

$$\bar{x} = \frac{W_1 x_1 + W_2 x_2 + W_3 x_3}{W_1 + W_2 + W_3}$$

The equation for the location of the center of gravity can be modified to apply to any number of particles. For n particles,

$$\bar{x} = \frac{W_1 x_1 + W_2 x_2 + \cdots + W_n x_n}{W_1 + W_2 + \cdots + W_n} \tag{8.2}$$

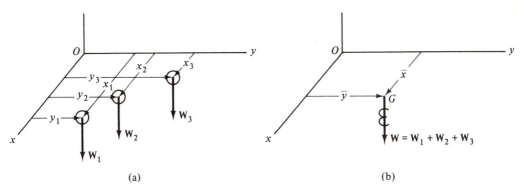

(a)
(b)

FIGURE 8.5

If the three particles lie in the xy plane, two coordinates are required to locate the center of gravity of the system. In Fig. 8.5(a) we show such a system. The particles have weights W_1, W_2, and W_3 and their locations are given by coordinates in the xy plane. The resultant or total weight is shown in Fig. 8.5(b). The magnitude of the total weight is found from the equation

$$\uparrow \Sigma\ F = -W_1 - W_2 - W_3 = -W$$

or

$$W = W_1 + W_2 + W_3$$

and the location of the total weight is found from the theorem of moments. The sum of the moments of the weights about the y axis is equal to the moment of the total weight about the y axis.

$$\Sigma\ M_y = +W_1 x_1 + W_2 x_2 + W_3 x_3 = +W\bar{x}$$

or

$$\bar{x} = \frac{W_1 x_1 + W_2 x_2 + W_3 x_3}{W_1 + W_2 + W_3} \tag{8.3}$$

Similarly for the x axis,

$$\Sigma\ M_x = -W_1 y_1 - W_2 y_2 - W_3 y_3 = -W\bar{y}$$

or

$$\bar{y} = \frac{W_1 y_1 + W_2 y_2 + W_3 y_3}{W_1 + W_2 + W_3} \tag{8.4}$$

For n particles in the xy plane, the location of the center of gravity is given by

$$\bar{x} = \frac{W_1 x_1 + W_2 x_2 + \cdots + W_n x_n}{W_1 + W_2 + \cdots + W_n}$$

and   (8.5)

$$\bar{y} = \frac{W_1 y_1 + W_2 y_2 + \cdots + W_n y_n}{W_1 + W_2 + \cdots + W_n}$$

EXAMPLE 8.2

Find the center of gravity for the system of four particles in the xy plane shown in Fig. 8.6.

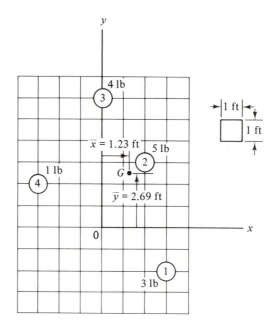

FIGURE 8.6

Solution

From Fig. 8.6 the weights and coordinates are as follows:

$$W_1 = 3 \text{ lb at } x_1 = 3 \text{ ft} \qquad y_1 = -2 \text{ ft}$$
$$W_2 = 5 \text{ lb at } x_2 = 2 \text{ ft} \qquad y_2 = 3 \text{ ft}$$

$$W_3 = 4 \text{ lb at } x_3 = 0 \qquad y_3 = 6 \text{ ft}$$

$$W_4 = 1 \text{ lb at } x_4 = -3 \text{ ft} \qquad y_4 = 2 \text{ ft}$$

From Eqs. (8.5),

$$\bar{x} = \frac{W_1 x_1 + W_2 x_2 + W_3 x_3 + W_4 x_4}{W_1 + W_2 + W_3 + W_4}$$

$$= \frac{3(3) + 5(2) + 4(0) + 1(-3)}{3 + 5 + 4 + 1} = \frac{16}{13}$$

$$= 1.23 \text{ ft}$$

and

$$\bar{y} = \frac{W_1 y_1 + W_2 y_2 + W_3 y_3 + W_4 y_4}{W_1 + W_2 + W_3 + W_4}$$

$$= \frac{3(-2) + 5(3) + 4(6) + 1(2)}{3 + 5 + 4 + 1} = \frac{35}{13}$$

$$= 2.69 \text{ ft}$$

The location of the center of gravity G is shown in Fig. 8.6.

PROBLEMS

8.1 through 8.4 Determine the moments of the forces shown about the x axis, y axis, and z axis.

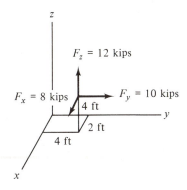

PROB. 8.1

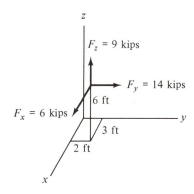

PROB. 8.2

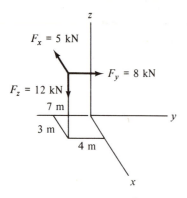

PROB. 8.3

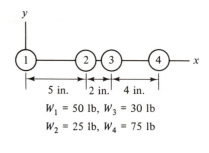

$W_1 = 50$ lb, $W_3 = 30$ lb

$W_2 = 25$ lb, $W_4 = 75$ lb

PROB. 8.6

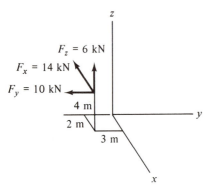

PROB. 8.4

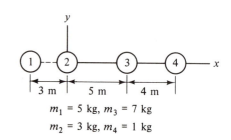

$m_1 = 5$ kg, $m_3 = 7$ kg

$m_2 = 3$ kg, $m_4 = 1$ kg

PROB. 8.7

8.5 through 8.16 Determine the center of gravity of the system of particles with weights or masses and locations as shown.

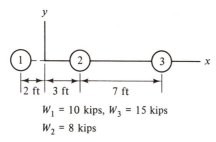

$W_1 = 10$ kips, $W_3 = 15$ kips

$W_2 = 8$ kips

PROB. 8.5

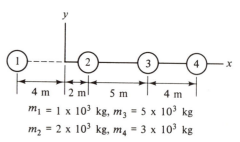

$m_1 = 1 \times 10^3$ kg, $m_3 = 5 \times 10^3$ kg

$m_2 = 2 \times 10^3$ kg, $m_4 = 3 \times 10^3$ kg

PROB. 8.8

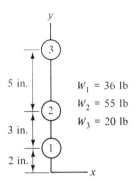

$W_1 = 36$ lb
$W_2 = 55$ lb
$W_3 = 20$ lb

PROB. 8.9

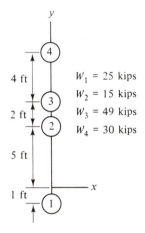

$W_1 = 25$ kips
$W_2 = 15$ kips
$W_3 = 49$ kips
$W_4 = 30$ kips

PROB. 8.10

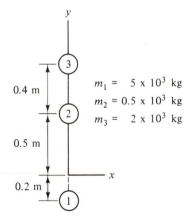

$m_1 = 5 \times 10^3$ kg
$m_2 = 0.5 \times 10^3$ kg
$m_3 = 2 \times 10^3$ kg

PROB. 8.11

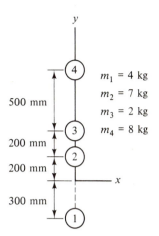

$m_1 = 4$ kg
$m_2 = 7$ kg
$m_3 = 2$ kg
$m_4 = 8$ kg

PROB. 8.12

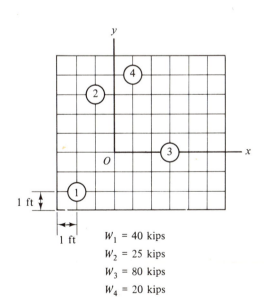

$W_1 = 40$ kips
$W_2 = 25$ kips
$W_3 = 80$ kips
$W_4 = 20$ kips

PROB. 8.13

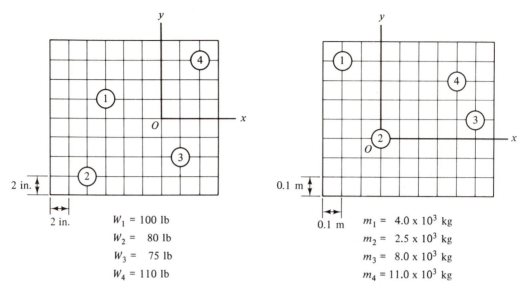

$W_1 = 100$ lb
$W_2 = 80$ lb
$W_3 = 75$ lb
$W_4 = 110$ lb

PROB. 8.14

$m_1 = 4.0 \times 10^3$ kg
$m_2 = 2.5 \times 10^3$ kg
$m_3 = 8.0 \times 10^3$ kg
$m_4 = 11.0 \times 10^3$ kg

PROB. 8.15

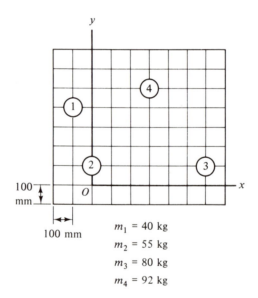

$m_1 = 40$ kg
$m_2 = 55$ kg
$m_3 = 80$ kg
$m_4 = 92$ kg

PROB. 8.16

8.4 CENTER OF GRAVITY FOR THIN PLATES AND CENTROIDS

In Fig. 8.7 a plane figure and a thin homogeneous plate of uniform thickness in the same shape as the plane figure are shown. The *center of gravity*

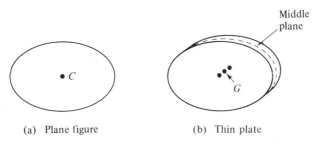

(a) Plane figure   (b) Thin plate

FIGURE 8.7

of the thin plate is on its middle plane. The *centroid* is at the same location in the plane of the thin homogeneous plate. The terms "centroid" and "center of gravity" are often used to indicate the same point. They are at the same location only if the plate is homogeneous.

There are many common plane figures for which the centroid can be located by inspection by making use of the following information. For any plane area with an axis of symmetry, the centroid lies on the axis of symmetry. For any plane area with two axes of symmetry, the centroid lies on the intersection of the two symmetric axes. Examples of plane areas with axes of symmetry are given in Fig. 8.8

Other plane figures whose centroid can be located by inspection are shown in Fig. 8.9. In these figures any point A in the figure has a corresponding point B so that the centroid C of the figure bisects the line AB.

The centroids for several plane figures are given in Table A.3 of the Appendix.

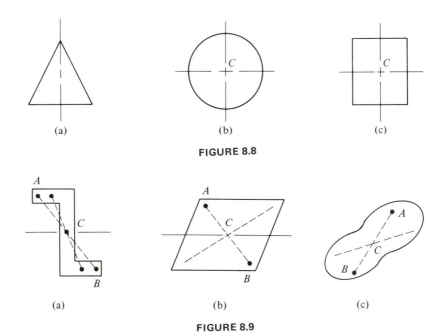

(a)   (b)   (c)

FIGURE 8.8

(a)   (b)   (c)

FIGURE 8.9

8.5 MOMENTS OF PLANE AREAS

Consider the plane area A whose centroid C is known in Fig. 8.10. The first moment of the area about the x axis is the product of the area and the distance from the centroid of the area to the x axis. That is,

$$M_x = A\overline{y} \qquad \text{(a)}$$

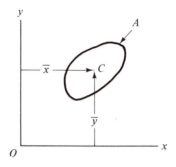

FIGURE 8.10

Similarily, the first moment of the area about the y axis is the product of the area and the distance from the centroid of the area to the y axis. That is,

$$M_y = A\overline{x} \qquad \text{(b)}$$

EXAMPLE 8.3

Find the moment about the x and y axis of the triangular area in Fig. 8.11.

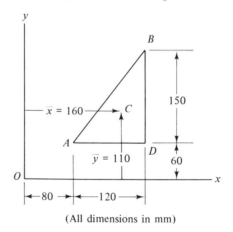

(All dimensions in mm)

FIGURE 8.11

Solution

From Table A.3 of the Appendix, the formulas for the area and centroidal location for a triangle are known. The area

$$A = \frac{bh}{2} = \frac{120(150)}{2} = 9000 \text{ mm}^2$$

The distance from the vertical side *BD* of the triangle to the centroid of the triangle is $b/3 = 120/3 = 40$ mm. The distance from the *y* axis to the centroid is $\bar{x} = 80 + 120 - 40 = 160$ mm and the moment of the triangular area about the *y* axis is given by Eq. (b).

$$M_y = A\bar{x} = 9000(160) = 1.44 \times 10^6 \text{ mm}^3 \qquad \text{Answer}$$

The distance from the horizontal side *AD* of the triangle to the centroid is $h/3 = 150/3 = 50$ mm. The distance from the *x* axis to the centroid is $\bar{y} = 60 + 50 = 110$ mm and the moment of the triangular area about the *x* axis is given by Eq. (a).

$$M_x = A\bar{y} = 9000(110) = 0.99 \times 10^6 \text{ mm}^3 \qquad \text{Answer}$$

EXAMPLE 8.4

Find the moment of the composite area shown in Fig. 8.12(a) about the *x* and *y* axes.

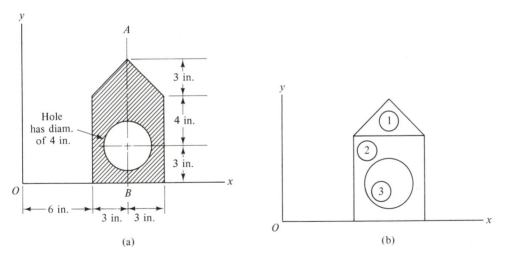

(a)   (b)

FIGURE 8.12

Solution

Let us visualize the composite area as made up of three simple areas, indicated in Fig. 8.12(b), where area ① is a triangle, area ② is a rectangle, and area ③ is a circular hole. The areas and centroidal distances are shown in tabular form on page 182.

	Shape	Area (in.²)	y (in.)	$M_x = Ay$ (in.³)
①	Triangle	$bh/2 = 6(3)/2 = 9$	$7 + 3/3 = 8$	72
②	Rectangle	$bh = 6(7) = 42$	$7/2 = 3.5$	147
③	Circle	$-\pi d^2/4 = -\pi(4)^2/4 = -12.57$	3	-37.7
		$\Sigma\ A = 38.43$ in.²		$\Sigma\ Ay = 181.3$ in.³

To set up this table, we have used the results given in Table A.3. We use a negative sign for area ③ because the circular hole has been removed from the rectangle. The moment of the composite areas about the x axis is equal to the sum of the moments of the simple areas about the x axis as shown in the table.

$$M_x = \Sigma\ A\bar{y} = 72.0 + 147.0 - 37.70 = 181.3 \text{ in.}^3 \qquad \text{Answer}$$

Since AB is an axis of symmetry for the composite area, the centroid lies on the line AB and the centroidal distance $\bar{x} = 6 + 3 = 9$ in. The total area is equal to the sum of the areas as shown in the table.

$$A = \Sigma\ A = 9.0 + 42.0 - 12.57 = 38.43 \text{ in.}^2$$

The moment about the y axis is given by

$$M_y = (\Sigma\ A)\bar{x} = 38.43(9) = 345.9 \text{ in.}^3 \qquad \text{Answer}$$

8.6 CENTROIDS OF COMPOSITE AREAS

When a composite area can be broken up into simple areas whose centroids are known, the centroid of the composite area can be determined. The method is similar to the method of finding the center of gravity of a system of particles.

In Fig. 8.13(a), we consider three simple areas A_1, A_2, and A_3 whose centroids C_1, C_2, and C_3 are known. We also consider, in Fig. 8.13(b), the composite system made up of the total area with an unknown centroid. By construction, the sum of the simple areas must be equal to the composite or total area.

$$A = \Sigma\ A = A_1 + A_2 + A_3$$

The sum of the moments of the simple areas about an axis must be equal to the moment of the composite area about the same axis. For the y axis,

$$M_y = A_1 x_1 + A_2 x_2 + A_3 x_3 = (A_1 + A_2 + A_3)\bar{x}$$

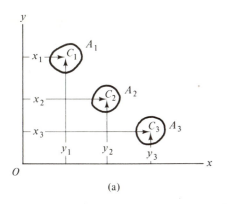

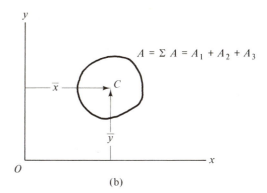

(a)   (b)

FIGURE 8.13

or

$$\bar{x} = \frac{A_1 x_1 + A_2 x_2 + A_3 x_3}{A_1 + A_2 + A_3} = \frac{\Sigma\ Ax}{\Sigma\ A} \qquad (8.6)$$

For the x axis,

$$M_x = A_1 y_1 + A_2 y_2 + A_3 y_3 = (A_1 + A_2 + A_3)\bar{y}$$

or

$$\bar{y} = \frac{A_1 y_1 + A_2 y_2 + A_3 y_3}{A_1 + A_2 + A_3} = \frac{\Sigma\ Ay}{\Sigma\ A} \qquad (8.7)$$

EXAMPLE 8.5

Find the centroid of the composite area shown in Fig. 8.12(a).

Solution

In Example 8.4 we found that

$$\Sigma\ Ax = 345.9\ \text{in.}^3, \quad \Sigma\ Ay = 181.3\ \text{in.}^3, \quad \text{and} \quad \Sigma\ A = 38.43\ \text{in.}^2$$

From Eq. (8.6),

$$\bar{x} = \frac{\Sigma\ Ax}{\Sigma\ A} = \frac{345.9}{38.43} = 9\ \text{in.} \qquad\qquad \text{Answer}$$

and from Eq. (8.7),

$$\bar{y} = \frac{\Sigma\ Ay}{\Sigma\ A} = \frac{181.3}{38.43} = 4.72\ \text{in.} \qquad\qquad \text{Answer}$$

The centroidal distance $\bar{x} = 9$ in. was already known because AB is an axis of symmetry.

EXAMPLE 8.6

Find the centroid of the composite area shown in Fig. 8.14(a).

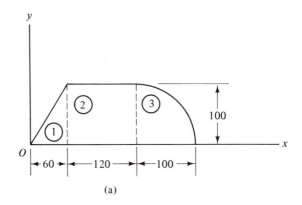

(a)

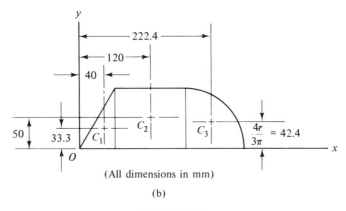

(All dimensions in mm)

(b)

FIGURE 8.14

Solution

Imagine the composite area as made up of three simple areas as shown in Fig. 8.14(b), where ① is the triangle, ② is the rectangle, and ③ is a quarter-circle. The areas and centroid distances are shown in Figs. 8.14(b) and tabulated as shown. To calculate these values we used formulas given in Table A.3 of the Appendix. From Eqs. (8.6) and (8.7) and the sums in the table,

$$\bar{x} = \frac{\Sigma\, Ax}{\Sigma\, A} = \frac{3.307 \times 10^6}{22.85 \times 10^3} = 144.7 \text{ mm} \qquad\qquad \text{Answer}$$

and

$$\bar{y} = \frac{\Sigma \, Ay}{\Sigma \, A} = \frac{1.033 \times 10^6}{22.85 \times 10^3} = 45.2 \text{ mm}$$ Answer

	Shape	A (mm²)	x (mm)	Ax (mm³)	y (mm)	Ay (mm³)
①	Triangle	3 000	40	0.120×10^6	33.3	0.100×10^6
②	Rectangle	12 000	120	1.440×10^6	50	0.600×10^6
③	Quarter-circle	7 850	222.4	1.747×10^6	42.4	0.333×10^6
		$\Sigma A =$ 22 850 mm²		$\Sigma Ax =$ 3.307×10^6 mm³		$\Sigma Ay =$ 1.033×10^6 mm³

PROBLEMS

8.17 through 8.20   Locate the centroid of the
   plane areas shown, by inspection.

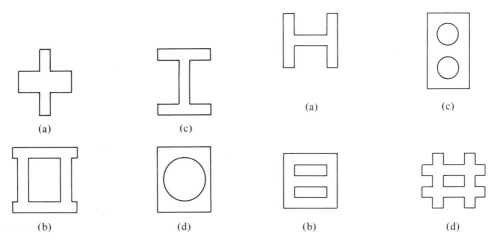

(a)   (c)   (a)   (c)

(b)   (d)   (b)   (d)

PROB. 8.17   PROB. 8.18

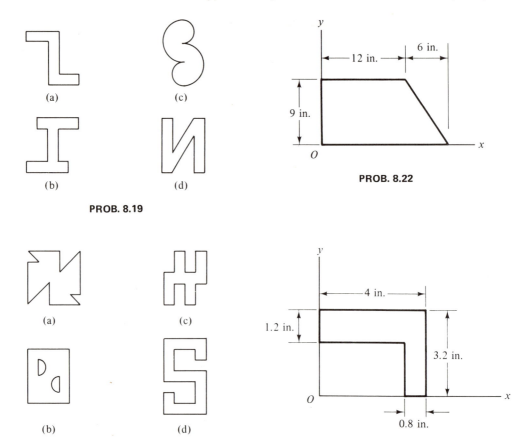

(a)

(c)

(b)

(d)

PROB. 8.19

PROB. 8.22

(a)

(c)

(b)

(d)

PROB. 8.20

PROB. 8.23

8.21 through 8.36 Locate the centroid of the
plane composite area shown.

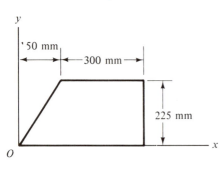

PROB. 8.21

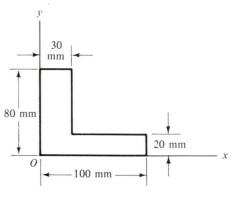

PROB. 8.24

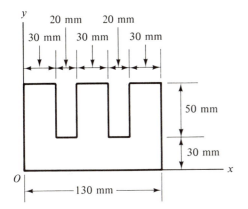

PROB. 8.25

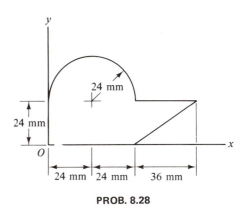

PROB. 8.28

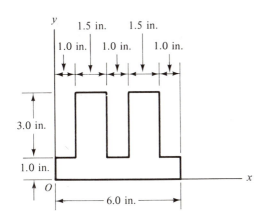

PROB. 8.26

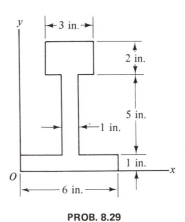

PROB. 8.29

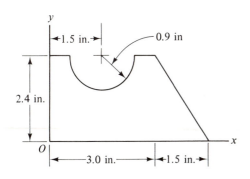

PROB. 8.27

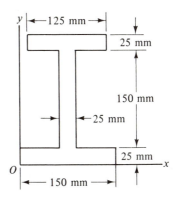

PROB. 8.30

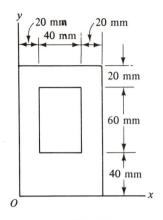

PROB. 8.31

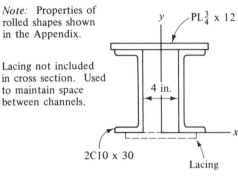

Note: Properties of rolled shapes shown in the Appendix.

Lacing not included in cross section. Used to maintain space between channels.

2C10 x 30

Lacing

PL$\frac{3}{4}$ x 12

4 in.

PROB. 8.34

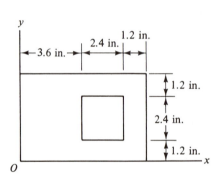

PROB. 8.32

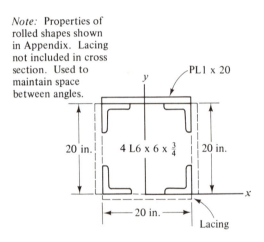

Note: Properties of rolled shapes shown in Appendix. Lacing not included in cross section. Used to maintain space between angles.

PL1 x 20

4 L6 x 6 x $\frac{3}{4}$

20 in.   20 in.

20 in.

Lacing

PROB. 8.35

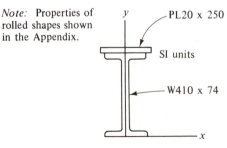

Note: Properties of rolled shapes shown in the Appendix.

PL20 x 250

SI units

W410 x 74

PROB. 8.33

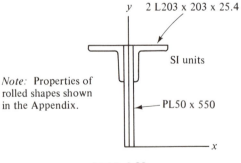

2 L203 x 203 x 25.4

SI units

Note: Properties of rolled shapes shown in the Appendix.

PL50 x 550

PROB. 8.36

8.7 MOMENT OF INERTIA OF A PLANE AREA

In the study of the strength of beams and columns and statics of fluids, the moment of inertia or second moment of a plane area is required. Consider the plane area shown in Fig. 8.15. The area is divided into n thin strips parallel to the x axis of areas $\Delta A_1, \Delta A_2, \ldots, \Delta A_n$ with centroidal distances $y_1, y_2, \ldots, y_n$ from the x axis to the centroid of the area.

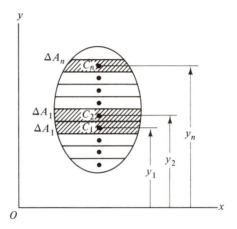

FIGURE 8.15

The moment of inertia of the area ΔA about the x axis is equal to the product of the square of the centroidal distance y and the area. With n areas we add up the moments of inertia for each area to find the total approximate moment of inertia. Thus the approximate moment of inertia with respect to the x axis I_x is given by

$$I_x = y_1^2 \Delta A_1 + y_2^2 \Delta A_2 + \cdots + y_n^2 \Delta A_n = \Sigma \, y_i^2 \, \Delta A_i \qquad (8.8)$$

The approximate moment of inertia with respect to the y axis I_y is given by

$$I_y = x_1^2 \Delta A_1 + x_2^2 \Delta A_2 + \cdots + x_n^2 \Delta A_n = \Sigma \, x_i^2 \, \Delta A_i \qquad (8.9)$$

where the thin strips of area ΔA_i are parallel to the y axis and x_i is the distance from the y axis to the centroid of the area ΔA_i.

If we let the number of areas increase without limit, we obtain the exact moment of inertia. This requires integral calculus and is beyond the scope of this text.

Formulas for the moment of inertia of simple areas are given in Table A.4 of the Appendix.

EXAMPLE 8.7

For the area shown in Fig. 8.16(a), determine the approximate value and exact value for I_x.

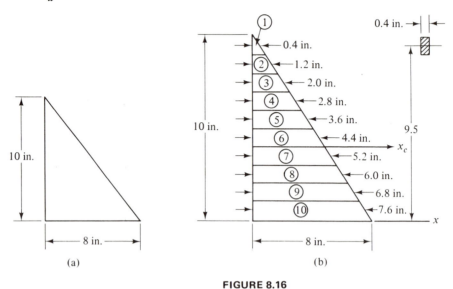

FIGURE 8.16

Solution

As shown in Fig. 8.16(b), divide the triangle into 10 horizontal strips 1 in. wide. Replace each strip with a small rectangle whose length is the average length of the strip and whose centroidal distance is the mean distance of the rectangle from the x axis.

The area of strip ①, $\Delta A_1 = 0.4$ in.^{2} and the centroid distance $y_1 = 9.5$ in. from the x axis. The moment of inertia I_{x_1} is

$$I_{x_1} = y_1^2 \Delta A_1 = (9.5)^2 (0.4) = 36.1 \text{ in.}^4$$

The remaining moments of inertia are calculated in the same way. The calculations are tabulated as shown. From the sum shown in the table,

$$I_x = 670 \text{ in.}^4 \qquad\qquad \text{Answer}$$

The exact value is calculated from the formula given in Table A.4 of the Appendix.

$$I_x = \frac{bh^3}{12} = \frac{8(10)^3}{12} = 666.7 \text{ in.}^4 \qquad\qquad \text{Answer}$$

The error in the approximation is only 0.5 percent. With additional strips the error will decrease.

This method for finding the approximate moment of inertia can be useful when the areas are irregular and formulas for finding their moments of inertia are not available.

	y_i (in.)	ΔA_i (in.^{2})	$I_{x_i} = y_i^2 \Delta A_i$ (in.^{4})
①	9.5	0.4	36.1
②	8.5	1.2	86.7
③	7.5	2.0	112.5
④	6.5	2.8	118.3
⑤	5.5	3.6	108.9
⑥	4.5	4.4	89.1
⑦	3.5	5.2	63.7
⑧	2.5	6.0	37.5
⑨	1.5	6.8	15.3
⑩	0.5	7.6	1.9
			$\Sigma I_{x_i} = 670$ in.^{4}

8.8 POLAR MOMENT OF INERTIA

The product of the area ΔA_i and the square of the distance r_i from the origin O (or z axis) to the centroid of the area is called the *polar moment of inertia* for the area ΔA_i (Fig. 8.17). Adding the polar moments of

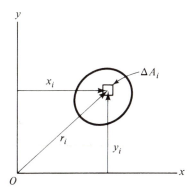

FIGURE 8.17

inertia for each area, we find the total approximate polar moment of inertia J, that is,

$$J = \Sigma \ r_i^2 \Delta A_i \qquad (8.10)$$

From the Pythagorean theorem, $r_i^2 = x_i^2 + y_i^2$. Therefore,

$$J = \Sigma \ (x_i^2 + y_i^2) \Delta A_i = \Sigma \ x_i^2 \Delta A_i + \Sigma \ y_i^2 \Delta A_i$$

From Eqs. (8.8) and (8.9),

$$J = I_x + I_y \qquad (8.11)$$

EXAMPLE 8.8

A plane area has moments of inertia $I_x = 89.2$ in.^{4} and $I_y = 143.5$ in.^{4}, find the polar moment of inertia.

Solution

From Eq. (8.11),

$$J = I_x + I_y = 89.2 + 143.5 = 232.7 \text{ in.}^4 \qquad \text{Answer}$$

8.9 RADIUS OF GYRATION

The radius of gyration r of an area with respect to a given axis is defined by the relationship

$$r = \sqrt{\frac{I}{A}} \quad \text{or} \quad I = Ar^2 \qquad (8.12)$$

where I is the moment of inertia with respect to the given axis and A the cross-sectional area. The radius of gyration is used in structural calculations and is tabulated for standard structural shapes in Tables A.5 through A.9 of the Appendix.

EXAMPLE 8.9

The 12-in. by 8-in. structural tubing shown in Fig. 8.18 has moments of inertia $I_x = 337$ in.^{4}, $I_y = 181$ in.^{4}, and cross-sectional area $A = 17.9$ in.^{2}. Determine the radius of gyration with respect to the x and y axes.

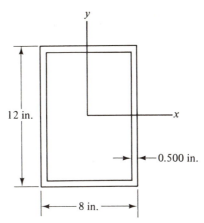

TS12 x 8 x 500 (structural tubing)

FIGURE 8.18

Solution

From Eq. (8.12),

$$r_x = \sqrt{\frac{I_x}{A}} = \sqrt{\frac{337}{17.9}} = 4.34 \text{ in.}$$ Answer

$$r_y = \sqrt{\frac{I_y}{A}} = \sqrt{\frac{181}{17.9}} = 3.18 \text{ in.}$$ Answer

PROBLEMS

8.37 through 8.40 For the area shown, determine (a) the exact value of the moment of inertia I_x by formula, (b) the approximate value of the moment of inertia I_x if the area is divided into 10 horizontal strips of equal width, and (c) the percent error in the approximation. Percent error equals 100 (approximate I_x - exact I_x)/exact I_x.

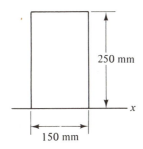

PROB. 8.37

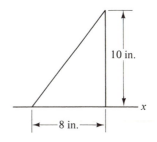

PROB. 8.38

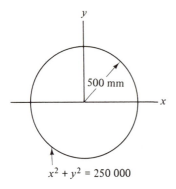

$x^2 + y^2 = 250\ 000$

PROB. 8.39

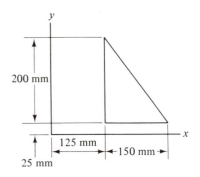

PROB. 8.42

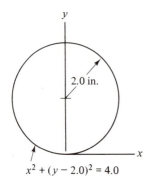

$x^2 + (y - 2.0)^2 = 4.0$

PROB. 8.40

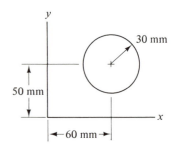

PROB. 8.43

8.41 through 8.44 For the areas shown, determine (a) the centroidal moment of inertia I_{x_c} and I_{y_c} and (b) the moment of inertia I_x and I_y.

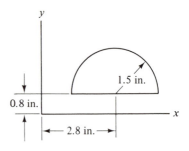

PROB. 8.44

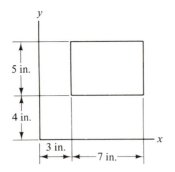

PROB. 8.41

In Probs. 8.45 through 8.48, determine the polar centroidal moment of inertia J_c and the polar moment of inertia J for the areas shown.

8.45 Use the figure for Prob. 8.41.

8.46 Use the figure for Prob. 8.42.

8.47 Use the figure for Prob. 8.43.

8.48 Use the figure for Prob. 8.44.

8.10 TRANSFER FORMULA

The *transfer formula* or parallel-axis theorem gives a relationship between the moment of inertia with respect to any axis and the moment of inertia with respect to a parallel axis through the centroid. Consider the area shown in Fig. 8.19. The bb' axis is a centroidal axis, that is, through the centroid. The aa' axis is any axis parallel to the centroidal axis.

The moment of inertia of the element of area ΔA_i with respect to the aa' axis is given by

$$I_i = (y_i + d)^2 \Delta A_i$$

The moment of inertia with respect to the aa' axis of the total area A is found by summing I_i for all the elements of area ΔA_i.

$$I = \Sigma\, I_i = \Sigma\, (y_i + d)^2 \Delta A_i$$

Expanding and distributing terms, we have

$$I = \Sigma\, (y_i^2 + 2\, dy_i + d^2)\Delta A_i$$
$$= \Sigma\, y_i^2 \Delta A_i + \Sigma\, 2\, dy_i \Delta A_i + \Sigma\, d^2 \Delta A_i$$

The first term on the right is the moment of inertia with respect to the centroidal axis I_c. The second term is equal to $2d \Sigma\, y_i \Delta A_i$. Since the moment of area with respect to the centroidal axis, $\Sigma\, y_i \Delta A_i$, is equal to zero, the second term is equal to zero. The last term is equal to $d^2 \Sigma\, \Delta A_i$ and $\Sigma\, \Delta A_i$ is the total area A. Therefore,

$$I = I_c + Ad^2 \tag{8.13}$$

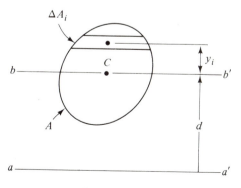

FIGURE 8.19

where I_c is the moment of inertia with respect to the centroidal axis, A the total area, and d the distance between the two parallel axes aa' and bb', the centroidal axis.

We notice that the term Ad^2 is *added* when the transfer is made from the centroidal axis but *subtracted* when the transfer is made to the centroidal axis. Therefore, the moment of inertia with respect to the centroidal axis is always smaller than with respect to any other parallel axis.

Similar relationships exist for the polar moment of inertia and the radius of gyration between a centroidal axis and any other parallel axis. Thus

$$J = J_c + Ad^2 \tag{8.14}$$

$$r^2 = r_c^2 + d^2 \tag{8.15}$$

EXAMPLE 8.10

Find the moment of inertia of the triangle shown in Fig. 8.20(a) with respect to the x and x_c axes.

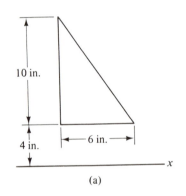

(a)

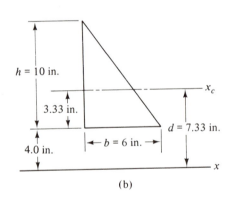

(b)

FIGURE 8.20

Solution

From Table A.3 of the Appendix the centroid of the triangle is $h/3 = 10/3 = 3.33$ in. above the base of the triangle [Fig. 8.20(b)]. The distance between the parallel axes is $d = 4.0 + 3.33 = 7.33$ in. The area of the triangle is $A = bh/2 = 6(10)/2 = 30$ in.^{2}, and from Table A.4 of the Appendix the moment of inertia with respect to the centroidal axis is

$$I_c = \frac{bh^3}{36} = \frac{6(10)^3}{36} = 166.7 \text{ in.}^4$$

From Eq. (8.13),

$$I_x = I_c + Ad^2 = 166.7 + 30\,(7.33)^2$$

$$= 1780 \text{ in.}^4$$

8.11 MOMENT OF INERTIA OF COMPOSITE AREAS

When a composite area can be broken up into simple areas whose centroid and moment of inertia with respect to the centroid are known or can be calculated, the moment of inertia of the composite area can be determined. The following example will illustrate the method.

EXAMPLE 8.11

Find the moment of inertia with respect to the y axis for the composite area shown in Fig. 8.21(a).

Solution

The composite area is divided into three simple shapes (a triangle, rectangle, and quarter-circle) as shown in Fig. 8.21(b)–(d). The centroidal distance, area, and centroidal moment of inertia are calculated for each shape from the formulas given in Table A.4 of the Appendix and displayed in Fig. 8.21. To find the moment of inertia with respect to the y axis for areas ①, ②, and ③, the transfer formula is used as follows:

$$(I_y)_1 = (I_c)_1 + (Ad^2)_1 = 0.6 \times 10^6 + 3 \times 10^3 (40)^2$$

$$= 5.4 \times 10^6 \text{ mm}^4$$

$$(I_y)_2 = (I_c)_2 + (Ad^2)_2 = 14.4 \times 10^6 + 12 \times 10^3 (120)^2$$

$$= 187.2 \times 10^6 \text{ mm}^4$$

$$(I_y)_3 = (I_c)_3 + (Ad^2)_3 = 5.5 \times 10^6 + 7.85 \times 10^3 (222.4)^2$$

$$= 394.1 \times 10^6 \text{ mm}^4$$

The total moment of inertia about the y axis for the composite area is found by adding the moment of inertia for the three areas.

$$I_y = (I_y)_1 + (I_y)_2 + (I_y)_3$$

$$= 5.4 \times 10^6 + 187.2 \times 10^6 + 394.1 \times 10^6$$

$$= 586.7 \times 10^6 \text{ mm}^4 \qquad\qquad \text{Answer}$$

For convenience, the calculations have also been tabulated as shown on page 199. The column for $I_y = I_c + Ad^2$ could have been omitted from the table and the moment of inertia obtained by adding the sums ΣI_c and $\Sigma (Ad^2)$. That is,

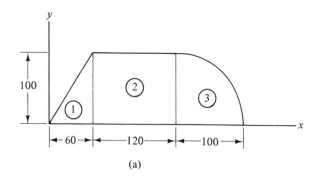

(a)

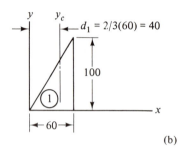

$A_1 = bh/2 = 60(100)/2 = 3 \times 10^3$ mm^2

$I_{c_1} = bh^3/36 = 100(60)^3/36 = 0.6 \times 10^6$ mm^4

(b)

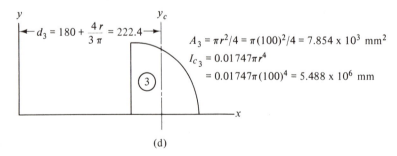

$A_2 = bh = 120(100) = 12 \times 10^3$ mm^2

$I_{c_2} = bh^3/12 = 100(120)^3/12 = 14.4 \times 10^6$ mm^4

(c)

$A_3 = \pi r^2/4 = \pi(100)^2/4 = 7.854 \times 10^3$ mm^2

$I_{c_3} = 0.01747\pi r^4$

$= 0.01747\pi(100)^4 = 5.488 \times 10^6$ mm

(d)

(All dimensions in mm)

FIGURE 8.21

$$I_y = \Sigma\, I_c + \Sigma\, (Ad^2) = 20.5 \times 10^6 + 566.2 \times 10^6$$

$$= 586.7 \times 10^6 \text{ mm}^4$$

which is the same result as before.

	Shape	$A\ (mm^2)$	$d\ (mm)$	$I_c\ (mm^4)$	$Ad^2\ (mm^4)$	$I_y = I_c + Ad^2\ (mm^4)$
①	Triangle	3×10^3	40	0.6×10^6	4.8×10^6	5.4×10^6
②	Rectangle	12×10^3	120	14.4×10^6	172.8×10^6	187.2×10^6
③	Quarter-circle	7.85×10^3	222.4	5.5×10^6	388.6×10^6	394.1×10^6
				$\Sigma\, I_c =$ 20.5×10^6 (mm^4)	$\Sigma\, (Ad^2) =$ 566.2×10^6 (mm^4)	$\Sigma\, I_y =$ 586.7×10^6 (mm^4)

EXAMPLE 8.12

Find the moment of inertia with respect to the centroidal axis y_c for the composite area shown in Fig. 8.21(a).

Solution

The centroid of the area was determined in Example 8.6 and the moment of inertia with respect to the y axis in Example 8.11. The values obtained are shown in Fig. 8.22. From the transfer formula, $I_y = I_c + Ad^2$ or $I_c = I_y - Ad^2$. Therefore,

$$I_c = 586.7 \times 10^6 - 22.85 \times 10^3\, (144.7)^2$$

$$= 108.3 \times 10^6 \text{ mm}^4 \qquad\qquad\qquad \text{Answer}$$

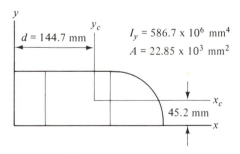

FIGURE 8.22

We observe that the moment of inertia with respect to the centroidal axis is smaller than with respect to any other parallel axis.

EXAMPLE 8.13

A structural member made up of two American Standard channels and a plate are fastened together as shown in Fig. 8.23(a). Determine (a) the moment of inertia of the composite section with respect to a centroidal axis parallel with the plate, and (b) the corresponding radius of gyration.

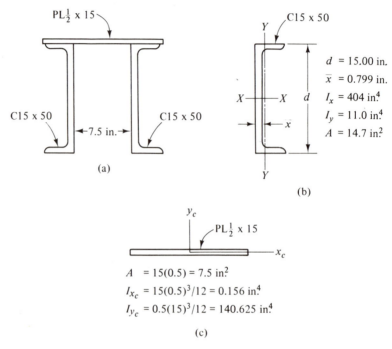

(a)

C15 x 50

d = 15.00 in.
$\bar{x}$ = 0.799 in.
I_x = 404 in⁴
I_y = 11.0 in⁴
A = 14.7 in²

(b)

A = 15(0.5) = 7.5 in²
I_{x_c} = 15(0.5)³/12 = 0.156 in⁴
I_{y_c} = 0.5(15)³/12 = 140.625 in⁴

(c)

FIGURE 8.23

Solution

Properties of the C15 × 50 channel are tabulated in Table A.7 of the Appendix and shown in Fig. 8.23(b). Properties of the plate (rectangle) are calculated and shown in Fig. 8.23(c). The origin of the coordinate axis for the composite cross section is located so that the x axis passes through the centroids of the two channel sections and the y axis is an axis of symmetry for the cross section (Fig. 8.24).

Since the y axis is an axis of symmetry, the composite centroid lies on the y axis and $\bar{x} = 0$. To find $\bar{y}$ we use the method of Sec. 8.6. The results are tabulated. From the sums shown in the table

$$\bar{y} = \frac{\Sigma Ay}{\Sigma A} = \frac{58.12}{36.9} = 1.575 \text{ in.}$$

	Shape	A (in.^{2})	y (in.)	Ay (in.^{3})
①	C15 × 50 (R)	14.7	0	0
②	C15 × 50 (L)	14.7	0	0
③	PL$\frac{1}{2}$ × 15	7.5	7.75	58.12
		$\Sigma A = 36.9$ in.^{2}		$\Sigma Ay = 58.12$ in.^{3}

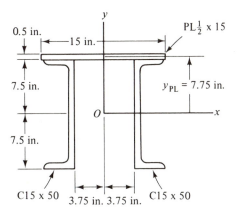

FIGURE 8.24

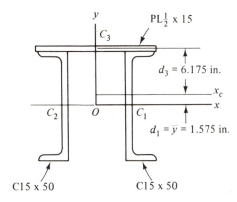

FIGURE 8.25

To find the moment of inertia with respect to the centroidal axis parallel to the plate, the methods of Sec. 8.11 will be used. Transfer distances for the channels and plate are shown in Fig. 8.25. Properties for the channel and plate were shown in

Fig. 8.23(b) and (c). The results are tabulated as shown. From the transfer formula and the sums shown in the table,

$$I_{x_c} = \Sigma\, I_c + \Sigma\, (Ad^2) = 808.2 + 358.9$$

$$I_{x_c} = 1167 \text{ in.}^4 \qquad\qquad \text{Answer}$$

	Shape	A (in.^{2})	d (in.)	I_c (in.^{4})	Ad^2 (in.^{4})
①	C15 × 50 (R)	14.7	−1.575	404	36.47
②	C15 × 50 (L)	14.7	−1.575	404	36.47
③	PL$\frac{1}{2}$ × 15	7.5	6.175	0.156	286.0
		$\Sigma\, A = 36.9$ in.^{2}		$\Sigma\, I_c =$ 808.2 (in.^{4})	$\Sigma\, (Ad^2) =$ 358.9 (in.^{4})

The radius of gyration is defined by Eq. (8.12); therefore,

$$r_{x_c} = \sqrt{\frac{I_{x_c}}{A}} = \sqrt{\frac{1167}{36.9}} = 5.62 \text{ in.} \qquad\qquad \text{Answer}$$

8.12 PROPERTIES OF STRUCTURAL STEEL SHAPES

Structural steel is rolled into a variety of shapes and sizes. The steel shapes are designated by the shape and size of their cross section. The properties of the cross sections of selected structural steel shapes are given in Tables A.5 through A.9 for common units. SI units are given in Tables (SI)A.5 through (SI)A.9.

Some of the typical shapes may be described as follows:

1. Wide-flange beam. This section has a shape similar to an I. The two parallel horizontal parts of the cross section are called *flanges* and the vertical part of the cross section is called the *web*. This beam is designated by the symbol W, the nominal depth in inches or mm, and weight in lb per ft or mass in kg per m. *Example:* W16 × 96; (SI units) W410 × 143.

2. American Standard beam. This section is commonly called an *I-beam*. The flanges are narrower than those of the wide-flange beam and the inner flange surface has a slope of approximately 2 to 12 in. It is designated by the symbol S, the nominal depth in inches or mm, and the weight in lb per ft or mass in kg per m. *Example:* S18 × 70; (SI units) S300 × 74.

3. American Standard channel. This section is similar to the American Standard beam with the flanges removed from one side. It is designated by the symbol C, the nominal depth in inches or mm, and the weight in lb per ft or mass in kg per m. *Example:* C15 × 50; (SI units) C375 × 74.4.

    4. Angles. This section has a shape like an L. The horizontal and vertical parts are called *legs.* It is designated by the letter L, the length of the long or equal leg, the length of the short or equal leg, and the thickness of the legs. The lengths and thickness are given in inches or mm. *Examples:* L8 × 8 × 1 and L5 × 3 × $\frac{1}{2}$; (SI units) L203 × 203 × 25.4 and L127 × 76 × 12.7.

    5. Plates. A plate has a rectangular cross section, usually 6 in. (152 mm) or more in width. It is designated by the symbol PL, the thickness of the plate, and the width of the plate. The thickness and width are given in inches or mm. *Example:* PL$\frac{1}{2}$ × 12; (SI units) PL12.7 × 305.

PROBLEMS

8.49 through 8.68 For the composite area shown, determine the moment of inertia with respect to the centroidal axes (a) I_{x_c}, and (b) I_{y_c}.

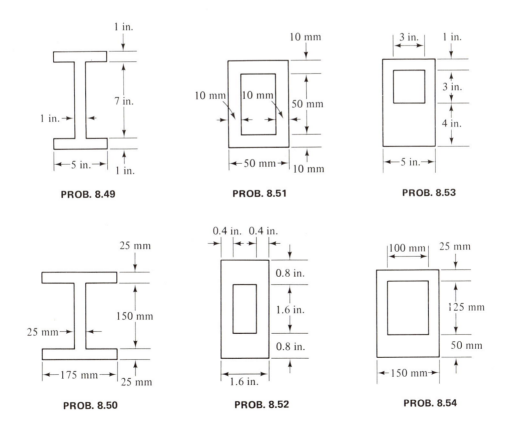

PROB. 8.49   PROB. 8.51   PROB. 8.53

PROB. 8.50   PROB. 8.52   PROB. 8.54

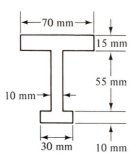

PROB. 8.55

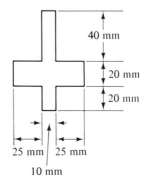

PROB. 8.59

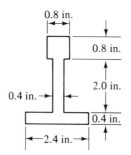

PROB. 8.56

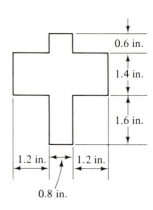

PROB. 8.60

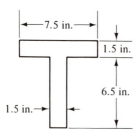

PROB. 8.57

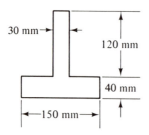

PROB. 8.58

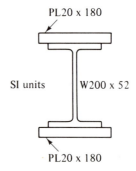

PROB. 8.61

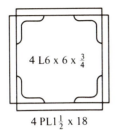

$4 L6 \times 6 \times \frac{3}{4}$

$4 PL1\frac{1}{2} \times 18$

PROB. 8.62

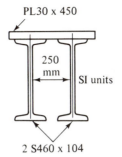

PL30 x 450

250 mm

SI units

2 S460 x 104

PROB. 8.66

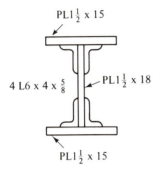

$PL1\frac{1}{2} \times 15$

$4 L6 \times 4 \times \frac{5}{8}$ $PL1\frac{1}{2} \times 18$

$PL1\frac{1}{2} \times 15$

PROB. 8.63

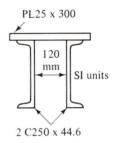

PL25 x 300

120 mm

SI units

2 C250 x 44.6

PROB. 8.67

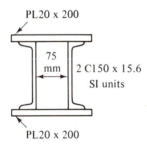

PL20 x 200

75 mm

2 C150 x 15.6
SI units

PL20 x 200

PROB. 8.64

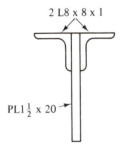

2 L8 x 8 x 1

$PL1\frac{1}{2} \times 20$

PROB. 8.68

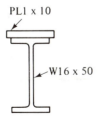

PL1 x 10

W16 x 50

PROB. 8.65

In Probs. 8.69 through 8.72, determine the radius of gyration with respect to the centroidal axes for the areas shown.

8.69   Use the figure for Prob. 8.50.

8.70   Use the figure for Prob. 8.52.

8.71   Use the figure for Prob. 8.62.

8.72   Use the figure for Prob. 8.64.

Internal Reactions—
Stress for Axial Loads

9.1 INTRODUCTION

Strength of materials or mechanics of materials is concerned with the stresses and deflections or deformations in a body. These are produced by loads on the body and by the dimensions of the body. We will be able to determine from our study of strength of materials if a body fulfills its intended purpose and to design the body so that it can safely support a given load.

Although not as old as statics, strength of materials dates back to the time of the Renaissance. Galileo in his book *Two New Sciences*, published in 1638, made reference to the properties of structural materials and discussed the strength of beams. In 1678 Hooke published the experimental relationship between force and displacement which is now called

Hooke's law. However, much of what is now called strength of materials was developed by French investigators in the late 1700s and the early 1800s. Most notable among them were Coulomb and Navier.

In statically determinate problems the internal reaction will be found by cutting through a body with a section, isolating the free body on either side of the section, and applying the equilibrium equations. Forces at the cut section are internal reactions and represent the resultant of forces that are distributed over the section within the body. These distributed forces per unit of area are called *stresses*. However, in statically indeterminate problems it will be necessary to consider changes in shape or size of the body in addition to the equilibrium equation.

9.2 METHOD OF SECTIONS AND INTERNAL REACTIONS

In our study of trusses we found internal forces in members of a truss by cutting the truss with a section and constructing a free-body diagram of

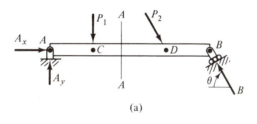

(a)

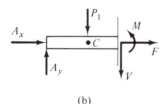

(b)

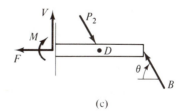

(c)

FIGURE 9.1

either part of the truss. The forces in the members cut became external forces and were found by the equilibrium equations. In the case of the truss the members cut were two-force members and the magnitude of the internal force was unknown, but its direction was the same as the direction of the member. In the more general case, when a member is cut by a section, both an unknown internal force and couple act at the section and both magnitude and direction are unknown.

Consider the free-body diagram of a member involving loads and reactions in a plane [Fig. 9.1(a)]. The three reactions A_x, A_y, and B are found from the equations of equilibrium. To find the internal reactions or stress resultants at a section A–A, we cut the member at right angles to the axis of the member, separate the member at that section, and draw a free-body diagram of either part of the member. The *axial force F*, the *shear force V*, and the bending couple M which act on the left-hand part of the member are shown in Fig. 9.1(b). The equal and opposite forces and couples that act on the right-hand part of the member are shown in Fig. 9.1(c). Notice that if the two parts of the member are put back together, the internal reactions add to zero and we have a free-body diagram of the entire member. For convenience, the axial force and shear force act through the centroid of the cross section.

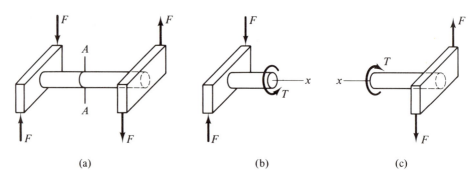

(a)   (b)   (c)

FIGURE 9.2

We will be interested in one other kind of internal reaction or stress resultant. It cannot be produced by forces in the plane of the member and is called the torque or twisting moment. The *torque T* is the moment about the axis of a member. Loads that produce a torque are shown in Fig. 9.2(a). The member is cut at section A–A and the internal torques that occur are shown in Fig. 9.2(b) and (c).

EXAMPLE 9.1

Calculate the internal reactions for the bent bar shown in Fig. 9.3(a) at sections A–A and B–B.

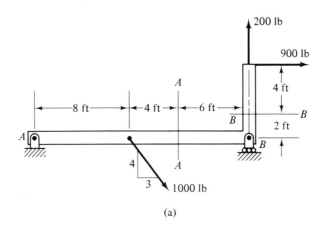

(a)

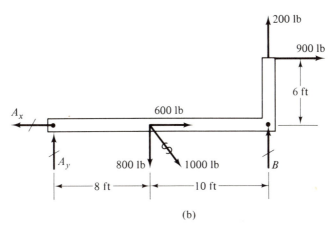

(b)

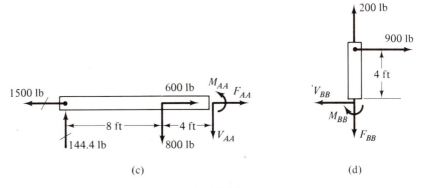

(c)   (d)

FIGURE 9.3

Solution

The free-body diagram of the bent bar is shown in Fig. 9.3(b). Components of the 1000-lb force are calculated and shown on the diagram. From equilibrium:

$$\circlearrowright \Sigma \, M_A = 0 \quad -800\,(8) + B\,(18) + 200\,(18) - 900\,(6) = 0$$

$$B = 455.6 \text{ lb}$$

$$\circlearrowright \Sigma \, M_B = 0 \quad -A_y\,(18) + 800\,(10) - 900\,(6) = 0$$

$$A_y = 144.4 \text{ lb}$$

$$\rightarrow \Sigma \, F_x = 0 \quad -A_x + 600 + 900 = 0$$

$$A_x = 1500 \text{ lb}$$

Check:

$$\uparrow \Sigma \, F_y = 0 \quad A_y + B - 800 + 200 = 0$$

$$144.4 + 455.6 - 800 + 200 = 0 \quad \text{OK}$$

For Section A–A

The free-body diagram for the bent bar to the left of the section A–A is shown in Fig. 9.3(c). The bar to the left of the section A–A is selected because it results in a simple free-body diagram. The axial force, shear force, and bending moment for section A–A are shown. For convenience the axial force and shear force act through the centroid C of the cross section. This portion of the beam is in equilibrium under the action of the forces, reactions, and internal reactions and couple. Therefore,

$$\rightarrow \Sigma \, F_x = 0 \quad 600 - 1500 + F_{AA} = 0$$

$$F_{AA} = 900 \text{ lb} \quad \text{(tension)} \qquad\qquad \text{Answer}$$

$$\uparrow \Sigma \, F_y = 0 \quad 144.4 - 800 - V_{AA} = 0$$

$$V_{AA} = -655.6 \text{ lb}$$

$$= 656 \text{ lb} \uparrow \qquad\qquad \text{Answer}$$

$$\circlearrowright \Sigma \, M_c = 0 \quad -144.4\,(12) + 800\,(4) + M_{AA} = 0$$

$$M_{AA} = -1467 \text{ lb-ft}$$

$$= 1467 \text{ lb-ft} \; \circlearrowleft \qquad\qquad \text{Answer}$$

For Section B–B

The free-body diagram for the bent bar above section B–B is shown in Fig. 9.3(d). The bar above section B–B is selected because it results in a simple free-body diagram. As before, the part of the beam shown is in equilibrium. Therefore,

$$\uparrow \Sigma \, F_y = 0 \quad 200 - F_{BB} = 0$$

$$F_{BB} = 200 \text{ lb} \quad \text{(tension)} \qquad\qquad \text{Answer}$$

$$\rightarrow \Sigma \, F_x = 0 \quad 900 - V_{BB} = 0$$

$$V_{BB} = 900 \text{ lb} \leftarrow \qquad\qquad \text{Answer}$$

$$\circlearrowright \Sigma \, M_c = 0 \qquad -900(4) - M_{BB} = 0$$

$$M_{BB} = -3600 \text{ lb-ft}$$

$$= 3600 \text{ lb-ft} \circlearrowleft \qquad\qquad \text{Answer}$$

PROBLEMS

9.1 through 9.16 Calculate the internal reactions for the member shown at the section(s) indicated.

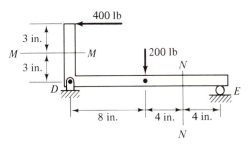

PROB. 9.4

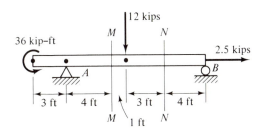

PROB. 9.1

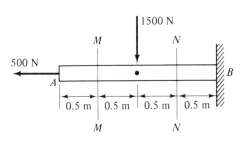

PROB. 9.2

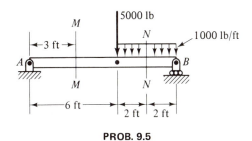

PROB. 9.5

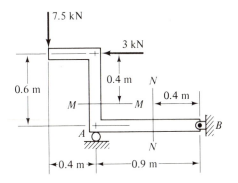

PROB. 9.3

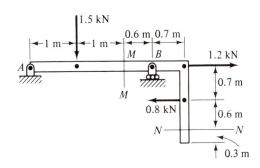

PROB. 9.6

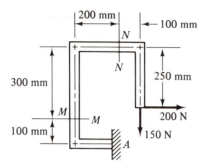

PROB. 9.7

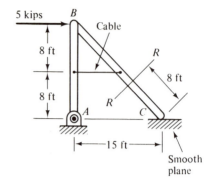

PROB. 9.10

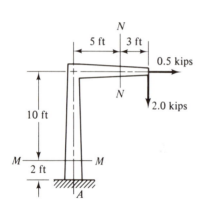

PROB. 9.8

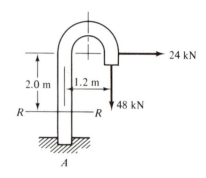

PROB. 9.11

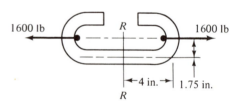

PROB. 9.12

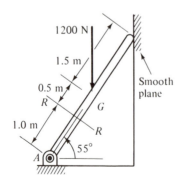

PROB. 9.9

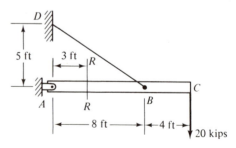

PROB. 9.13

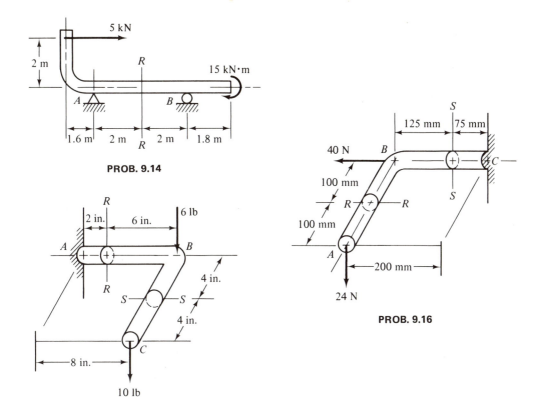

PROB. 9.14

PROB. 9.16

PROB. 9.15

9.3 STRESS

The internal reactions for a section are the resultant of distributed forces ΔT, which act on each small element of the cross-sectional area. In general, they vary in direction and magnitude, as shown in Fig. 9.4(a), for each small element of area. We are interested in intensity or force per unit area of these forces at various points of the cross section because their inten-

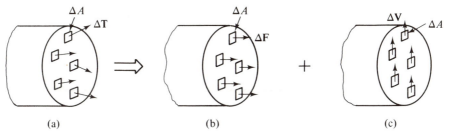

FIGURE 9.4

sity affects the ability of the material to support loads and resist changes in shape. We shall resolve the forces into components perpendicular, or normal, and parallel, or tangent, to the area on which they act, as shown in Fig. 9.4(b) and (c).

The intensity of the force normal to the area is called the *normal stress* and will be represented by the Greek letter σ (sigma). The average value of the normal stress over the area ΔA is defined by the equation

$$\sigma = \frac{\Delta F}{\Delta A}$$

where ΔF is the component of the force normal to the area ΔA. To find the stress at a point, the area ΔA must be decreased in size until it approaches a point. A normal stress is called *tensile stress* when it stretches the material on which it acts, and *compressive stress* when it shortens the material on which it acts.

The intensity of the force parallel to the area is called the *shear stress* and will be represented by the Greek letter τ (tau). The average value of the shear stress over the area ΔA is defined by the equation

$$\tau = \frac{\Delta V}{\Delta A}$$

where ΔV is the component of the force parallel to the area ΔA. The shear stress at a point is found by decreasing the area ΔA in size until it approaches a point.

In cases where we have *uniform stress*, that is, stress which does not vary over the cross section or we wish to find an *average stress*, the normal stress can be found from the equation

$$\sigma = \frac{F}{A} \tag{9.1}$$

and the shear stress from the equation

$$\tau = \frac{V}{A} \tag{9.2}$$

The force F is the sum of the normal forces ΔF, the force V the sum of the parallel (tangential) forces ΔV and A the area of the cross section or the sum of the elements of area ΔA of the cross section.

The normal stress is caused by force components normal or

perpendicular to the area of the cross section, and the shear stress is caused by force components parallel or tangent to the cross section.

Units

Stresses are measured by units of force divided by units of area. The usual units for stress in the U.S. common system are pounds per square inch, abbreviated as *psi*, or kips per square inch, abbreviated *ksi*. In metric, or SI units, stress is in newtons per square meter, abbreviated as N/m^2, or also designated as a pascal (Pa). The pascal is a small unit of stress (1 psi = 6892.4682 Pa) and it may be more convenient to use the kilopascal (1 kPa = 10^3 Pa) or megapascal (1 MPa = 10^6 Pa).

9.4 STRESS IN AN AXIALLY LOADED MEMBER

Consider a straight two-force member of uniform cross section. The line of action of the loads passes through the centroid of the cross section as shown in Fig. 9.5(a). Near the middle of the bar at section B-B, a plane

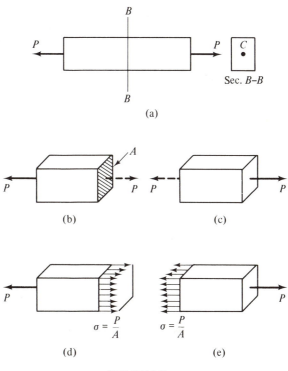

FIGURE 9.5

perpendicular to the line of action of the loads (axis of the bar), the internal reaction is a single axial force equal to P as shown in Fig. 9.5(b) and (c). The normal stress distributed on the cross section is a *uniform* tension. From Eq. (9.1), the tensile stress

$$\sigma = \frac{P}{A} \quad \text{or} \quad \frac{\text{force}}{\text{area}} \quad \left[\frac{\text{lb}}{\text{in.}^2} = \text{psi}\right] \quad \text{or} \quad \left[\frac{\text{N}}{\text{m}^2} = \text{Pa}\right] \tag{9.3}$$

where P is the axial force and A the cross-sectional area. Since the stress distribution is uniform, Eq. (9.3) gives the actual stress for an axially loaded member. The distribution of stress is shown in Fig. 9.5(d) and (e). Notice that the stress multiplied by the cross-sectional area is equal to the axial force or stress resultant ($\sigma A = P$). No shear stress acts on the cross section.

If the direction of the forces in Fig. 9.5(a) are reversed, the member will be in compression. For a short member, where the compressive forces do not produce buckling or collapse, the compressive stress can also be calculated by Eq. (9.3).

Bearing Stress

In certain structural and mechanical problems, one body is supported by another as shown in Fig. 9.6. The force intensity or normal stress between the two bodies in contact can be calculated by Eq. (9.1), if the resultant

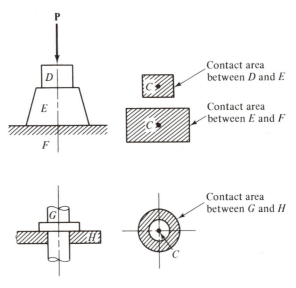

FIGURE 9.6

of the applied loads acts through the centroid C of the contact area. This normal stress—called *bearing stress*—is found by dividing the resultant of the applied loads by the contact area. That is,

$$\sigma_{bearing} = \frac{P}{A_{contact}} \tag{9.4}$$

9.5 AVERAGE SHEAR STRESS

There are many problems in which applied forces are transmitted from one body to another by developing internal reactions on planes that are parallel to the applied force. The planes are in shear and the average shear

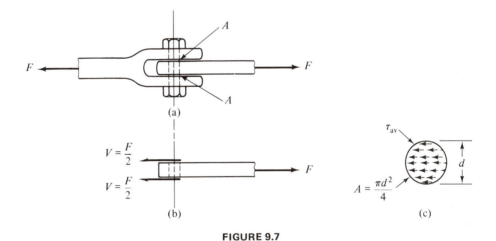

FIGURE 9.7

stress can be found by dividing the shear force on the plane by the shear area of the plane. That is, applying Eq. (9.2), we have

$$\tau_{av} = \frac{V}{A_s} \quad \text{or} \quad \frac{\text{force}}{\text{area}} \quad \left[\frac{\text{lb}}{\text{in.}^2} = \text{psi}\right] \quad \text{or} \quad \left[\frac{\text{N}}{\text{m}^2} = \text{Pa}\right] \tag{9.5}$$

An application of the concept of the average shear stress is shown in Fig. 9.7(a). Two I-bars are connected by a shear pin. We separate the bars by cutting the shear pin along two separate planes of area $A = \pi d^2/4$ as shown in Fig. 9.7(b) and (c). From equilibrium each plane must have an internal reaction equal to $F/2$, which is parallel to the plane and therefore in shear. The shear stress for each plane is far from uniform but an average value can be calculated from Eq. (9.5):

$$\tau_{av} = \frac{V}{A} = \frac{F/2}{\pi d^2/4}$$

Since the shear pin has two planes that resist the force, it is said to be in *double shear*.

In another example, shown in Fig. 9.8(a), two plates are joined by a rivet. Separation of the plates requires that the rivet be cut along a single plane as shown in Fig. 9.8(b) and (c). The internal shear reaction from the equilibrium equation is equal to F and the *average* value of the shear stress is

$$\tau_{av} = \frac{V}{A} = \frac{F}{\pi d^2/4}$$

The rivet is said to be in single shear.

In the third example a key is used to secure a crank to a shaft [Fig. 9.9(a)]. The force on the key is found from a free-body diagram of the crank shown in Fig. 9.9(b). Summing moments about O, we obtain

$$\Sigma M_O = F'_{key}\, r - Pa = 0$$

$$F_{key} = \frac{Pa}{r}$$

The key must be sheared along the area $A = bL$ as shown in Fig. 9.9(c).

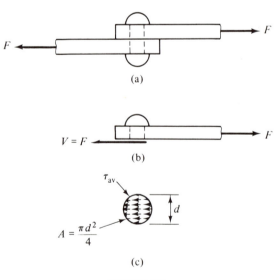

(a)

(b)

(c)

FIGURE 9.8

Therefore, the average shear stress is given by

$$\tau_{av} = \frac{F_{key}}{A} = \frac{Pa/r}{bL}$$

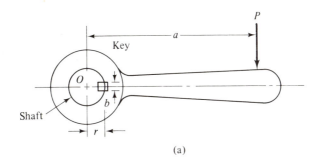

(a)

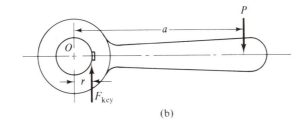

(b)

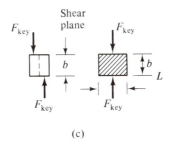

(c)

FIGURE 9.9

9.6 PROBLEMS INVOLVING NORMAL AND SHEAR STRESS

The equations for normal stress and shear stress are quite simple. Difficulty may arise in applying the equations either in finding the internal reaction from the equation of equilibrium or in visualizing the area that is acted on by normal or shear stress. We will consider several examples that will help to clarify the problem areas.

EXAMPLE 9.2

A timber frame or truss shown in Fig. 9.10(a) supports a load P of 50 kN. Find (a) the normal stress in all the members, (b) the horizontal shearing stress in timber AC, and (c) the bearing stress of AC on plate D, which measures 0.25 m $\times$ 0.25 m $\times$ 0.025 m.

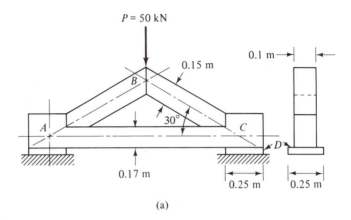

(a)

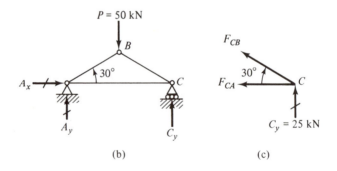

(b)   (c)

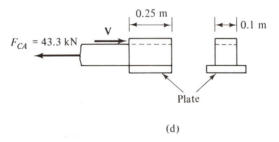

(d)

FIGURE 9.10

Solution

We idealize the problem by replacing the timber frame with a pin-connected simple truss as shown in Fig. 9.10(b). From the equilibrium equations, $A_x = 0$, $A_y = 25.0$ kN, and $C_y = 25.0$ kN. The truss will be solved by the method of joints. A free-body diagram of joint C is shown in Fig. 9.10(c). For equilibrium of concurrent forces

$$\uparrow \Sigma F_y = 0 \qquad 25.0 + F_{CB} \sin 30° = 0$$

$$F_{CB} = -\frac{25}{\sin 30} = -50 \text{ kN}$$

and

$$\rightarrow \Sigma F_x = 0 \qquad -F_{CA} - F_{CB} \cos 30° = 0$$

$$F_{CA} = +50 \cos 30° = 43.3 \text{ kN}$$

The force in AB is the same as the force in BC from symmetry. Therefore,

$$F_{AB} = -50 \text{ kN}$$

(a) The cross-sectional area of members AB and BC are $0.15 \times 0.1 = 0.015 \text{ m}^2$ and member AC is $0.17 \times 0.1 = 0.017 \text{ m}^2$. The normal stress in members AB and BC is

$$\sigma = \frac{F_{AB}}{A} = \frac{-50 \text{ kN}}{0.015 \text{ m}^2} = -3333 \text{ kPa}$$

or

$$\sigma = 3.33 \text{ MPa} \quad \text{(compression)} \qquad\qquad \text{Answer}$$

The normal stress in member AC:

$$\sigma = \frac{F_{AC}}{A} = \frac{43.3 \text{ kN}}{0.017 \text{ m}^2} = 2547 \text{ kPa}$$

or

$$\sigma = 2.55 \text{ MPa} \quad \text{(tension)} \qquad\qquad \text{Answer}$$

(b) The horizontal shear force acting on AC must be equal to 43.3 kN and the shear area $0.25 \text{ m} \times 0.1 \text{ m} = 0.025 \text{ m}^2$ as shown in Fig. 9.10(d). Therefore,

$$\tau_{av} = \frac{V}{A} = \frac{43.3 \text{ kN}}{0.025 \text{ m}^2} = 1732 \text{ kPa}$$

or

$$\tau_{av} = 1.73 \text{ MPa} \qquad\qquad \text{Answer}$$

(c) The force on the plate is equal to the reaction $C_y = 25$ kN and the area in contact between the plate and member AC is $0.25 \text{ m} \times 0.1 \text{ m} = 0.025 \text{ m}^2$. Therefore, the bearing stress

$$\sigma_{\text{bearing}} = \frac{P}{A} = \frac{25 \text{ kN}}{0.025 \text{ m}^2} = 1000 \text{ kPa}$$

or

$$\sigma_{\text{bearing}} = 1.0 \text{ MPa} \qquad\qquad \text{Answer}$$

EXAMPLE 9.3

For the pin-connected frame of Fig. 9.11(a), determine (a) the axial normal stress in the 0.65-in.-diameter rod AB, (b) the average shear stress in the 0.75-in.-diameter pin at D (pin D is in single shear), and (c) the average shear stress in the 0.50-in.-diameter pin at B (pin B is in double shear).

Solution

The free-body diagram for the entire frame is shown in Fig. 9.11(b). From the equations of equilibrium

$$\circlearrowright \Sigma M_E = 0 \qquad 1.2(20) + 1.5(12) - C_x(15) = 0$$
$$C_x = 2.8 \text{ kips}$$

$$\rightarrow \Sigma F_x = 0 \qquad C_x - E_x = 0$$
$$E_x = C_x = 2.8 \text{ kips}$$

$$\uparrow \Sigma F_y = -1.2 - 1.5 + E_y = 0$$
$$E_y = 2.7 \text{ kips}$$

Check:

$$\circlearrowright \Sigma M_C = 0 \qquad 1.2(16) + 1.5(8) + E_y(4) - E_x(15) = 0$$
$$1.2(16) + 1.5(8) + 2.7(4) - 2.8(15) = 0 \qquad \text{OK}$$

The free-body diagrams for members AE, AB, and DC are shown in Fig. 9.11(c)–(e). From the free-body diagram of AE:

$$\circlearrowright \Sigma M_A = 0 \qquad -1.5(8) + D_y(16) + 2.7(20) = 0$$
$$D_y = -2.625 \text{ kips}$$

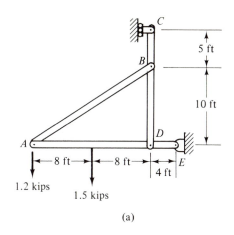

(a)

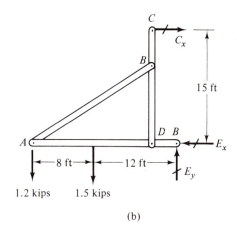

(b)

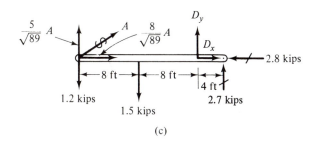

(c)

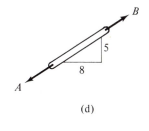

(d)

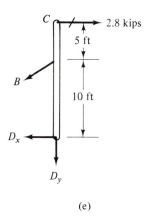

(e)

FIGURE 9.11

$$\uparrow \Sigma F_y = 0 \qquad \frac{5}{\sqrt{89}} A - 1.5 - 1.2 + D_y + 2.7 = 0$$

$$\frac{5}{\sqrt{89}} A = 1.5 + 1.2 - (-2.625) - 2.7$$

$$A = 4.953 \text{ kips}$$

Member AB is a two-force member; therefore, $A = B = 4.953$ kips.

$$\rightarrow \Sigma \, F_x = 0 \qquad \frac{8}{\sqrt{89}} A + D_x - 2.8 = 0$$

$$D_x = 2.8 - \frac{8}{\sqrt{89}} (4.953) = -1.400 \text{ kips}$$

We check our answer with the free-body diagram of member CD.

$$\circlearrowright \Sigma \, M_B = 0 \qquad -2.8(5) - D_x(10) = 0$$

$$-2.8(5) - (-1.40)(10) = 0 \qquad \text{OK}$$

$$\rightarrow \Sigma \, F_x = 0 \qquad \frac{-8}{\sqrt{89}} B - D_x + 2.8 = 0$$

$$\frac{-8}{\sqrt{89}} (4.953) - (-1.4) + 2.8 = 0.000 \qquad \text{OK}$$

$$\uparrow \Sigma \, F_y = 0 \qquad \frac{-5}{\sqrt{89}} B - D_y = 0$$

$$\frac{-5}{\sqrt{89}} (4.953) - (-2.625) = 0.000 \qquad \text{OK}$$

(a) The cross-sectional area of rod AB is $A = \pi d^2/4 = A = \pi (0.65)^2/4 = 0.3318$ m^2 and the axial force is $F_{AB} = 4.953$ kips; therefore, the tensile stress is

$$\sigma_{AB} = \frac{F_{AB}}{A} = \frac{4.953}{0.3318} = 14.93 \text{ ksi} \qquad\qquad \text{Answer}$$

(b) The cross-sectional area of the pin in single shear at D is $A = \pi d^2/4 = \pi(0.75)^2/4 = 0.4418$ m^2 and the force on the pin at D is $F_D = (D_x^2 + D_y^2)^{1/2} = [(1.4)^2 + (2.626)^2]^{1/2} = 2.975$ kips; therefore, the average shear stress is

$$\tau_D = \frac{F_D}{A} = \frac{2.975}{0.4418} = 6.73 \text{ ksi} \qquad\qquad \text{Answer}$$

(c) The cross-sectional area of the pin in double shear at B is $A = 2\pi d^2/4 = 2\pi(0.5)^2/4 = 0.3927$ in.^{2} and the force on the pin at B is 4.953 kips; therefore, the average shear stress is

$$\tau_B = \frac{F_B}{A} = \frac{4.953}{0.3927} = 12.61 \text{ ksi} \qquad\qquad \text{Answer}$$

EXAMPLE 9.4

The collar bearing supports a load $P = 200$ kN as shown in Fig. 9.12(a). Find the tensile stress in the shaft, the shearing stress between the collar and the shaft, and the bearing stress between the collar and the support.

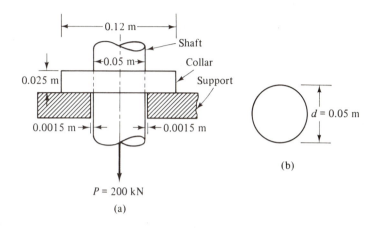

(a)

(b)

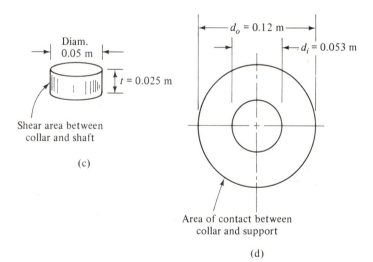

(c)

(d)

FIGURE 9.12

Solution

Tensile Stress

The area of the shaft in tension is shown in Fig. 9.12(b). The area $A = \pi d^2/4 = \pi(0.05)^2/4 = 0.001\ 963$ m^2. The *tensile stress*

$$\sigma = \frac{P}{A} = \frac{200 \text{ kN}}{0.001\ 963 \text{ m}^2} = 101\ 900 \text{ kPa}$$

$$= 101.9 \text{ MPa} \qquad\qquad\qquad \text{Answer}$$

Shear Stress

The shear area between the collar and shaft is shown in Fig. 9.12(c). The area $A = \pi\, dt = \pi(0.05)(0.025) = 0.003\ 927 \text{ m}^2$. The average shear stress

$$\tau_{av} = \frac{V}{A} = \frac{200 \text{ kN}}{0.003\ 927 \text{ m}^2} = 50\ 930 \text{ kPa}$$

$$= 50.9 \text{ MPa} \qquad\qquad\qquad \text{Answer}$$

Bearing Stress

The bearing area between the collar and support is shown in Fig. 9.12(d). The area $A = \pi d_o^2/4 - \pi d_i^2/4 = \pi[(0.12)^2 - (0.053)^2]/4 = 0.009\ 104 \text{ m}^2$. The bearing stress

$$\sigma_{Br} = \frac{P}{A} = \frac{200 \text{ kN}}{0.009\ 104 \text{ m}^2} = 21\ 970 \text{ kPa}$$

$$= 21.97 \text{ MPa} \qquad\qquad\qquad \text{Answer}$$

EXAMPLE 9.5

A steel column rests on a steel base plate which is supported by a concrete pier that is supported by a concrete footing as shown in Fig. 9.13(a). Find the bearing stress between (a) the steel base plate and the pier, (b) the pier and footing, and (c) the footing and soil. Concrete weighs 150 lb/ft^3.

Solution

The free-body diagram is shown in Fig. 9.13(b). The weight of the pier and footing and the contact areas have been calculated and shown on the figure.

(a) The force between the base plate and pier is $P = 350$ kips. The bearing stress

$$\sigma_{Br} = \frac{P}{A} = \frac{350}{528} = 0.663 \text{ ksi} \qquad\qquad \text{Answer}$$

(b) The force between the pier and footing is $P = P + W_1 = 350 + 5 = 355$ kips. The bearing stress

$$\sigma_{Br} = \frac{P}{A} = \frac{355}{960} = 0.370 \text{ ksi} \qquad\qquad \text{Answer}$$

(c) The force between the footing and the soil is $R = P + W_1 + W_2 = 350 + 5 + 5.4 = 360.4$ kips. The bearing stress

$$\sigma_{Br} = \frac{P}{A} = \frac{360.4}{2400} = 0.150 \text{ ksi}$$ Answer

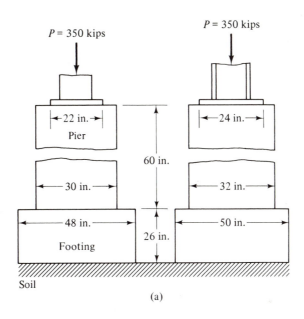

(a)

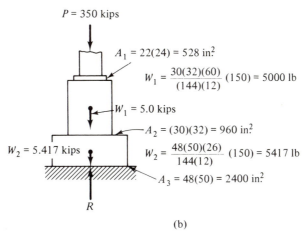

(b)

FIGURE 9.13

PROBLEMS

9.17 A round aluminum tube has a wall thickness of 0.150 in. and an outside diameter of 1.5 in. The tube is used as a short compression member. If the axial load on the member is 15.5 kips, determine the axial stress.

9.18 A wooden post 220 mm by 220 mm supports a compressive load of 350 kN. Determine the compressive stress in the post.

9.19 The failure tensile load for a steel wire 1.0 mm in diameter is 1600 N. What is the failure tensile stress?

9.20 A round tie rod has a diameter of 0.5 in. The axial load on the member is 20,000 lb tension. What is the axial stress?

9.21 A steel column with a cross-sectional area of 2.51 in.² supports an axial compressive load of $P = 45$ kips and bears on a 9-in. by 11-in. steel base plate as shown. Determine (a) the compressive stress in the column and (b) the bearing stress between the plate and the concrete footing.

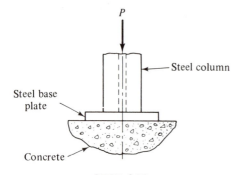

PROB. 9.21

9.22 A 400-N weight is supported by three wires joined together at C as shown. If the diameter of the wire is 5 mm, determine the tensile stress in each wire.

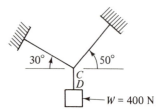

PROB. 9.22

9.23 The bracket shown supports a load of $P = 500$ lb. Members AB and BC each have cross-sectional areas of 0.5 in.² Determine the axial stress in each member.

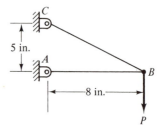

PROB. 9.23

9.24 A force of 54,000 lb on a $\frac{3}{4}$-in.-diameter punch, punches a hole in a $\frac{3}{8}$-in.-thick steel plate. Determine (a) the average shear stress in the plate and (b) the compressive stress in the punch.

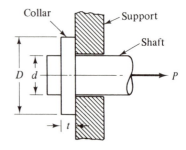

PROB. 9.25

9.25 A thrust bearing supports a load P as shown. If $P = 45$ kN, $D = 30$ mm, $d = 15$ mm, and $t = 9.5$ mm, determine (a) the tensile stress in the shaft, (b) the shearing stress between the collar and the shaft, and (c) the bearing stress between the collar and the support.

9.26 The timber truss shown supports a load of $P = 10$ kips. Find (a) the normal stress in all the members, (b) the horizontal shear stress in timber AC at C, and (c) the bearing stress of AC on plate D, which measures 5 in. by 5 in. and is 0.25 in. thick.

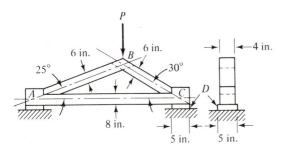

PROB. 9.26

9.27 A plate is supported by a $\frac{1}{4}$-in.-diameter bolt and bracket shown. If the load on the plate is 2000 lb, determine the average shear stress in the bolt.

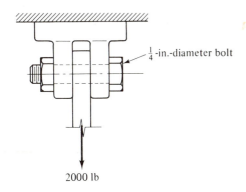

2000 lb

PROB. 9.27

9.28 The hand crank illustrated is keyed to a round shaft. Determine the shear stress in the key if $P = 300$ N, $L = 400$ mm, $d = 40$ mm, $w = 8$ mm, $t = 6$ mm, and the key is 50 mm long.

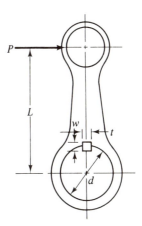

PROB. 9.28

9.29 A hoist, consisting of a 6-in. by 8-in. wooden post AB and a 1.125-in.-diameter steel bar BC, supports a load $P = 9500$ lb as shown. Determine (a) the axial stresses in AB and BC and (b) the average shear stress in the 1.5-in.-diameter pin at C.

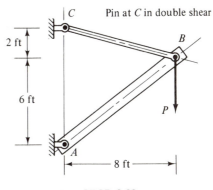

PROB. 9.29

9.30 The round hanger rod illustrated is supported by a floor plank. If the load on the rod is 12,000 lb, find (a) the tensile stress in the rod and (b) the bearing stress between the washer and the floor plank.

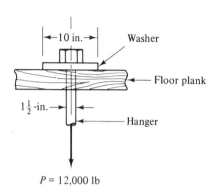

$P = 12,000$ lb

PROB. 9.30

9.31 The loads shown are applied axially on the round steel bar which has a diameter of 60 mm. Determine the axial stress between (a) A and B, (b) B and C, and (c) C and D.

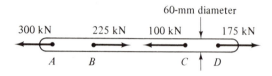

PROB. 9.31

9.32 An aluminum bar with a diameter of 2.2 in. has axial loads as illustrated. Determine the axial stress between (a) S and T, (b) T and U, and (c) U and V.

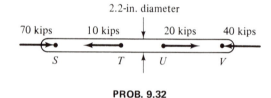

PROB. 9.32

9.7 ALLOWABLE WORKING STRESS

The allowable working stress is the maximum stress that is considered safe for a material to support under certain loading conditions. It is the stress used in the design of load-supporting members of a structure or machine. The allowable working stress values are determined by tests on the material with various loading conditions and by experience gained from the performance of previous designs under service conditions.

With the allowable stress σ_a known, we can solve Eq. (9.3) for the required area in tension or compression.

$$A_{\text{req}} = \frac{P}{\sigma_a} \tag{9.6}$$

If the allowable shear stress τ_a is known, we can solve Eq. (9.4) for the required area.

$$A_{\text{req}} = \frac{V}{\tau_a} \tag{9.7}$$

EXAMPLE 9.6

A hollow cylinder is to be designed to support a compressive load of 650 kN. The allowable compressive stress $\sigma_a = 69\ 200$ kN/m^2. Compute the outside diameter of the cylinder if the wall thickness is 50 mm.

Solution

Solving for the required area from Eq. (9.6), we have

$$A_{\text{req}} = \frac{P}{\sigma_a} = \frac{650}{69\ 200} = 0.009\ 393\ \text{m}^2$$

$$= 9393\ \text{mm}^2$$

In Fig. 9.14 the compression area is shown. The formula for the area is shown in the figure.

$$A = \frac{\pi}{4}(d_o^2 - d_i^2) = 9393\ \text{mm}^2$$

$$d_o^2 - d_i^2 = \frac{4}{\pi}(9393) = 11\ 960$$

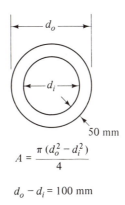

$$A = \frac{\pi\,(d_o^2 - d_i^2)}{4}$$

$$d_o - d_i = 100\ \text{mm}$$

FIGURE 9.14

Factoring the left-hand side of this equation yields

$$(d_o - d_i)(d_o + d_i) = 11\ 960 \tag{a}$$

Since the wall thickness is 50 mm,

$$d_o - d_i = 100.0 \tag{b}$$

Dividing Eq. (a) by Eq. (b), we have

$$d_o + d_i = \frac{11\ 960}{100} = 119.60 \tag{c}$$

Adding Eqs. (b) and (c) yields

$$2d_o = 219.60$$

$$d_o = 109.8 \text{ mm} \hspace{3cm} \text{Answer}$$

EXAMPLE 9.7

A punch press is used to punch a 1.5-in.-diameter hole in a 0.35-in.-thick steel plate. Determine the force exerted by the press on the plate when the average shear resistance to punching of the steel plate is 58,000 psi.

Solution

The punch shears out a cylinder. The shear area is the surface area of the cylinder, which has a diameter of 1.5 in. and a height equal to the thickness of the plate, which is 0.35 m, as shown in Fig. 9.15.

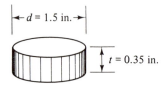

FIGURE 9.15

The area is given by

$$A = \pi\, dt = \pi(1.5)(0.35) = 1.650 \text{ in.}^2$$

From Eq. (9.5),

$$P_{\text{punch}} = V = A\sigma = 1.650\,(58,000)$$

$$= 95,700 \hspace{3cm} \text{Answer}$$

EXAMPLE 9.8

The bell crank shown in Fig. 9.16(a) is in equilibrium with loads P and Q. The allowable stress in tension is 140 000 kN/m^2 and the allowable stress in shear is 90 000 kN/m^2.

The pins at A, B, and C are in double shear. Determine (a) the allowable load P, (b) the allowable load Q, and (c) the required diameter of the rod d_2 at C and the diameter of the pin at A, B, and C.

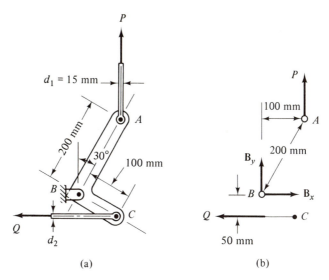

(a)   (b)

FIGURE 9.16

Solution

(a) The area of the rod at A is

$$A = \pi d_1^2 / 4 = \pi (0.015)^2 / 4 = 176.7 \times 10^{-6} \text{ m}^2$$

The allowable load P from Eq. (9.3) is given by

$$P = \sigma A = 140 \times 10^3 (176.7 \times 10^{-6}) = 24.74$$

$$= 24.7 \text{ kN} \qquad\qquad\qquad\qquad \text{Answer}$$

(b) From the free-body diagram of the bell crank shown in Fig. 9.16(b):

$$\circlearrowright \Sigma M_B = 0 \qquad 100P - 50Q = 0$$

$$Q = 2P = 2(24.74) = 49.48$$

$$= 49.5 \text{ kN} \qquad\qquad\qquad \text{Answer}$$

$$\rightarrow \Sigma F_x = 0 \qquad B_x - Q = 0$$

$$B_x = 49.48 \text{ kN}$$

$$\uparrow \Sigma F_y = 0 \qquad B_y = P = 24.74 \text{ kN}$$

$$B = (B_x^2 + B_y^2)^{1/2} = [(24.74)^2 + (49.48)^2]^{1/2}$$

$$= 55.32 \text{ kN}$$

(c) From Eq. (9.6):

$$A_{\text{req}} = \frac{P}{\sigma_a}$$

The area of the rod at C is given by

$$A_{req} = \frac{\pi d_2^2}{4} = \frac{Q}{\sigma_a} = \frac{49.48}{140 \times 10^3} = 3.534 \times 10^{-4} \text{ m}^2$$

$$d_2^2 = \frac{4}{\pi} (3.534 \times 10^{-4}) = 4.50 \times 10^{-4} \text{ m}^2$$

$$d_2 = 0.0212 \text{ m}$$

$$= 21.2 \text{ mm} \qquad\qquad \text{Answer}$$

For the diameter of the pins, from Eq. (9.7):

$$A_{req} = \frac{2\pi d^2}{4} = \frac{V}{\tau_a} = \frac{V}{90 \times 10^3}$$

or

$$d^2 = 7.074 \times 10^{-6} V$$

At A:

$$d_A^2 = 7.074 \times 10^{-6} (24.74) = 1.750 \times 10^{-4} \text{ m}^2$$

$$d_A = 1.322 \times 10^{-2} \text{ m}$$

$$= 13.22 \text{ mm} \qquad\qquad \text{Answer}$$

At C:

$$d_C^2 = 7.074 \times 10^{-6} (49.48) = 3.500 \times 10^{-4} \text{ m}^2$$

$$d_C = 1.870 \times 10^{-2} \text{ m}$$

$$= 18.70 \text{ mm} \qquad\qquad \text{Answer}$$

At B:

$$d_B^2 = 7.074 \times 10^{-6} (55.32) = 3.913 \times 10^{-4} \text{ m}^2$$

$$d_B = 1.978 \times 10^{-2} \text{ m}$$

$$= 19.78 \text{ mm} \qquad\qquad \text{Answer}$$

PROBLEMS

9.33 A rectangular bar with a cross section 2.25 in. by 1.75 in. is acted on by an axial force that causes a tensile stress of 22,000 psi. What is the axial force?

9.34 A circular bar supports an axial force of 35 kN. If the compressive stress in the bar is not to exceed 75.8 MN/m^2, determine the diameter of the bar.

9.35 A circular tube has an outer diameter of 1.2 in. and an inner diameter of 1.0 in. The allowable stress is 24,000 psi in tension and 16,000 psi in compression. Determine the allowable tensile and compressive loads.

9.36 A steel wire must support a tensile load of 2000 N. The allowable stress $\sigma_a = 165$ MN/m^2. Determine the required diameter of the wire.

9.37 The steel column illustrated in the figure for Prob. 9.21 supports an axial compressive load of $P = 78$ kips. The allowable compressive stress in the steel is 15.5 ksi and the allowable bearing stress in the concrete footing is 750 psi. Determine the required cross-sectional area of the column and the length of the steel base plate if the width of the plate is 10 in.

9.38 The bracket illustrated in the figure for Prob. 9.23 supports a load $P = 4.5$ kN. The allowable stress is 165 MN/m^2 in tension and 110 MN/m^2 in compression. Determine the required cross-sectional areas of members AB and BC.

9.39 A thrust bearing as shown in the figure for Prob. 9.25 supports a load $P = 5000$ lb. The allowable stress is 22 ksi in tension, 14.5 ksi for shear between the collar and shaft, and 3.5 ksi for bearing between the collar and support. Determine the required values for d, t, and D.

9.40 The pin-connected truss supports loads of 10 kN and 35 kN as shown. The

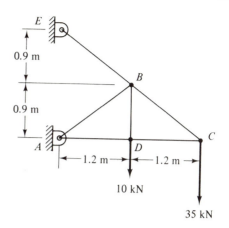

PROB. 9.40

allowable stress is 150 MN/m^2 in tension and 75 MN/m^2 in compression. Determine the required cross-sectional area for each member.

9.41 The hand crank illustrated in the figure for Prob. 9.28 supports a load $P = 50$ lb. If $L = 18$ in., $d = 1.5$ in., the key is 3 in. long, and the allowable shear stress in the key is 15,000 psi, determine the required thickness, t, of the key.

9.42 The hoist shown in the figure for Prob. 9.29 supports a load of $P = 10$ kips. Member BC is a round steel rod and member AB a square wooden post. The allowable stress is 20 ksi for steel in tension and 800 psi for wood in compression parallel to the grain. Determine the diameter of the steel rod and the dimensions of the square wooden post.

9.8 FURTHER ANALYSIS OF AXIAL LOADS—STRESSES ON OBLIQUE SECTIONS

In Sec. 9.4 we considered the stress on a cross section perpendicular to the axis of the bar. In this section we will consider the stress on a plane that is perpendicular to the plane of the bar but oblique (inclined) with the axis. A straight two-force member of uniform cross-sectional area A is shown in Fig. 9.17(a).

To find the internal reactions on the oblique section *B–B*, we isolate the bar to the left of the section [Fig. 9.17(b)]. The resultant force on the oblique section must be equal to *P* for equilibrium. The force is inclined at an angle of θ with the normal to the oblique section. Therefore, the normal force $F = P \cos \theta$ and the shear force $V = P \sin \theta$. The area of the oblique cross-sectional area is equal to the cross-sectional area divided by the cos θ. That is,

$$A_\theta = \frac{A}{\cos \theta}$$

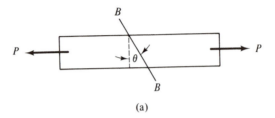

(a)

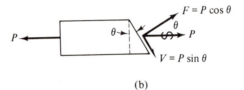

(b)

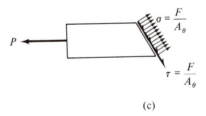

(c)

FIGURE 9.17

Therefore, the normal stress on the oblique plane

$$\sigma_\theta = \frac{F}{A_\theta} = \frac{P \cos \theta}{A/\cos \theta}$$

or

$$\sigma_\theta = \frac{P \cos^2 \theta}{A} \tag{9.8}$$

and the shear stress on the oblique plane

$$\tau_\theta = \frac{V}{A_\theta} = \frac{P \sin \theta}{A/\cos \theta}$$

or

$$\tau_\theta = \frac{P \sin \theta \cos \theta}{A}$$

Since $\sin \theta \cos \theta = (\sin 2\theta)/2$,

$$\tau_\theta = \frac{P \sin 2\theta}{2A} \tag{9.9}$$

The maximum normal stress from Eq. (9.8) occurs when $\theta = 0°$. That is, $\sigma_0 = P/A$. The maximum value of the shear stress from Eq. (9.9) occurs when $\theta = 45°$ and is equal to

$$\tau_{45°} = \tau_{max} = \frac{P}{2A}$$

that is, *half* the value of the *maximum tensile* stress. Notice that the tensile stress on a plane when $\theta = 45°$ has the same value as the shear stress on the same plane.

EXAMPLE 9.9

A steel bar 0.75 in. in diameter has an axial load of 12,000 lb. Determine (a) the normal and shearing stresses on a plane through the bar which forms an angle of 30° with the cross section and (b) the maximum normal and shearing stresses in the bar.

Solution

The bar is shown in Fig. 9.18(a). A plane whose normal makes an angle of 30° with the axis of the bar cuts through the bar and a free-body diagram of the bar to the left of the cut is shown in Fig. 9.18(b). Included are the normal and shear components of the internal reactions. The cross-sectional area perpendicular to the axis of the bar $A = \pi d^2/4 = \pi(0.75)^2/4 = 0.4418$ in.^{2}. The area along an oblique plane of 30° is

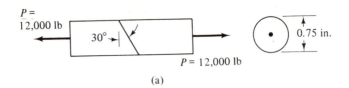

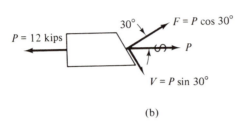

(b)

FIGURE 9.18

$A_\theta = A/\cos \theta$ or $A_{30} = 0.4418/\cos 30° = 0.5101$ in.^{2}. The normal internal reactions $F = P \cos 30° = 12,000 \cos 30° = 10,392$ lb and $V = P \sin 30° = 12,000 \sin 30° = 6000$ lb. Therefore:

(a) The normal stress

$$\sigma_{30} = \frac{F}{A_{30}} = \frac{10,390}{0.5101} = 20,400 \text{ psi} \qquad \text{Answer}$$

$$\tau_{30} = \frac{V}{A_{30}} = \frac{6000}{0.5101} = 11,760 \text{ psi} \qquad \text{Answer}$$

(b) The maximum normal and shearing stresses can be calculated by formula.

$$\sigma_{max} = \sigma_0 = \frac{P}{A} = \frac{12,000}{0.4418} = 27,200 \text{ psi} \qquad \text{Answer}$$

$$\tau_{max} = \tau_{45°} = \frac{P}{2A} = \frac{12,000}{2(0.4418)} = 13,580 \text{ psi} \qquad \text{Answer}$$

EXAMPLE 9.10

A short compression block with a cross section that measures 0.15 m $\times$ 0.20 m is loaded with an axial load P. The allowable compressive stress is $\sigma = 10\ 000$ kN/m^2 and the allowable shear stress is $\tau = 830$ kN/m^2. Find the maximum allowable load.

Solution

The formula for the maximum normal and shear stress may be used. The cross-sectional area

$$A = 0.15(0.20) = 0.03 \text{ m}^2$$

For compression,

$$\sigma_{max} = \frac{P}{A} \qquad P_a = \sigma_a A = 10\ 000\,(0.03) = 300\ \text{kN}$$

For shear,

$$\tau_{max} = \frac{P}{2A} \qquad P_a = \sigma_a\, 2A = 830\,(2)\,(0.03) = 49.8\ \text{kN}$$

The allowable load is therefore based on the maximum shear stress. The allowable load is

$$P_a = 49.8\ \text{kN} \qquad\qquad \text{Answer}$$

If the load on the compression block is increased until failure occurs, it will occur in shear along a plane that makes an angle of $45°$ with the axis of the compression block.

PROBLEMS

9.43 A steel bar 0.875 in. in diameter has an axial tensile load of 20,000 lb. Determine (a) the normal and shearing stresses on a plane that makes an angle of $35°$ with the direction of the load, and (b) the maximum normal and shearing stresses in the bar.

9.44 A short compression block with a cross section that measures 6 in. by 8 in. is loaded with an axial load P. The al-

lowable compressive stress is 880 psi and the allowable shear stress is 225 psi. Find the maximum allowable load.

9.45 The $\frac{3}{4}$-in.-thick plate shown supports loads of $P = 800$ lb and $Q = 400$ lb. Determine (a) the normal and shear forces on the oblique plane from A to C, and (b) the normal and shear stresses on the oblique plane from B to D.

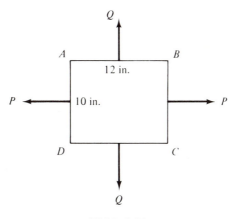

PROB. 9.45

9.46 The uniform plate 25 mm thick is acted on by the stresses illustrated in the figure. Determine the normal and shear stresses on the oblique plane from A to C.

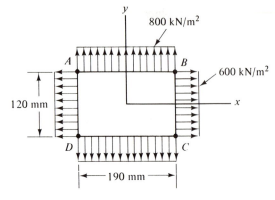

PROB. 9.46

Strain for Axial Loads—
Hooke's Law _____

10.1 AXIAL STRAIN

We have seen that when an axial load is applied to a bar, normal stresses are produced on a cross section perpendicular to the axis of the bar. In addition, the bar increases in length as shown in Fig. 10.1. The increase in

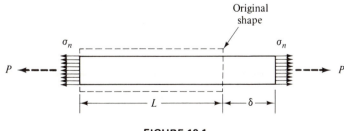

FIGURE 10.1

length, or change in length, represented by the Greek letter δ (delta) is called *deformation*. The change in length δ is for a bar of length L. The change in length per unit of length, represented by the Greek letter ε (epsilon), is called *strain*. Strain is defined by the equation

$$\epsilon = \frac{\delta}{L} \tag{10.1}$$

Strain is usually expressed dimensionally as inches per inch or meters per meter, even though it is a dimensionless quantity. Equation (10.1) gives the average strain over a length L. To obtain strain at a point, we let the length L approach zero. Lengths as small as 3 mm ($\frac{1}{8}$ in.) can be realized with electrical strain gauges.

In the case of an axial load the stress in the direction of the load is called *axial stress* and the strain in the direction of the load is called *axial strain*. With axial strain we also have a smaller normal or lateral strain perpendicular to the load. When the axial stress is tensile, the axial strain is associated with an increase in length and the lateral strain is associated with a decrease in width. The reverse is true for a compression stress. Tensile strain is called *positive strain* and compressive strain *negative strain*.

EXAMPLE 10.1

A concrete test cylinder 200 mm in diameter by 400 mm high is tested to failure. The compressive strain at failure is 0.0012. Determine the total amount the cylinder shortens before failure.

Solution

From the definition for strain,

$$\epsilon = \frac{\delta}{L} \quad \text{or} \quad \delta = \epsilon L$$

The total shortening or deformation of the cylinder is given by

$$\delta = 0.0012(400) = 0.48 \text{ mm} \qquad \text{Answer}$$

10.2 TENSION TEST AND STRESS-STRAIN DIAGRAM

One of the most common tests of material is the tension test. In the usual tension test the cross section of the specimen is round, square, or rectangular. If a large enough piece of the material to be tested is available and can be machined, a round cross section could be used. For thin plate,

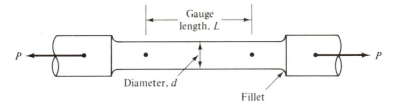

FIGURE 10.2

a rectangular or square section would be used. The profile for a typical round test specimen is shown in Fig. 10.2. The fillets are provided to reduce the stress concentration caused by the abrupt change in section. The deformation or change in length of the specimen is measured for the

FIGURE 10.3 Universal Testing Machine. (*Courtesy Tinius Olsen Testing Machine Company.*)

FIGURE 10.4 Extensometer which separates when its range has been exceeded. (*Courtesy Tinius Olsen Testing Machine Company.*)

gauge length. The strain is therefore the deformation divided by the gauge length. The ends of the specimen must be properly shaped to fit the gripping device on the testing machine used to apply the deformation or load.

The tension specimen is placed in a testing machine such as the one shown in Fig. 10.3. A device, called an extensometer (Fig. 10.4), for measuring the deformation over the gauge length is attached. Deformations or loads are gradually applied to the specimen and simultaneous readings of the load and deformation are taken at specified intervals. Values of stress are found by dividing the load by the original cross-sectional area and the corresponding value of strain by dividing the deformation by the gauge length. The values obtained can then be plotted in a stress–strain curve. The *shape* of the curve will depend on the kind of material tested. (The temperature and speed at which the test was performed may also affect the results.)

The stress–strain curves shown in Fig. 10.5 are for three different kinds of material:

1. Low-carbon steel, a ductile material with a yield point [Fig. 10.5(a)].
2. A ductile material, such as aluminum alloy, which does not have a yield point [Fig. 10.5(b)].
3. A brittle material, such as cast iron or concrete in compression [Fig. 10.5(c)].

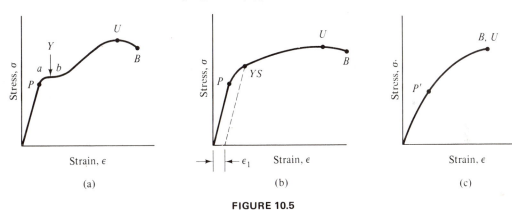

FIGURE 10.5

Ductile materials have the capacity to deform considerably under a tensile load before breaking; a brittle material has a limited capacity to deform before breaking.

We are concerned here with the shape of the curves only. The actual magnitude of the stresses and strains for various materials would differ widely. For example, the stress at point U for concrete and cast iron could be 4000 psi or 20,000 psi, respectively.

There are several points of interest which can be identified on the curves as follows:

1. *Proportional limit:* the maximum stress for which stress is proportional to strain. (Stress at point P.)

2. *Yield point:* stress for which the strain increases without an increase in stress. (Horizontal portion of the curve *ab*. Stress at point *Y*.)

3. *Yield strength:* the stress that will cause the material to undergo a certain specified amount of permanent strain after unloading. (Usual permanent strain ϵ_1 = 0.2 percent. Stress at point *YS*.)

4. *Ultimate strength:* maximum stress material can support up to failure. (Stress at point *U*.)

5. *Breaking strength:* stress in the material based on original cross-sectional area at the time it breaks. Also called fracture or rupture strength. (Stress at point *B*.)

Compression tests are made in a manner similar to the tension test just described. The cross section of the compression specimen is preferably of a uniform circular shape, although a rectangular or square shape is often used. The deformations or loads are applied directly by bearing on the specimen and the ratio of length to diameter is limited by the tendency of the specimen to bend.

10.3 HOOKE'S LAW

We can see in Fig. 10.5(a) and (b) and to a lesser degree in Fig. 10.5(c) that stress is directly proportional to strain (the curve is a straight line) on the lower end of the stress–strain curve. Based on tests of various materials and on the idealized behavior of those materials, *Hooke's law states that stress is proportional to strain.* In Fig. 10.6 we show a stress–strain curve for a material that follows Hooke's law. The slope of the stress–strain curve is the *elastic modulus* or modulus of elasticity, *E*.

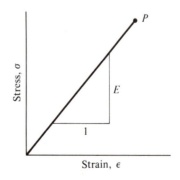

FIGURE 10.6

The elastic modulus is also sometimes called Young's modulus. The elastic modulus, E, is equal to the slope of the stress–strain curve.

$$E = \frac{\text{stress}}{\text{strain}} = \frac{\sigma}{\epsilon} \quad \text{or} \quad \sigma = E\epsilon \qquad (10.2)$$

This is the mathematical statement of Hooke's law. Hooke's law only applies up to the proportional limit of the material.

Since strain is dimensionless, the elastic modulus, E, has the same units as stress. The modulus is a measure of the stiffness or resistance of a material to loads. Steel has an approximate value of $E = 30 \times 10^6$ psi (200×10^6 kN/m^2), while wood has an approximate value of $E = 1.5 \times 10^6$ psi (10×10^6 kN/m^2). Average values of the elastic modulus are given in Table A.11 of the Appendix.

EXAMPLE 10.2

The axial load at the proportional limit of a 0.505-in.-diameter bar was 4900 lb. If a 2-in. gauge length on the bar has increased by 0.0032 in. at the proportional limit, find (a) the stress at the proportional limit, (b) the strain at the proportional limit, and (c) the elastic modulus.

Solution

(a) The cross-sectional area

$$A = \pi d^2/4 = \pi(0.505)^2/4 = 0.2003 \text{ in.}^2$$

and the stress at the proportional limit

$$\sigma = \frac{P_P}{A} = \frac{4900}{0.200} = 24{,}500 \text{ psi} \qquad \text{Answer}$$

(b) The strain at the proportional limit

$$\epsilon_P = \frac{\delta}{L} = \frac{0.0032 \text{ in.}}{2.0 \text{ in.}} = 0.0016 \qquad \text{Answer}$$

(c) The elastic modulus

$$E = \frac{\sigma}{\epsilon} = \frac{24{,}500}{0.0016} = 15.3 \times 10^6 \text{ psi} \qquad \text{Answer}$$

EXAMPLE 10.3

A flat bar with a cross section 5.0 mm by 50 mm elongates 2.1 mm in a length of 1.5 m as a result of an axial load of 45 kN. The proportional limit of the material is 240 000 kN/m^2. Determine (a) the axial stress in the bar, and (b) the elastic modulus.

Solution

(a) Stress is given by

$$\sigma = \frac{P}{A} = \frac{45}{5\,(50)} = 0.18 \ \frac{\text{kN}}{\text{mm}^2}$$

$$= 180\ 000 \ \text{kN/m}^2 \qquad\qquad\qquad \text{Answer}$$

(b) The length 1.5 m = 1500 mm. Therefore, the strain

$$\epsilon = \frac{\delta}{L} = \frac{2.1 \ \text{mm}}{1500 \ \text{mm}} = 0.0014$$

Since the stress is below the proportional limit, the elastic modulus can be found from

$$E = \frac{\sigma}{\epsilon} = \frac{180\ 000}{0.0014} = 128.6 \times 10^6 \ \frac{\text{kN}}{\text{m}^2} \qquad\qquad \text{Answer}$$

10.4 AXIALLY LOADED MEMBERS

From Hooke's law, Eq. (10.2),

$$\sigma = E\epsilon$$

When the stress and strain are caused by axial loads, we have

$$\frac{P}{A} = E\,\frac{\delta}{L}$$

or

$$\delta = \frac{PL}{AE} \qquad\qquad\qquad\qquad (10.3)$$

EXAMPLE 10.4

Determine the total change in length of the 0.025-m-diameter rod when it is acted on by three forces shown in Fig. 10.7(a). Let $E = 70 \times 10^6 \ \text{kN/m}^2$.

Solution

The total change in length must be equal to the sum of the changes in length for each section of the rod.

Cut the bar at any section *D–D* between points *A* and *B* and draw the free-body diagram of the bar to the left of the section as shown in Fig. 10.7(b). We see from the equilibrium equation that the internal reaction must be a tensile force of

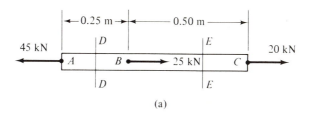

(a)

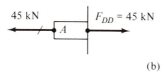

(b)

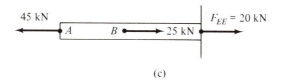

(c)

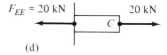

(d)

FIGURE 10.7

45 kN. Similarly, cut the bar at any section E–E between points B and C and draw a free-body diagram of the bar either to the left as shown in Fig. 10.7(c) or to the right as shown in Fig. 10.7(d) of the section. The internal reaction, as given by the equilibrium equation, must be a tensile force of 20 kN.

The cross-sectional area $A = \pi d^2/4 = \pi(0.025)^2/4 = 0.000\ 49\ \text{m}^2$. The change in length for the section of the rod from A to B is found from Eq. (10.3).

$$\delta_1 = \frac{PL}{AE} = \frac{45\,(0.25)}{0.000\ 49\,(70)(10^6)} = 0.000\ 328\ \text{m}$$

$$= 0.328\ \text{mm}$$

Similarly, the change in length for the section of the rod from B to C is

$$\delta_2 = \frac{PL}{AE} = \frac{20\,(0.50)}{0.000\ 49\,(70)(10^6)} = 0.000\ 292\ \text{m}$$

$$= 0.292\ \text{mm}$$

Therefore, the total change in length

$$\delta = \delta_1 + \delta_2 = 0.328 + 0.292 = 0.620\ \text{mm} \qquad \text{Answer}$$

249

PROBLEMS

10.1 A concrete test cylinder 6 in. in diameter and 12 in. high is loaded in compression. As a result of the load, the cylinder shortened by 0.01 in. Determine the strain.

10.2 A steel rod 2.5 mm in diameter is tested in tension. The strain at the proportional limit is 0.0016. Determine the deformation at the proportional limit for a gauge length 20 mm long.

10.3 An axial load is applied to a bar. The length $L = 16$ in. increases by 0.022 in. The width $w = 3$ in. decreases by 0.00112 in. and the thickness $t = 0.75$ in. decreases by 0.00028 in. Determine the three components of strain.

10.4 A steel cable is used on a hoist. If the maximum strain in the cable is 0.008, what is the deformation for 30 m of this cable?

In Probs. 10.5 and 10.6, data from a tension test of steel are as tabulated. Calculate stress and the corresponding strain and draw a complete stress–strain curve. Draw a second stress–strain curve up to the yield point on the same graph. Determine the (a) proportional limit, (b) modulus of elasticity, (c) upper and lower yield point, (d) ultimate strength, (e) breaking strength, (f) percent reduction in area (see Sec. 10.8), and (g) percent elongation (see Sec. 10.8).

10.5 For the complete stress–strain curve, scales of 1 in. = 10 ksi and 1 in. = 0.05 in./in. may be used. For the stress–strain curve up to the yield point, use a scale of 1 in. = 0.0005 in./in.

10.6 For the complete stress–strain curve, scales of 10 mm = 25 MPa and 10 mm = 0.020 mm/mm may be used. For the stress–strain curve up to the yield point, use a scale of 10 mm = 0.000 20 mm/mm.

(Data for Prob. 10.5)
Tension Test of Steel (Gauge Length = 2 in.; Original Diameter = 0.501 in.)

Load (kips)	Deformation (in.)	Load (kips)	Deformation (in.)
0	0.0000	8.4	0.0050
1.0	0.0003	7.8	0.0095
2.0	0.0007	9.65	0.1000
3.0	0.0011	11.8	0.2000
4.0	0.0014	12.25	0.3000
5.0	0.0018	12.5	0.5000
6.0	0.0021	12.2	0.6000
7.0	0.0025	10.2	0.7000
8.0	0.0028	8.4	0.7200
8.6	0.0035		

Final diameter = 0.331 in.

(Data for Prob. 10.6)
Tension Test of Steel (Gauge Length = 40 mm; Original Diameter = 10 mm)

Load (kN)	Deformation (mm)	Load (kN)	Deformation (mm)
0	0.0000	20.8	0.0920
2.0	0.0050	19.5	0.3400
4.0	0.0090	20.0	0.7600
6.0	0.0148	23.0	1.6000
8.0	0.0197	25.5	2.0000
10.0	0.0247	28.4	4.0000
12.0	0.0296	29.4	6.0000
14.0	0.0346	29.8	8.0000
16.0	0.0395	29.6	10.0000
18.0	0.0444	29.0	12.0000
20.0	0.0512	27.9	14.0000
21.36	0.0610	23.2	15.2000

Final diameter = 5.52 mm

In Probs. 10.7 and 10.8, data from a tension test of cast iron are as tabulated. Calculate stress and the corresponding strain and draw a complete stress–strain curve. Determine the (a) proportional limit, (b) modulus of elasticity, (c) yield strength at 0.05 percent offset, and (d) ultimate strength. [The yield strength at 0.05 percent offset is found by drawing a line parallel to the modulus line through the strain $\epsilon_1 = 0.0005$ on the stress–strain curve as shown in Fig. 10.5(b).]

line parallel to the modulus line through the strain $\epsilon_1 = 0.002$ on the stress–strain curve as shown in Fig. 10.5(b).]

10.7 Scales of 1 in. = 5 ksi and 1 in. = 0.0005 in./in. may be used.

10.9 For the complete stress–strain curve, scales of 1 in. = 10 ksi and 1 in. = 0.05 in./in. may be used. For the stress–strain curve up to the yield strength, use a scale of 1 in. = 0.002 in./in.

Tension Test of Cast Iron (Gauge Length = 2 in.; Original Diameter = 0.505 in.)

Load (kips)	Deformation (in.)	Load (kips)	Deformation (in.)
0	0.00000	3.5	0.00250
0.5	0.00034	4.0	0.00310
1.0	0.00064	4.5	0.00400
1.5	0.00101	4.8	0.00500
2.0	0.00136	5.0	0.00600
2.5	0.00170	5.2	(Failure)
3.0	0.00200		

Tension Test of Aluminum Alloy (Gauge Length = 2 in.; Original Diameter = 0.505 in.)

Load (kips)	Deformation (in.)	Load (kips)	Deformation (in.)
0	0.0000	8.0	0.0118
1.0	0.0010	9.0	0.0264
2.0	0.0020	10.0	0.0600
3.0	0.0030	11.0	0.1408
4.0	0.0040	12.0	0.3220
5.0	0.0050	13.0	0.7030
6.0	0.0063	13.2	(Failure)
7.0	0.0084		

10.8 Scales of 10 mm = 10 MPa and 10 mm = 0.0002 mm/mm may be used.

Tension Test of Cast Iron (Gauge Length = 40 mm; Original Diameter = 10 mm)

Load (kN)	Deformation (mm)	Load (kN)	Deformation (mm)
0	0.000 00	8.0	0.040 76
1.0	0.005 08	9.0	0.046 40
2.0	0.010 20	10.0	0.055 60
3.0	0.015 28	11.0	0.064 00
4.0	0.020 36	12.0	0.074 00
5.0	0.025 48	13.0	0.100 00
6.0	0.030 56	14.0	0.148 00
7.0	0.036 68	14.1	(Failure)

10.10 For the complete stress–strain curve, scales of 10 mm = 25 MPa and 10 mm = 0.0125 mm/mm may be used. For the stress–strain curve up to the yield strength, use a scale of 10 mm = 0.0005 mm/mm.

Tension Test of Aluminum Alloy (Gauge Length = 40 mm; Original Diameter = 10 mm)

Load (kN)	Deformation (mm)	Load (kN)	Deformation (mm)
0	0.0000	20.0	0.1976
2.0	0.0148	22.0	0.2888
4.0	0.0296	24.0	0.4844
6.0	0.0444	26.0	0.8652
8.0	0.0592	28.0	1.6408
10.0	0.0740	30.0	3.0428
12.0	0.0892	32.0	5.6880
14.0	0.1048	34.0	10.1200
16.0	0.1236	35.2	(Failure)
18.0	0.1500		

In Probs. 10.9 and 10.10, data from a tension test of an aluminum alloy are as tabulated. Calculate stress and the corresponding strain and draw the complete stress–strain curve. Draw a second stress–strain curve up to the yield strength at 0.2 percent offset on the same graph. Determine the (a) proportional limit, (b) modulus of elasticity, (c) yield strength at 0.2 percent offset, and (d) ultimate strength. [The yield strength at 0.2 percent offset is found by drawing a

10.11 A copper bar with a cross-sectional area of 1.2 in.^{2} is loaded as shown.

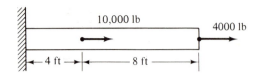

PROB. 10.11

What is the total change in length of the bar? $E_C = 16 \times 10^6$ psi.

10.12 An aluminum bar with a cross-sectional area of 650 mm² is loaded as shown. What is the total change in length of the bar? $E_A = 70 \times 10^6$ kN/m².

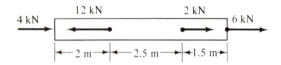

PROB. 10.12

10.13 The aluminum bar AB, shown in the figure, has a rectangular cross section 6 in. by 1.5 in. Determine (a) the axial stress in the bar and (b) the lengthening of the bar due to the loads. $E_A = 10 \times 10^6$ psi.

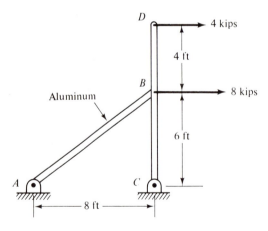

PROB. 10.13

10.14 The round steel rod shown in the figure has a diameter of 30 mm. Determine (a) the axial stress in the rod and

(b) the lengthening of the rod due to the load. $E_S = 200 \times 10^6$ kN/m².

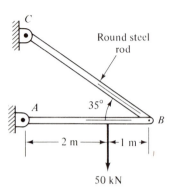

PROB. 10.14

10.15 A short square steel structural tube is used as a compression member as shown. The allowable compressive stress is 24,000 psi and the allowable axial deformation 0.006 in. Determine the allowable load. $E_S = 30 \times 10^6$ psi.

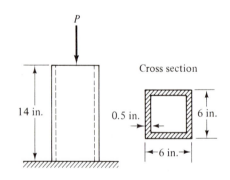

PROB. 10.15

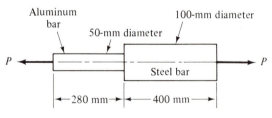

PROB. 10.16

10.16 Two round bars are joined and used as a tension member as illustrated. Determine the magnitude of the force P that will lengthen the two bars 0.33 mm. $E_S = 200 \times 10^6$ kN/m^2, $E_A = 70 \times 10^6$ kN/m^2.

10.17 The axial load at the proportional limit of an 0.8-in.-diameter bar was 12,300 lb. If a 4-in.-gauge length on the bar has increased by 0.0060 in. at the proportional limit, determine

(a) the stress at the proportional limit, (b) the strain at the proportional limit, and (c) the elastic modulus.

10.18 A flat bar of cross section 15 mm by 70 mm elongates 2.59 mm in a length of 1.3 m as the result of an axial load of 230 kN. The proportional limit of the material is 240 kN/m^2. Determine the axial stress in the bar and the elastic modulus.

10.5 STATICALLY INDETERMINATE PROBLEMS

If a machine or structure is made up of one or more axially loaded members, the equations of statics may not be sufficient to find the internal reaction in the members. In such cases, equations of geometric fit of the members may be required. Equation (10.3) will be used to write such equations.

EXAMPLE 10.5

Two steel plates 1 in. by 4 in. by 15 in. are attached to a pine block 4 in. by 4 in. by 15 in. as shown in Fig. 10.8(a). A rigid bearing plate on top transmits a load of 200 kips to the compression member. Find the internal reaction and stress in the pine and steel. The elastic modulus for steel is $E_S = 30 \times 10^6$ psi and for pine $E_P = 1.5 \times 10^6$ psi.

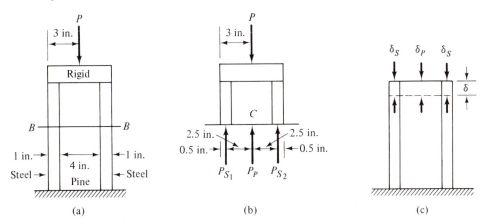

FIGURE 10.8

Solution

A section B–B is cut through the member and a free-body diagram is drawn for the member above the section as shown in Fig. 10.8(b). The forces in the steel plates are

P_{S_1} and P_{S_2}. The force in the pine block is P_P. From equilibrium equations:

$$\circlearrowright \Sigma M_C = 0 \quad P_{S_2}(2.5) - P_{S_1}(2.5) = 0$$

$$P_{S_2} = P_{S_1} = P_S$$

$$\uparrow \Sigma F_y = 0 \quad P_S + P_S + P_P - P = 0$$

$$2P_S + P_P = P = 200 \text{ kips}$$

Figure 10.8(c) shows the deformations for the steel plates and the pine block. Since the force on each steel plate is the same, the deformations of the steel plates are equal. The rigid bearing plate under the load ensures that the deformation of the pine block must equal the deformation of the steel plate. That is,

$$\delta_S = \delta_P$$

From Eq. (10.3),

$$\frac{P_S L_S}{A_S E_S} = \frac{P_P L_P}{A_P E_P} \quad \text{or} \quad P_S = \frac{A_S E_S L_P}{A_P E_P L_S} P_P$$

where the subscript, S, refers to the steel plate and the subscript, P, refers to the pine block. The length of both the steel and pine are equal. Substituting values of the cross-sectional area and elastic modules, we have

$$P_S = \frac{4(30)(10^6)L_P}{16(1.5)(10^6)L_S} P_P = 5P_P$$

Substituting $P_S = 5P_P$ in the equilibrium equation yields

$$2(5P_P) + P_P = 200$$

$$11P_P = 200$$

$$P_P = 18.18 \text{ kips}$$

Then

$$P_S = 5P_P = 5(18.18) = 90.9 \text{ kips}$$

$$P_P = 18.2 \text{ kips} \qquad P_S = 90.9 \text{ kips} \qquad\qquad \text{Answer}$$

The stress in the steel and pine is given by

$$\sigma_S = \frac{P_S}{A_S} = \frac{90.9}{4} = 22.7 \text{ ksi} \qquad\qquad \text{Answer}$$

$$\sigma_P = \frac{P_P}{A_P} = \frac{18.2}{16} = 1.14 \text{ ksi} \qquad\qquad \text{Answer}$$

EXAMPLE 10.6

The bar shown in Fig. 10.9(a) is attached at its ends to unyielding supports. There is no initial stress in the bar. A force P is then applied as shown. (a) What are the forces in the steel and aluminum parts of the bar? (b) If the allowable working stress in the steel is 150 MN/m^2 and in the aluminum 60 MN/m^2, what safe load P can the bar support? $E_S = 200 \times 10^6$ kN/m^2, $E_A = 70 \times 10^6$ kN/m^2.

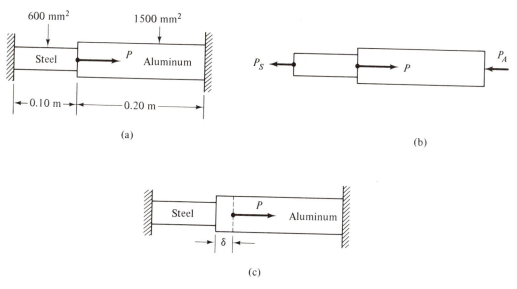

(a)

(b)

(c)

FIGURE 10.9

Solution

(a) In Fig. 10.9(b) a free-body diagram of the bar is shown. From equilibrium:

$$\rightarrow \Sigma F_x = 0 \qquad -P_S + P - P_A = 0$$

$$P_S + P_A = P$$

Figure 10.9(c) shows the deformation for the steel and aluminum. The tensile deformation for the steel is equal to the compression deformation of the aluminum. That is,

$$\delta_S = \delta_A$$

From Eq. (10.3),

$$\frac{P_S L_S}{A_S E_S} = \frac{P_A L_A}{A_A E_A} \quad \text{or} \quad P_S = \frac{A_S E_S L_A}{A_A E_A L_S} P_A$$

Substituting numerical values, we have

$$P_S = \frac{600\,(200)\,(10^9)\,(0.20)}{1500\,(70)\,(10^9)\,(0.10)}\,P_A = 2.286 P_A$$

Substituting $P_S = 2.286\,P_A$ in the equilibrium equation,

$$2.286 P_A + P_A = P$$

$$3.286 P_A = P \qquad P_A = 0.304 P \qquad\qquad \text{Answer}$$

$$P_S = 2.286 P_A = 2.286\,(0.304 P)$$

$$= 0.695 P \qquad\qquad \text{Answer}$$

(b) The allowable load in the steel and aluminum part of the bar is

$$P_A = \sigma_A A_A = 60\,(1500) = 90\ 000\ \text{N}$$

$$P_S = \sigma_S A_S = 150\,(600) = 90\ 000\ \text{N}$$

The *total* allowable load based on the allowable steel stress is

$$P = \frac{P_S}{0.695} = \frac{90\ 000}{0.695} = 129\ 500\ \text{N}$$

and on the allowable aluminum stress

$$P = \frac{P_A}{0.304} = \frac{90\ 000}{0.304} = 296\ 000\ \text{N}$$

Therefore, the maximum allowable load is

$$P = 129\ 500\ \text{N} \qquad\qquad \text{Answer}$$

PROBLEMS

10.19 A bar is supported at B and C as illustrated. A force of 10,000 lb is applied at A. Determine the reactions at the supports.

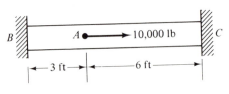

PROB. 10.19

10.20 As shown in the figure, the bar is supported at C and D. A 240-kip force is applied at E. Determine the reactions at the supports and the stress and strain in each part of the bar. Assume that $E = 30 \times 10^6$ psi.

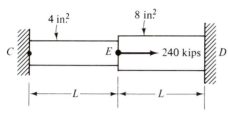

PROB. 10.20

10.21 A steel pipe is filled with concrete and used as a column with a cross section as shown. The column is loaded in compression. If the total compressive force is 200 kips, determine the stress in each material. $E_S = 30 \times 10^6$ psi and $E_C = 3 \times 10^6$ psi.

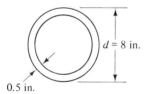

PROB. 10.21

10.22 A circular aluminum rod $1\frac{1}{2}$ in. in diameter is fitted inside a circular steel tube with an inside diameter of $1\frac{1}{2}$ in. and an outside diameter of 3 in. The rod and the tube have the same length and are acted on by a compressive axial load which causes a stress in the steel tube of 15,000 psi. Determine the axial load and the stress in the aluminum rod. $E_S = 30 \times 10^6$ psi and $E_A = 10 \times 10^6$ psi.

10.23 A rigid bar AB is supported by three rods as shown. If a load $P = 200$ kN is applied to the bar, determine the deformation and stress in each rod. $E_S = 200 \times 10^6$ kN/m^2 and $E_C = 110 \times 10^6$ kN/m^2.

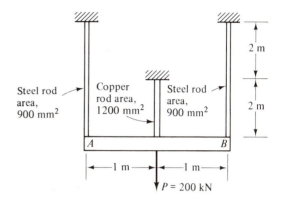

PROB. 10.23

10.24 In the figure shown, the steel bar AB has a cross section 0.75 in. by 0.5 in. and the aluminum bar CD has a cross section 1.0 in. by 0.5 in. Assume that bar BC is rigid. A load $P = 4000$ lb is applied to bar BC so that it remains horizontal. Determine (a) the axial stress in the steel and aluminum bar, (b) the vertical displacement of BC, and (c) the location of the load measured from point B. $E_S = 30 \times 10^6$ psi and $E_A = 10 \times 10^6$ psi.

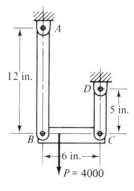

PROB. 10.24

10.25   Bars *M* and *O* are made of brass and bar *N* is made of steel as illustrated. Each bar has a cross-sectional area of 2000 mm². What load *P* will cause axial stress in the steel to be one-half of the axial stress in the brass? $E_S =$

200×10^6 kN/m² and $E_B = 100 \times 10^6$ kN/m².

10.26   Two wires are used to lift a 10-kip weight as shown. One wire is 60.0 ft long and the other wire is 60.08 ft long. If the cross-sectional area of each wire is 0.15 in.², determine the load supported by each wire. $E = 30 \times 10^6$ psi.

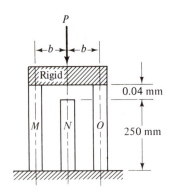

PROB. 10.25

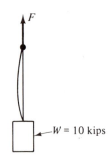

PROB. 10.26

10.6 POISSON'S RATIO

When a load is applied along the axis of a bar, axial strain is produced. At the same time, a lateral (perpendicular to the axis) strain is also produced. If the axial force is in tension, the length of the bar increases and the cross section contracts or decreases. That is, a *positive axial stress produces a positive axial strain and a negative lateral strain*. For a negative axial stress, the axial strain is negative and the lateral strain is positive.

The ratio of lateral strain to axial strain is called *Poisson's ratio*. It is constant for a given material provided that the material is not stressed above the proportional limit, is homogeneous, and has the same physical properties in all directions. Poisson's ratio, represented by the Greek letter ν (nu), is defined by the equation

$$\nu = \frac{-\text{lateral strain}}{\text{axial strain}} \tag{10.4}$$

The negative sign ensures that Poisson's ratio is a positive number.

The value of Poisson's ratio, ν, varies from 0.25 to 0.35 for different metals. For concrete it may be as low as $\nu = 0.1$ and for rubber as high as $\nu = 0.5$. Values of Poisson's ratio are given in Table A.11 of the Appendix.

EXAMPLE 10.7

A test conducted on a steel bar whose cross section is 0.5 in. by 4 in. is performed. An axial load of $P = 30,500$ lb produces a deformation of 0.00103 in. over a gauge length of 2 in. and a decrease of 0.000078 in. in the 0.5-in. thickness of the bar. Determine Poisson's ratio, ν, the elastic modulus, E, and the decrease in the 4.0-in. cross-sectional dimension.

Solution

The lateral strain is

$$\epsilon_t = \frac{\delta_t}{d} = \frac{-0.000078}{0.50} = -0.000156$$

Since the thickness decreases, the lateral strain is negative. The axial strain is

$$\epsilon_a = \frac{\delta_a}{L} = \frac{0.00103}{2.0} = 0.000515$$

Therefore, Poisson's ratio is given by

$$\nu = \frac{-\epsilon_t}{\epsilon_a} = \frac{-(-0.000156)}{0.000515} = 0.303 \qquad \text{Answer}$$

The cross-sectional area of the rod $A = (0.5)(4.0) = 2.0$ in.^{2} From Eq. (10.3),

$$\delta = \frac{PL}{AE} \qquad E = \frac{PL}{A\delta} = \frac{30,500}{2.0(0.00103)} = 29.6 \times 10^6 \text{ psi} \qquad \text{Answer}$$

The lateral strain is $\epsilon_t = -0.000156$. Therefore, $\delta_t = \epsilon_t L_t = -0.000156(4.0) = -0.000624$ in. The axial stress will be calculated to verify that the stress is below the proportional limit.

$$\sigma = \frac{P}{A} = \frac{30,500}{2.0} = 15,250 \text{ psi} < \sigma_P \qquad \text{OK}$$

10.7 THERMAL STRAIN

When the temperature of a material is raised (or lowered), the material expands (or contracts) unless it is prevented from doing so by external forces acting on the material. If the temperature of a body of length L is changed by a temperature Δt, the linear deformation is given by

$$\delta_t = L\alpha(\Delta t) \tag{10.5}$$

where α is the coefficient of linear expansion and has units of per Fahrenheit (Celsius) degree. Average values of α for selected materials are given in Table A.11 of the Appendix.

Indeterminate problems involving temperature changes follow the same methods as the problems in Sec. 10.4.

EXAMPLE 10.8

A steel rod having a cross-sectional area of 0.002 m² and length 1.5 m is attached to supports. Find (a) the force exerted by the rod on the supports if the temperature falls 60 C° and the supports are unyielding, and (b) the force exerted by the rod on the supports if they yield 0.0003 m while the temperature raises 50 C°.

$$E_S = 200 \times 10^9 \text{ N/m}^2 \qquad \alpha_S = 11.5 \times 10^{-6} \text{ per C}°$$

Solution

(a) Remove one end support, decrease the temperature of the steel rod by 60 C°, and let it shorten. Calculate the decrease in length as shown in Fig. 10.10(a). Then apply a load that will restore the bar to its original length as shown in Fig. 10.10(b). From Eq. (10.5),

$$\delta_t = L\alpha(\Delta t) = 1.5\,(11.5)\,(10^{-6})\,(60)$$

$$= -0.001\ 035 \text{ m}$$

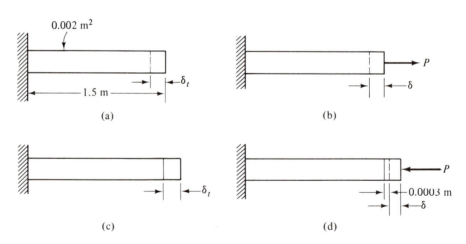

FIGURE 10.10

From Eq. (10.3),

$$\delta = \frac{PL}{AE}$$

or

$$P = \frac{\delta AE}{L} = \frac{0.001\ 035\,(0.002)\,(200)\,(10^9)}{1.5}$$

$$= 276\ 000\ \text{N} = 276\ \text{kN} \qquad \text{Answer}$$

The rod is in tension. Thus the force of the rod on the supports will tend to bring the supports closer together.

(b) Remove one support and increase the temperature of the steel rod. Calculate the increase in length due to a temperature rise of 50 C° as shown in Fig. 10.10(c). Then apply a force that will restore the rod to its original length less 0.0003 m, which the supports are permitted to yield, as shown in Fig. 10.10(d). From Eq. (10.5),

$$\delta_t = L\alpha(\Delta t) = 1.5\,(11.5)\,(10^{-6})\,(50)$$

$$= 0.000\ 862\ 5\ \text{m}$$

Since

$$\delta = \delta_t - 0.0003 = 0.000\ 862\ 5 - 0.0003$$

$$= 0.000\ 562\ 5\ \text{m}$$

From Eq. (10.3),

$$P = \frac{\delta AE}{L} = \frac{0.000\ 562\ 5\,(0.002)\,(200)\,(10^9)}{1.5}$$

$$= 150\ 000\ \text{N} = 150\ \text{kN} \qquad \text{Answer}$$

The rod is in compression. Thus the force of the rod on the supports will tend to move the supports farther apart.

EXAMPLE 10.9

A steel bolt through a 25-in.-long aluminum tube has the nut turned on until it just touches the end of the tube. No stress is introduced in bolt or tube. The temperature of the assembly is raised from 60°F to 70°F. If the cross-sectional area of the bolt is 0.08 in.^{2} and the cross-sectional area of the tube is 0.25 in.^{2}, find the stress induced into the bolt and tube by the temperature change. $E_S = 30 \times 10^6$ psi, $E_A = 10 \times 10^6$ psi, $\alpha_S = 6.5 \times 10^{-6}$ per F°, $\alpha_A = 13.0 \times 10^{-6}$ per F°.

Solution

Remove the steel bolt from the aluminum tube; increase their temperature. Let them expand freely and calculate the thermal expansion of the steel bolt, δ_{t_S}, and the thermal expansion of the aluminum tube, δ_{t_A}, as shown in Fig. 10.11(a).

$$\delta_{t_S} = L\alpha(\Delta t) = 25\,(6.5)\,(10^{-6})\,(10) = 0.001625 \text{ in.}$$

$$\delta_{t_A} = L\alpha(\Delta t) = 25\,(13.0)\,(10^{-6})\,(10) = 0.00325 \text{ in.}$$

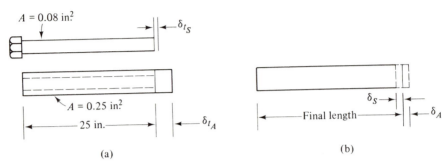

FIGURE 10.11

To place the steel bolt back into the aluminum tube, the steel bolt must be stretched δ_S by a force P_S and the aluminum tube must be compressed δ_A by a force P_A so that the tube and the bolt are the same length, as shown in Fig. 10.11(b). That is,

$$\delta_A + \delta_S = \delta_{t_A} - \delta_{t_S}$$

$$\frac{P_A L_A}{A_A E_A} + \frac{P_S L_S}{A_S E_S} = 0.00325 - 0.001625$$

$$\frac{P_A\,(25)}{0.25\,(10)\,(10^6)} + \frac{P_S\,(25)}{0.08\,(30)\,(10^6)} = 0.001625$$

$$10.0 P_A + 10.416 P_S = 1625$$

From equilibrium,

$$P_A = P_S = P$$

Therefore,

$$20.416 P = 1625$$

$$P = 79.6$$

and the stresses in the steel bolt and the aluminum tube are

$$\sigma_S = \frac{P}{A_S} = \frac{79.6}{0.08} = 995 \text{ psi} \quad \text{(tension)} \qquad \text{Answer}$$

$$\sigma_A = \frac{P}{A_A} = \frac{79.6}{0.25} = 318 \text{ psi} \quad \text{(compression)} \qquad \text{Answer}$$

10.8 ADDITIONAL MECHANICAL PROPERTIES OF MATERIALS

In addition to the properties discussed in Sec. 10.2, the following are of interest.

Elastic Limit: The highest stress that can be applied without permanent strain when the stress is removed. To determine the elastic limit would require the application of larger and larger loadings and unloadings of the material until permanent strain is detected. The test would be difficult. It is rarely done. Numerical values of the proportional limit are often used in its place.

Elastic Range: Response of the material as shown on the stress–strain curve from the origin up to the proportional limit P (Fig. 10.12).

Plastic Range: Response of the material as shown on the stress–strain curve from the proportional limit P to the breaking strength B (Fig. 10.12).

Percentage Reduction in Area: When a ductile material is stretched beyond its ultimate strength, the cross section "necks down" and the area reduces appreciably. It is defined by the equation

$$\text{Percentage reduction in area} = \frac{A_o - A_f}{A_o} \, 100$$

where A_o is the original and A_f the final minimum cross-sectional area. It is a measure of ductility.

Percentage Elongation: The percentage elongation represents a comparison of the increase in the length of the gauge length to the original gauge

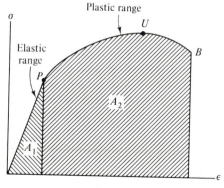

FIGURE 10.12

length. It is defined by the equation

$$\text{Percentage elongation} = \frac{L_f - L_o}{L_o}\ 100$$

where L_o is the original and L_f the final gauge length. It is also a measure of ductility.

Modulus of Resilience: The work done on a unit volume of material from a zero force up to the force at the proportional limit. This is equal to the area under the stress–strain curve from zero to the proportional limit. (Area A_1 of Fig. 10.12. Units of in.-lb/in.^{3} or N·m/m^3.)

Modules of Toughness: The work done on a unit volume of material from a zero force up to the force at the breaking point. This is equal to the area under the stress–strain curve from zero to the breaking strength. (Areas A_1 and A_2 of Fig. 10.12. Units of in.-lb/in.^{3} or N·m/m^3.)

10.9 FACTOR OF SAFETY

In Sec. 9.7 we defined the allowable working stress as the maximum stress that is considered safe for a material to support. The allowable working stress may be specified in terms of the factor of safety. The safety factor may be defined as the ratio of some stress that represents the strength of the material to the allowable stress. The factor of safety based on ultimate stress is defined as the ratio of the ultimate stress to the allowable stress.

$$\text{F.S.} = \frac{\sigma_{ult}}{\sigma_a} \tag{10.6}$$

Similarly, the factor of safety based on the yield stress is defined as the ratio of the yield stress to the allowable stress.

$$\text{F.S.} = \frac{\sigma_y}{\sigma_a} \tag{10.7}$$

For a ductile material such as structural steel, it would be logical to use the factor of safety based on the yield stress, while the factor of safety based on the ultimate stress would be used for brittle materials such as cast iron or concrete. In the case of the long column, the factor of safety is based on buckling stress.

Values of the factor of safety depend on the nature of the loads and the shape and kind of member involved. They may range in value from over 1 to 20 with values between 3 and 15 common.

PROBLEMS

10.27 Determine the Poisson's ratio for the bar material in Prob. 10.3.

10.28 A member 1.5 m long has a cross section 75 mm by 75 mm. The member becomes 0.7 mm longer and 0.01 mm narrower after loading. Determine Poisson's ratio for the material.

10.29 A load of 210 kips is applied to a 20-ft-long rod with a diameter of 2.25 in. The rod stretches 0.42 in. and the diameter decreases by 0.001 in. Determine Poisson's ratio and the elastic modulus for the material.

10.30 A flat bar of cross section 20 mm by 70 mm elongates 4.0 mm in a length of 2.0 m as a result of axial load. If the elastic modulus is 199×10^6 kN/m^2 and Poisson's ratio is 0.25 for the material, find the axial load and the total change in each cross-sectional dimension.

10.31 An aluminum rod with a diameter of 2.0 in. and a length of 1.5 ft is attached to supports at its ends. Determine (a) the force exerted by the supports on the rod if the temperature drops $50 F°$ and the supports are unyielding, and (b) the force exerted by the supports on the rod if they yield 0.0001 in. while the temperature increases 60 F°. $E_A = 10 \times 10^6$ psi and $\alpha_A = 13.2 \times 10^{-6} (F°)^{-1}$.

10.32 The composite aluminum and steel member illustrated is attached to supports at its ends. Determine (a) the force exerted by the supports on the member if the temperature drops $60 F°$ and the supports are unyielding, and (b) the force exerted by the supports on the member if they yield 0.01

in. while the temperature increases 70 F°. $E_S = 30 \times 10^6$ psi, $E_A = 10 \times 10^6$ psi, $\alpha_S = 6.5 \times 10^{-6} (F°)^{-1}$, and $\alpha_A = 13.2 \times 10^{-6} (F°)^{-1}$.

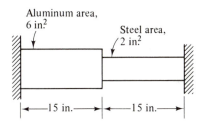

Aluminum area, 6 in²

Steel area, 2 in²

|←——15 in.——→|←——15 in.——→|

PROB. 10.32

10.33 A steel bar is attached at its ends to rigid supports. The tensile stress in the bar is 80 000 kN/m^2 when the temperature is 40°C. What is the temperature in the bar when the stress is 60 000 kN/m^2? $E_S = 200 \times 10^6$ kN/m^2 and $\alpha_S = 11.7 \times 10^{-6} (C°)^{-1}$.

10.34 An aluminum bar is attached at its ends to rigid supports. The compressive stress in the bar is 2500 psi when the temperature is 100°F. What is the stress in the bar when the temperature is lowered to 50°F? $E_A = 10 \times 10^6$ psi and $\alpha_A = 13.2 \times 10^{-6} (F°)^{-1}$.

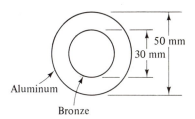

50 mm

30 mm

Aluminum

Bronze

PROB. 10.35

10.35 A bar with a composite cross section as shown has its temperature raised by $30C°$. If no slippage occurs between the bronze and aluminum, determine the axial stress in each material. $E_A = 70 \times 10^6$ kN/m^2, $E_B = 110 \times 10^6$ kN/m^2, $\alpha_A = 23.8 \times 10^{-6}$ $(C°)^{-1}$, and $\alpha_B = 18.0 \times 10^{-6} (C°)^{-1}$.

10.36 The cross section of a concrete column consisting of steel reinforcing bars embedded in a concrete cylinder is shown in the figure. If no slippage occurs between the concrete and steel, determine the axial stress introduced into the steel and concrete due to a temperature increase of $100F°$. $E_S = 30 \times 10^6$ psi, $E_C = 3 \times 10^6$ psi, $\alpha_S = 6.5 \times 10^{-6}$ $(F°)^{-1}$, and $\alpha_C = 6.2 \times 10^{-6} (F°)^{-1}$.

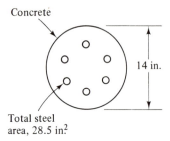

Concrete

14 in.

Total steel area, 28.5 in.^{2}

PROB. 10.36

Shear Stresses and Strains—
Torsion _____

11.1 INTRODUCTION

In this chapter we begin by discussing shearing stresses and strains and Hooke's law for shearing stresses and strains. The torsion of a circular shaft follows logically since the shaft is in a state of pure shear.

11.2 SHEARING STRESS ON PLANES AT RIGHT ANGLES

Consider the shear stress acting tangent to a plane edge of a flat plate of uniform thickness, t [Fig. 11.1(a)]. The average shear stress over a part of the edge from a to d is τ_1. The average shear stress is shown on the four sides of an element of the plate $abcd$ in Fig. 11.1(b). The plate is in equilibrium; therefore, the element $abcd$ of the plate is in equilibrium. We draw

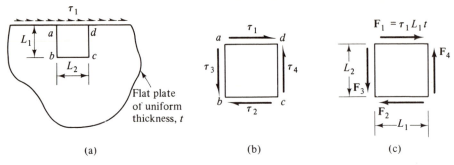

FIGURE 11.1

a free-body diagram of the element in Fig. 11.1(c). A force $F_1 = \tau_1 A = \tau_1 L_1 t$ acts to the right on the top of the element. The shear force on the bottom of the element is equal to the average shear stress τ_2 multiplied by the shear area $F_2 = \tau_2 A = \tau_2 L_1 t$. Forces on the left and right side of the element are $F_3 = \tau_3 L_2 t$ and $F_4 = \tau_4 L_2 t$, respectively. From equilibrium:

$$\rightarrow \Sigma F_x = 0 \qquad F_1 - F_2 = 0$$

or

$$\tau_1 L_1 t - \tau_2 L_1 t = 0$$

Therefore $\tau_1 = \tau_2$.

$$\uparrow \Sigma F_y = 0 \qquad -F_3 + F_4 = 0$$

or

$$-\tau_3 L_2 t + \tau_4 L_2 t = 0$$

Therefore $\tau_3 = \tau_4$.

$$\circlearrowright \Sigma M_b = 0 \qquad -F_1 L_2 + F_4 L_1 = 0$$
$$-\tau_1 L_1 t L_2 + \tau_4 L_2 t L_1 = 0$$

Therefore $\tau_1 = \tau_4$. We see that

$$\tau_1 = \tau_2 = \tau_3 = \tau_4 = \tau$$

Thus all four average shear stresses acting on the element are equal. Notice that the shear stresses meet tip to tip and tail to tail at the corners of the element. If we let the element shrink to a point, the average stresses become stresses at a point. Hence the *shearing stresses at a point acting on mutually perpendicular planes are equal.*

11.3 SHEARING STRAINS

The shearing stresses shown in Fig. 11.1(b) cause distortion of the element *abcd*. The diagonal *b-d* is lengthened and the diagonal *a-c* is shortened. The distortion is illustrated in Fig. 11.2. The shearing strain is equal to the

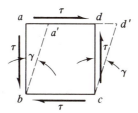

FIGURE 11.2

distance *aa'* divided by the distance *ad*. Since the angle represented by the Greek letter γ (gamma) is small,

$$\tan \gamma = \gamma \text{ (radians)} = \frac{aa'}{ab}$$

Hence the shearing strain is numerically equal to γ measured in radians.

11.4 HOOKE'S LAW FOR SHEAR

Experiments show that as for normal stress and strain, shearing stress is proportional to shearing strain as long as stress does not exceed the proportional limit. Hooke's law for shear may be expressed as

$$\tau = G\gamma \tag{11.1}$$

where G is the *shear modulus* or the modulus of rigidity. The shearing modulus, G, is a constant for a given material. It is expressed in the same units as stress since γ is measured in radians, which are dimensionless quantities. The shear modulus, G, has a value about 40 percent of the value of the elastic modulus E.

It can be shown that the three elastic constants E, G, and ν are not independent of each other for an isotropic material. An isotropic material is the same in all directions. The relationship between elastic constants is given by the equation

$$G = \frac{E}{2(1 + \nu)} \tag{11.2}$$

EXAMPLE 11.1

In the test on a steel bar, $E = 29.57 \times 10^6$ psi and $\nu = 0.303$. Find the shear modulus, G.

Solution

From Eq. (11.2),

$$G = \frac{29.57 \, (10^6)}{2 \, (1 + 0.303)} = 11.35 \times 10^6 \text{ psi} \qquad \text{Answer}$$

PROBLEMS

11.1 Determine the shear modulus for an aluminum alloy that has a modulus $E = 10.6 \times 10^6$ psi and Poisson's ratio $\nu = 0.325$.

11.2 Determine the shear modulus for a steel that has a modulus $E = 206 \times 10^6$ kN/m^2 (kPa) and Poisson's ratio $\nu = 0.25$.

11.3 Determine Poisson's ratio for a magnesium alloy that has a modulus $E = 44.8 \times 10^6$ kN/m^2 (kPa) and a shear modulus $G = 16.5 \times 10^6$ kN/m^2 (kPa).

11.4 Determine Poisson's ratio for a cast iron that has a modulus $E = 12 \times 10^6$ psi and a shear modulus $G = 4.8 \times 10^6$ psi.

11.5 TORSION OF A CIRCULAR SHAFT

We consider here the torsion of a circular shaft. Let the bottom end of the shaft be fixed and a torque T be applied at the top end as shown in Fig. 11.3(a).

Geometry of Deformation

Cutting through the shaft we draw a free-body diagram of the part of the shaft between plane cross sections J–J and K–K in Fig. 11.3(b). Both cross sections are normal to the axis of the member. We assume that these plane cross sections remain plane after the torque is applied. The plane $OABD$, which was parallel to the axis of the shaft, moves to a new position $OEBD$ as the top cross section rotates through an angle, φ (phi) after application of the torque T. The line OA remains straight as it rotates through the angle φ to the new position OE. The magnitude of the shearing strain at a distance r from the center of the shaft is given by the angle $FHG = \gamma$ expressed in radians. The angle γ is small and can be expressed closely in radian measure as

$$\gamma = \frac{FG}{FH} = \frac{FG}{L} \qquad \text{(a)}$$

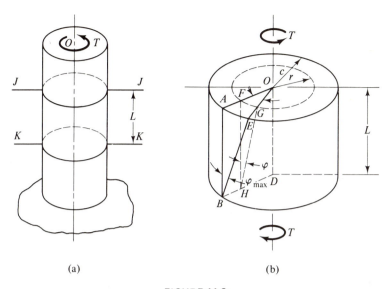

(a)   (b)

FIGURE 11.3

The angle $FOG = \varphi$ can be expressed in radians as $\varphi = FG/FO = FG/r$ or $FG = r\varphi$. Combining this with Eq. (a), we obtain

$$\gamma = \frac{r\varphi}{L} \tag{11.3}$$

Similarly, the magnitude of the maximum shearing strain is given by the angle $ABE = \gamma_{max}$. The angle γ_{max} is also small and can be expressed closely in radians measure as

$$\gamma_{max} = \frac{AE}{AB} = \frac{AE}{L} \tag{ }$$

The angle AOE or φ can be expressed in radians as $\varphi = AE/AO = AE/c$ or $AE = c\varphi$. Combining this with Eq. (b) yields

$$\gamma_{max} = \frac{c\varphi}{L} \tag{11.4}$$

Hooke's Law

In our consideration of the geometry of deformation, no restrictions have been placed on the material from which the shaft was made except that it must be isotropic (have the same physical properties in any direction). We assume now that the material follows Hooke's law for shear. Combining

271

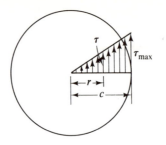

FIGURE 11.4

Hooke's law, Eq. (11.1), with Eqs. (11.3) and (11.4), we have

$$\tau = G\gamma = \frac{G\varphi}{L} r \tag{11.5}$$

and

$$\tau_{max} = G\gamma_{max} = \frac{G\varphi}{L} c \tag{11.6}$$

Dividing Eq. (11.6) by Eq. (11.5),

$$\frac{\tau_{max}}{\tau} = \frac{c}{r} \tag{c}$$

Thus we see that the shear stress on the cross section of the shaft is proportional to the distance from the center of the cross section as shown in Fig. 11.4.

Equilibrium

Consider (Fig. 11.5) a narrow ring of mean radius r and area ΔA concentric with the center of the cross section of the shaft. The magnitude of the shearing stress at every point in the ring is equal to τ, and it is directed normal to a radius drawn from the center of the cross section O to that

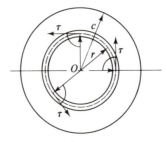

FIGURE 11.5

point. The moment about O or torque of the stresses on the ring are, therefore,

$$\underset{\substack{\text{(torque} \\ \text{for} \\ \text{ring)}}}{\Delta T} = \underset{\substack{\text{(stress)} \\ \text{(force)}}}{\tau} \; \underset{\text{(area)}}{\Delta A} \; \underset{\text{(moment arm)}}{r}$$

Adding up the torque for all the rings to include the entire cross-sectional area, we obtain the total torque.

$$T = \underset{\substack{\text{total} \\ \text{area}}}{\Sigma \tau} \; r \Delta A$$

Deformation of a Circular Shaft

Substituting the value of the shear stress, τ, from Eq. (11.5), we have

$$T = \underset{\substack{\text{total} \\ \text{area}}}{\Sigma} \frac{G\varphi}{L} r^2 \Delta A$$

Since the shear modulus G, angle of twist φ, and the length L are constant for each ring,

$$T = \frac{G\varphi}{L} \underset{\substack{\text{total} \\ \text{area}}}{\Sigma} r^2 \Delta A \qquad \text{(d)}$$

However, $\Sigma r^2 \Delta A$ is also a constant for a particular cross section called the *polar moment of inertia*, J. The polar moment of inertia was discussed in Sec. 8.8. The formula for the polar moment of inertia for a circle of radius c was given as

$$J = \frac{\pi c^4}{2} = \frac{\pi d^4}{32}$$

Substituting the polar moment of inertia, J, in Eq. (d), we obtain

$$T = \frac{G\varphi J}{L}$$

or

$$\varphi = \frac{TL}{GJ} \qquad (11.7)$$

where φ is the angle of twist in radians, a dimensionless quantity. The internal torque T is expressed in pound-inches or newton·meters (N·m), L in inches or meters (m), G in psi or newtons/meter^2 (N/m^2), and J in in.^{4} or m^4.

Stresses in a Circular Shaft

The relationship of shear stress to torque may be obtained by substituting Eq. (11.7) in (11.5), that is,

$$\tau = \frac{Gr\varphi}{L} = \frac{Gr}{L}\frac{TL}{GJ}$$

or

$$\tau = \frac{Tr}{J} \tag{11.8}$$

The maximum value of the shear stress occurs when r has its maximum value, $r = c$.

$$\tau_{max} = \frac{Tc}{J} \tag{11.9}$$

This equation is known as the *torsion formula*. It gives the maximum stress in terms of the torque and the dimensions of the member. Torque, T, will be expressed in inch-pounds or newton·meters (N·m), c in inches or meters, and J in in.^{4} or m^4. The torsion shearing stress will have units of

$$\frac{(\text{in.-lb})(\text{in.})}{(\text{in.}^4)} = \frac{\text{lb}}{\text{in.}^2} = \text{psi}$$

or in SI units

$$\frac{(\text{N·m})(\text{m})}{(\text{m}^4)} = \frac{\text{N}}{\text{m}^2} \text{ or Pa (pascals)}$$

11.6 FURTHER COMMENTS ON THE TORSION OF A CIRCULAR SHAFT

In our analysis of the torsion of a circular shaft we have only considered the shear stress acting on a cross section of the shaft.

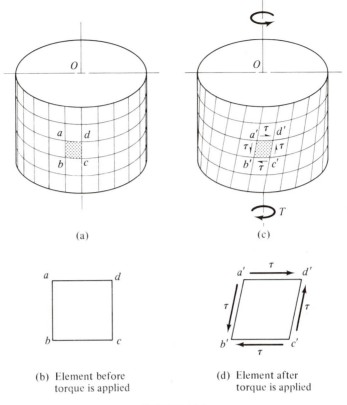

(a)

(c)

(b) Element before
    torque is applied

(d) Element after
    torque is applied

FIGURE 11.6

In Fig. 11.6 we consider an element *abcd* on the outside of a circular shaft before and after torque is applied. Side *ad* of the element lies along a cross section and side *ab* is normal to the cross section. Thus the shear stress on side *ad* and *ab* must be equal, from Sec. 11.2. The value of the stress is given by Eq. (11.9).

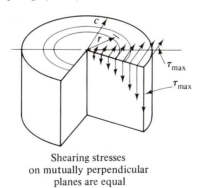

Shearing stresses
on mutually perpendicular
planes are equal

FIGURE 11.7

We may also apply the same argument to an element on any concentric cylinder of radius r inside the circular shaft. Therefore, the distribution of shear stress on the cross section and on a plane normal to the cross section are the same (Fig. 11.7).

11.7 PROBLEMS INVOLVING DEFORMATION AND STRESS IN A CIRCULAR SHAFT

In this section we solve several problems involving the deformation and stress in a circular shaft.

EXAMPLE 11.2

Find the torque that a solid circular shaft with a diameter of 0.1 m can transmit if the maximum shearing stress is 50 000 kN/m². What is the angle of twist per meter if $G = 80 \times 10^6$ kN/m²?

Solution

The polar moment of inertia, J, is given by

$$J = \frac{\pi d^4}{32} = \frac{\pi (0.1)^4}{32} = 9.82 \times 10^{-6} \text{ m}^4$$

From Eq. (11.9), $\tau_{max} = Tc/J$; therefore,

$$T = \frac{J\tau_{max}}{c} = \frac{(9.82)(10^{-6})(50)(10^3)}{0.05} = 9.82 \text{ kN} \cdot \text{m}$$

$$= 9820 \text{ N} \cdot \text{m} \qquad\qquad \text{Answer}$$

The angle of twist is given by Eq. (11.7):

$$\varphi = \frac{TL}{GJ} = \frac{9.82(1.0)}{80(10^6)(9.82)(10^{-6})} = 0.0125 \text{ rad}$$

or

$$\varphi = \frac{180}{\pi}(0.0125) = 0.72° \qquad\qquad \text{Answer}$$

EXAMPLE 11.3

Find the shearing stress on the outside and inside of a long hollow circular shaft with an outside diameter of 6 in. and an inside diameter of 4 in. if a torque of 20,000 lb-ft is applied. What will be the angle of twist in a length of 15 ft? $G = 12 \times 10^6$ psi.

Solution

The polar moment of inertia for a hollow circular shaft with cross section shown in Fig. 11.8 is given by

$$J = \frac{\pi(d_o^4 - d_i^4)}{32} = \frac{\pi[(6)^4 - (4)^4]}{32} = 102 \text{ in.}^4$$

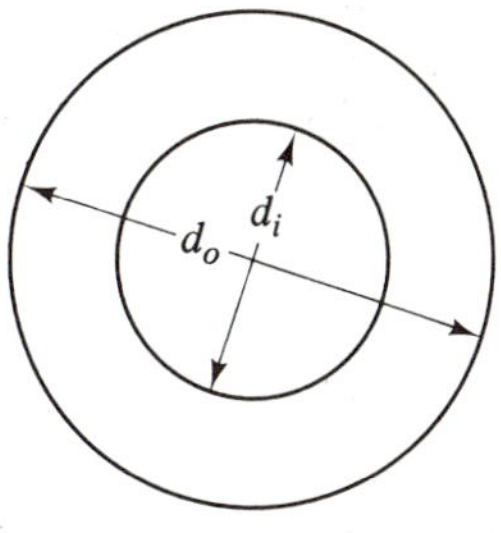

$$J = \frac{\pi}{32}(d_o^4 - d_i^4)$$

FIGURE 11.8

From Eq. (11.8), the stress on the outside of the shaft is

$$\tau_o = \frac{Tr_o}{J} = \frac{20,000(12)(3)}{102} = 7060 \text{ psi} \qquad \text{Answer}$$

while the stress on the inside of the shaft is

$$\tau_i = \frac{Tr_i}{J} = \frac{20,000(12)(2)}{102} = 4700 \text{ psi} \qquad \text{Answer}$$

The angle of twist, from Eq. (11.7), is

$$\varphi = \frac{TL}{GJ} = \frac{20,000(12)(15)(12)}{12(10^6)(102)} = 0.0353 \text{ rad}$$

or

$$\varphi = \frac{180}{\pi}(0.0353) = 2.02^\circ \qquad \text{Answer}$$

EXAMPLE 11.4

A solid shaft with a 20-mm diameter has built-in ends as shown in Fig. 11.9(a). The shaft is subjected to a torque $T = 40$ N·m applied to an intermediate point as shown. Determine the maximum shearing stress in the shaft and the angle of twist between A and B. $G = 80 \times 10^6$ kN/m^2.

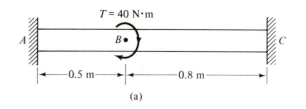

$T = 40$ N·m

(a)

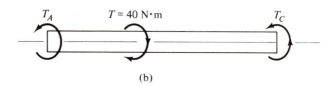

T_A $T = 40$ N·m T_C

(b)

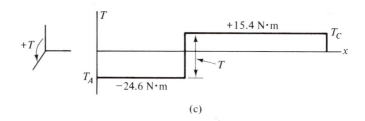

(c)

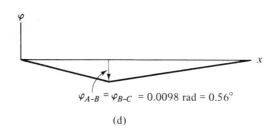

$\varphi_{A-B} = \varphi_{B-C} = 0.0098$ rad $= 0.56°$

(d)

FIGURE 11.9

Solution

This is a statically indeterminate problem in torsion. It is similar to the axially loaded problem discussed in Example 10.6 of Sec. 10.5.

In Fig. 11.9(b) we show the free-body diagram for the shaft. The reactive torques at A and C must oppose the applied torque at B. $T = 40$ N·m.

From equilibrium:

$$\searrow \Sigma T = 0 \qquad -T_A - T_C + 40 = 0 \tag{a}$$

From geometric fit, the angle of twist φ_{A-B} from A to B must be equal to the angle of twist φ_{B-C} from B to C.

$$\varphi_{A-B} = \varphi_{B-C} \tag{b}$$

From Eq. (11.7), $\varphi = TL/GJ$. Substituting the value of φ into Eq. (b) yields

$$\frac{T_A L_{AB}}{GJ} = \frac{T_C L_{BC}}{GJ}$$

or

$$T_A = \frac{L_{BC}}{L_{AB}} T_C = \frac{0.8}{0.5} T_C$$

$$= 1.6 T_C$$

Substituting the value of T_A into Eq. (a), we obtain

$$1.6 T_C + T_C = 40$$

$$T_C = 15.38 \text{ N·m}$$

and

$$T_A = 1.6 T_C = 1.6(15.38) = 24.62 \text{ N·m}$$

The polar moment of inertia is given by

$$J = \frac{\pi d^4}{32} = \frac{\pi (0.02)^4}{32} = 15.7 \times 10^{-9} \text{ m}^4$$

The maximum shearing stress will occur between A and B. From Eq. (11.9),

$$\tau_{max} = \frac{T_A c}{J} = \frac{24.6(0.01)}{15.7 \times 10^{-9}} = 15.67 \text{ MN/m}^2 \qquad \text{Answer}$$

The angle of twist between A and B is given by Eq. (11.7):

$$\varphi_{A-B} = \frac{T_A L_{AB}}{GJ} = \frac{24.62(0.5)}{80(10^9)(15.7)(10^{-9})} = 0.0098 \text{ rad}$$

or

$$\varphi_{A-B} = \frac{180}{\pi} (0.0098) = 0.56° \qquad \text{Answer}$$

The torque and angle of twist diagram is shown in Fig. 11.9(c) and (d).

EXAMPLE 11.5

Determine the diameter of a solid shaft required to deliver 15 horsepower at 25 cycles per second if the maximum shearing stress is limited to 50 000 kN/m^2.

Solution

It can be shown that

$$T = \frac{118.7 \text{ hp}}{n} \tag{11.10}$$

where

T = torque, N·m
n = revolutions per second
hp = horsepower

From Eq. (11.10),

$$T = \frac{118.7\,(15)}{25} = 71.4 \text{ N} \cdot \text{m}$$

From Eq. (11.9), $\tau_{\max} = Tc/J$; therefore,

$$\frac{J}{c} = \frac{T}{\tau_{\max}} = \frac{71.4}{50\,(10^6)} = 1.428 \times 10^{-6} \text{ m}^3$$

$$= \frac{\pi d^4/32}{d/2} = \frac{\pi d^3}{16} = 1.428\,(10^{-6}) \text{ m}^3$$

$$d^3 = 7.27 \times 10^{-6} \text{ m}^3$$

$$d = 19.37 \times 10^{-3} \text{ m}$$

$$= 19.37 \text{ mm} \qquad\qquad \text{Answer}$$

In U.S. common units,

$$T = \frac{63,000 \text{ hp}}{N} \tag{11.11}$$

where

T = torque, in.-lb
N = rpm (revolutions per minute)
hp = horsepower

11.8 TORSION TEST

In a test of a solid circular shaft in torsion, the outside of the shaft is more highly stressed than the interior of the shaft, and therefore the outside of the shaft reaches the proportional limit or yield point before it is reached in the interior of the shaft. As a result, the onset of yielding is covered up by the fact that the inside of the shaft continues to support loads with stresses proportional to strain. Only after considerable yielding has taken place does the effect show in the test data. To overcome this difficulty, a

FIGURE 11.10 Bench-type torsion testing machine. (*Courtesy Tinius Olsen Testing Machine Company.*)

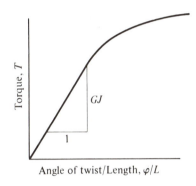

FIGURE 11.11

test on a thin hollow circular specimen can be performed. The stress across the thin wall of the specimen will have approximately the same value and the onset of yielding will show in the test data soon after it first occurs. From Eq. (11.7), $\varphi = TL/GJ$ or

$$\frac{T}{\varphi/L} = GJ \tag{a}$$

In the torsion test a torque is applied to a solid or hollow circular shaft by a torsion testing machine as shown in Fig. 11.10. For each specified value of the torque, we measure the angle of twist with a mechanical device called a troptometer. The values obtained can then be plotted in a torque angle of twist per unit of length curve. A typical curve is shown in Fig. 11.11.

From Eq. (a) the slope of such a curve will be equal to the shear modulus multiplied by the polar moment of inertia. The shear modulus is approximately 40 percent of the elastic modulus ($G \approx 0.40E$).

PROBLEMS

Where appropriate, use the following values of the shear modulus: $G_{steel} = 12 \times 10^6$ psi $(83 \times 10^6$ kPa$)$ and $G_{aluminum} = 4 \times 10^6$ psi $(28 \times 10^6$ kPa$)$.

11.5 Determine the maximum shearing stress for a circular steel shaft with a diameter of 50 mm if the torque applied is 2250 N·m.

11.6 Determine the maximum shearing stress for a circular steel shaft with a diameter of 2 in. if the torque applied is 5000 lb-in.

11.7 Determine the maximum shearing stress for a circular aluminum alloy tube with an outside diameter of 3 in. and an inside diameter of 2.5 in. if the torque applied is 25,000 lb-in.

11.8 Determine the maximum shearing stress for a circular aluminum alloy tube with an outside diameter of 75 mm and an inside diameter of 60 mm if the torque applied is 1695 N·m.

11.9 Determine the torque required to produce a maximum shearing stress of 140 000 kN/m² (kPa) in a solid circular shaft with a diameter of 15 mm.

11.10 Determine the torque required to produce a maximum shearing stress of 20,000 psi in a solid circular steel shaft with a diameter of 0.6 in.

11.11 Determine the torque required to produce a maximum shearing stress of 15,000 psi in a circular tube with an outside diameter of 6 in. and an inside diameter of 4 in.

11.12 Determine the torque required to produce a maximum shearing stress of 100 MN/m² (MPa) in a circular tube with an outside diameter of 150 mm and an inside diameter of 100 mm.

11.13 Determine the angle of twist in degrees for a solid circular steel shaft with a diameter of 100 mm if the torque applied is 20 kN·m and the shaft is 1.5 m long.

11.14 Determine the angle of twist in degrees for a solid circular steel shaft with a diameter of 4 in. if the torque applied is 15,000 lb-ft and the shaft is 5 ft long.

11.15 Determine the angle of twist in degrees for a circular aluminum alloy tube with an outside diameter of 4 in. and an inside diameter of 3.4 in. if the torque applied is 30,000 lb-in. and the shaft is 4.5 ft long.

11.16 Determine the angle of twist in degrees for a circular steel tube with an outside diameter of 100 mm and an inside diameter of 85 mm if the torque applied is 26 kN·m and the shaft is 2.2 m long.

11.17 A solid circular steel shaft has a diameter of 50 mm. Determine the torque required to produce an angle of twist of 3 degrees in a length of 1.8 m.

11.18 A solid circular aluminum alloy shaft has a diameter of 0.85 in. Determine the torque required to produce an angle of twist of 3.5 degrees in a length of 2.5 ft.

11.19 A circular steel tube has an outside diameter of 2.5 in. and an inside diameter of 1.75 in. Determine the torque required to produce an angle of twist of 2.2 degrees in a length of 4 ft.

11.20 A circular aluminum alloy tube has an outside diameter of 120 mm and an inside diameter of 90 mm. Determine the torque required to produce an angle of twist of 2 degrees in a length of 0.9 m.

11.21 A solid round steel shaft 0.75 m long transmits a torque $T = 600$ N·m. If the maximum shear stress in the shaft is 150×10^3 kN/m² (kPa), determine the diameter of the shaft and the angle of twist in degrees.

11.22 A solid, round aluminum alloy shaft 40 in. long transmits a torque $T = 3000$ lb-in. If the maximum shear stress in the shaft is 18 ksi, determine the diameter of the shaft and the angle of twist in degrees.

11.23 A circular steel tube 4 ft long transmits a torque $T = 5500$ lb-in. If the maximum shear stress in the tube is 25 ksi and the inside diameter of the tube is nine-tenths of the outside diameter of the tube, determine the outside diameter of the shaft and the angle of twist in degrees.

11.24 A circular aluminum alloy tube 1.2 m long transmits a torque $T = 7000$ N·m. If the maximum shear stress in the tube is 140×10^3 kN/m² (kPa) and the tube thickness is one-tenth of the outside diameter of the tube, determine the outside diameter of the tube and the angle of twist in degrees.

11.25 The stepped shaft shown is made of steel. Determine the maximum shear stress and the angle of twist in degrees if $d_1 = 60$ mm, $d_2 = 40$ mm, $L_1 = 0.5$ m, $L_2 = 0.25$ m, and the shaft transmits a torque $T = 400$ N·m.

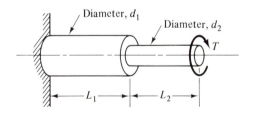

PROB. 11.25 and PROB. 11.26

11.26 The stepped shaft shown is made of aluminum alloy. Determine the maximum shear stress and the angle of twist in degrees if $d_1 = 2.5$ in., $d_2 = 2$ in., $L_1 = 2$ ft, $L_2 = 1$ ft, and the shaft transmits a torque $T = 450$ lb-ft.

11.27 Pipes A and B are joined together as shown. Pipe A, which is made of steel, has an outside diameter of 3.25 in., an inside diameter of 2.75 in., and a length $a = 2.5$ ft. Pipe B, which is made of aluminum alloy, has an outside diameter of 2.75 in., an inside diameter of 2.0 in., and a length $b = 1.25$ ft. If a torque $T = 800$ lb-in. is applied, determine the maximum shear stress and the angle of twist in degrees.

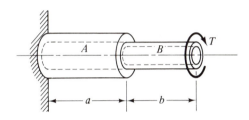

PROB. 11.27 and PROB. 11.28

11.28 Pipes A and B are joined as shown. Aluminum pipe A has an outside diameter of 80 mm, an inside diameter of 70 mm, and a length $a = 0.75$ m. Steel pipe B has an outside diameter of 70 mm, an inside diameter of 60 mm, and a length $b = 1.25$ m. If a torque $T = 100$ N·m is applied, determine the maximum shear stress and the angle of twist in degrees.

11.29 A solid steel shaft with a diameter $d = 60$ mm has built-in ends at A and C. The shaft is subjected to a torque $T = 306$ N·m applied to an intermediate point B so that $L_1 = 0.5$ m and $L_2 = 1.0$ m, as shown. Determine

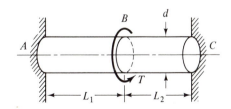

PROB. 11.29 and PROB. 11.30

the reactions at A and C and the maximum shearing stress in the shaft.

11.30 A solid aluminum alloy shaft with a diameter $d = 2.4$ in. has built-in ends at A and C. The shaft is subjected to a torque $T = 2100$ lb-in. applied to an intermediate point B so that $L_1 = 1.5$ ft and $L_2 = 2$ ft, as shown. Determine the reactions at A and C and the maximum shearing stress in the shaft.

11.31 A stepped shaft with diameters $D = 3.0$ in. and $d = 1.7$ in. has built-in ends at P and R. The shaft is subjected to a torque $T = 2500$ lb-in. applied to an intermediate point Q so that $a = 22$ in. and $b = 15$ in. as shown. Determine the reactions at P and R and the maximum shearing stress in the shaft.

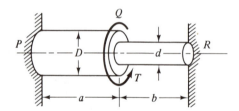

PROB. 11.31 and PROB. 11.32

11.32 A stepped shaft with diameters $D = 70$ mm and $d = 35$ mm has built-in ends at P and R. The shaft is subjected to a torque $T = 350$ N·m applied to an intermediate point Q so that $a = 0.6$ m and $b = 0.35$ m as shown. Determine

the reactions at P and R and the maximum shearing stress in the shaft.

11.33 Show that the torque transmitting capacity of a solid shaft is reduced by one-sixteenth by boring an axial hole through the center if the diameter of the hole is one-half the diameter of the shaft.

11.34 Show that the torque transmitting capacity of a solid shaft is reduced by one-ninth by boring an axial hole through the center if the area of the hole is one-third of the original shaft area.

11.35 Determine the diameter of a solid steel shaft required to deliver 5 hp at 60 cycles per second if the maximum shearing stress is limited to 95 000 kN/m^3 (kPa). (See Example 11.5.)

11.36 Determine the diameter of a solid steel shaft required to deliver 12 horsepower at 1800 rpm if the maximum shearing stress is limited to 14,000 psi. (See Example 11.5.)

11.37 Determine the horsepower a solid steel shaft with a diameter of 1.5 in. can deliver at 2500 rpm if the maximum allowable stress is 12,000 psi. (See Example 11.5.)

11.38 Determine the horsepower a solid steel shaft with a diameter of 20 mm can deliver at 60 cycles per second if the maximum allowable shearing stress is 90 000 kN/m^2 (kPa). (See Example 11.5.)

12

Beams—
Shear Forces and Bending Moments _____

12.1 INTRODUCTION

We previously considered four kinds of internal reactions in members. They were axial forces, shear forces, bending moments, and torsions. In Chapters 9 and 10 we examined the axial forces and the stresses and strains they produce. The ability of torque to produce stress and strain was explored in Chapter 11. In this chapter we examine the beam, that is, a member acted on by loads that produce bending.

The internal reactions in the beam are shear forces and bending moments and to a lesser extent axial forces. In addition to the values of axial forces, shear forces, and bending moments at selected cross sections of a beam, we may also be interested in their value at cross sections throughout the beam. Such information is represented graphically on an axial force diagram, shear force diagram, or bending moment diagram. The horizontal axis of the diagram represents the location of the beam cross section. The vertical axis represents the value of the particular internal reaction at each cross-section location.

12.2 TYPES OF BEAMS

As previously indicated, beams are structural and machine members acted on by loads that produce bending. In order to produce bending, the loads must be applied transversely, that is, perpendicular to the axis of the beam.

  Beams may be classified according to the kinds of supports and loading. If a beam is freely supported at its ends with either pins or rollers, it is called a *simply supported beam* or *simple beam*. Simple beams are shown in Fig. 12.1(a) and (b). The beam that is fixed at one end and free at the other is called a *cantilever beam*. Figure 12.1(c) and (d) show canti-lever beams. Beams with overhanging ends are shown in Fig. 12.1(e) and

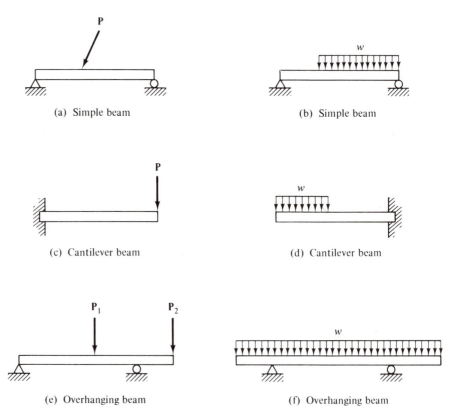

FIGURE 12.1 Statically determinate beams.

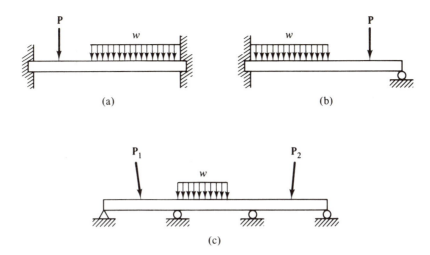

FIGURE 12.2 Statically indeterminate beams.

(f). The supports for such beams are either pins or rollers. The simple beam, cantilever beam, and beam with overhanging ends are all *determinate*, since the three unknown reactions for each beam can be determined by the equations of static equilibrium. Examples of *statically indeterminate* beams are shown in Fig. 12.2, where we have a *fixed beam* in which both ends are fixed, a beam in which one end is fixed and the other end supported by a roller, and a *continuous beam*, which is supported by a pin and two or more rollers.

12.3 BEAM REACTIONS

Before an analysis of the internal reactions can be made, the beam reactions must be calculated. Since the internal reactions depend upon the beam reactions, care should be taken to ensure their accuracy. In the following examples the procedure for finding beam reactions is reviewed.

EXAMPLE 12.1

Find the reactions for the simply supported beam [Fig. 12.3(a)] loaded as shown. Neglect the weight of the beam.

Solution

A free-body diagram is shown in Fig. 12.3(b). The uniformly distributed load has been replaced by its resultant and the inclined 5-kip force has been replaced by horizontal and vertical components. From equilibrium:

$$\Sigma\, M_A = 0 \quad -3.830(3) - 7.2(9) + D_y 12 = 0$$

$$D_y = 6.36 \text{ kips} \qquad \text{Answer}$$

287

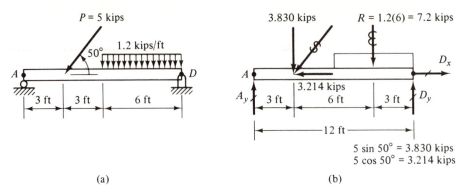

(a)

(b)

FIGURE 12.3

$$\circlearrowright \Sigma\, M_D = 0 \quad -A_y 12 + 3.830\,(9) + 7.2\,(3) = 0$$

$$A_y = 4.67 \text{ kips} \hspace{5cm} \text{Answer}$$

$$\rightarrow \Sigma\, F_x = 0 \quad -3.214 + D_x = 0$$

$$D_x = 3.214 \text{ kips} \hspace{5cm} \text{Answer}$$

Check:

$$\uparrow \Sigma\, F_y = 0 \quad A_y - 3.830 - 7.2 + D_y = 0$$

$$4.67 - 3.830 - 7.2 + 6.36 = 0 \quad \text{OK}$$

EXAMPLE 12.2

Find the reaction for the cantilever beam shown in Fig. 12.4(a). The uniform load has been replaced in Fig. 12.4(b), by its resultant. (Note that the moment of the couple is the same about any point in the plane of the couple.)

Solution

From equilibrium:

$$\rightarrow \Sigma\, F_x = 0 \quad A_x = 0 \hspace{4cm} \text{Answer}$$

$$\uparrow \Sigma\, F_y = 0 \quad A_y - 4 - 1.53 = 0$$

$$A_y = 5.53 \text{ kN} \hspace{4.5cm} \text{Answer}$$

$$\circlearrowright \Sigma\, M_A = 0 \quad M_A - 4\,(0.9) + 3.0 - 1.53\,(3.15) = 0$$

$$M_A = 5.42 \text{ kN} \cdot \text{m} \hspace{4cm} \text{Answer}$$

Check:

$$\circlearrowright \Sigma\, M_B = 0 \quad . \quad M_A - A_y 3.60 + 4\,(2.70) + 3.0 + 1.53\,(0.45) = 0$$

$$5.42 - 5.53\,(3.60) + 4\,(2.70) + 3.0 + 1.53\,(0.45) = 0.005 \quad \text{OK}$$

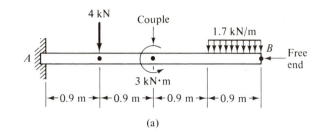

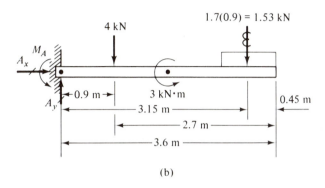

FIGURE 12.4

12.4 AXIAL FORCE, SHEARING FORCE, AND BENDING MOMENT

We use the method of sections to cut a beam at various sections and determine the internal reactions for that section. Consider a simply supported beam with a concentrated load P and uniformly distributed load w [Fig. 12.5(a)]. In a preliminary calculation the reactions were determined. To find the axial force, shearing force, or bending moment at a given section, we cut the beam at that section and draw a free-body diagram of either end of the beam.

Free-body diagrams are shown in Fig. 12.5(b) and (c) for section A–A and in Fig. 12.5(d) and (e) for section B–B. Either end of the beam must be in equilibrium under the external loads and internal reactions. The internal reactions play an important part in the study of beams and we will consider each separately.

Axial Force

The axial force has already been considered in Chapter 9. It is important to remember that for convenience the axial force acts through the centroid of the cross section. Recall that the axial force is in tension and therefore positive if it acts away from the beam. Axial forces will be shown in tension on the free-body diagram. If, from the equilibrium

289

equation, the sign of the axial force is negative, it means that the forces are in compression. Frequently, no loads or reactions act in the direction of the axis of the beam and therefore no axial force exists in the beam.

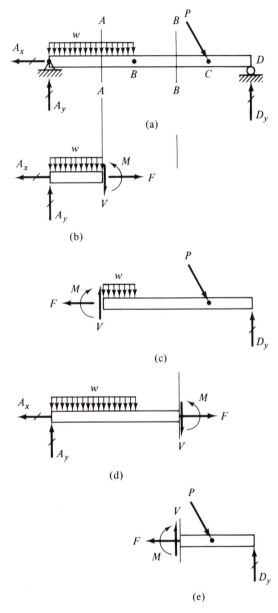

FIGURE 12.5

Shearing Force

Forces that act normal (perpendicular) to the axis of the beam produce a shearing force or shear in the plane of the cross section as shown in Fig. 12.5. The shear forces are all shown in a positive direction. See Fig. 12.6

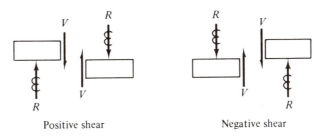

Positive shear   Negative shear

FIGURE 12.6

for the positive and negative sign convention for shear forces. We will assume a positive shear on our free-body diagrams. If the equilibrium equation gives a negative value for shear, it means that our assumption was incorrect and the shear is negative.

Bending Moments

The forces that act normal to the axis of the beam and forces in the axial direction that are not directed along the axis of the beam cause bending moments or couples as shown in Fig. 12.5. The bending moments are shown in a positive direction. See Fig. 12.7 for the positive and negative

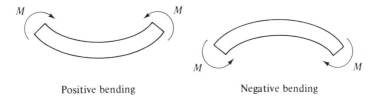

Positive bending   Negative bending

FIGURE 12.7

sign convention. Physically, a positive bending moment means compression on the top and tension on the bottom of the beam. It also means that the beam is bent concave upward. To find the bending moment, we write a moment equilibrium equation about a point at the centroid of the cross section.

In the following examples we calculate the axial force, shearing force, and bending moment at selected cross sections of beams.

EXAMPLE 12.3

For the beam shown in Fig. 12.8(a), find the shearing force at $x = 2$ ft, 6 ft, and 10 ft from A.

Kip = 1,000 #

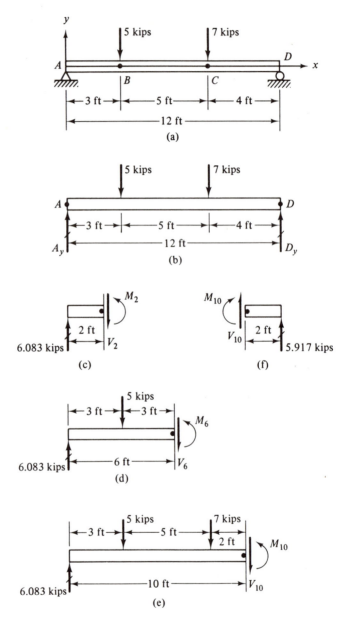

FIGURE 12.8

Solution

A free-body diagram for the entire beam is shown in Fig. 12.8(b). From equilibrium the reactions are

$$A_y = 6.083 \text{ kips} \qquad D_y = 5.917 \text{ kips}$$

The free-body diagram for the beam to the left of various sections are shown in the figure. We write the equation for vertical equilibrium in each case.

For the section at $x = 2$ ft [Fig. 12.8c)]:

$$\uparrow \Sigma F_y = 0 \qquad 6.083 - V_2 = 0$$
$$V_2 = 6.083 \text{ kips} \qquad \qquad \text{Answer (a)}$$

For $x = 6$ ft [Fig. 12.8(d)]:

$$\uparrow \Sigma F_y = 0 \qquad 6.083 - 5 - V_6 = 0$$
$$V_6 = 6.083 - 5.0 \qquad \qquad \text{(b)}$$
$$= 1.083 \text{ kips} \qquad \qquad \text{Answer}$$

For $x = 10$ ft [Fig. 12.8(e)]:

$$\uparrow \Sigma F_y = 0 \qquad 6.083 - 5 - 7 - V_{10} = 0$$
$$V_{10} = 6.083 - 5 - 7 \qquad \qquad \text{(c)}$$
$$= -5.917 \text{ kips} \qquad \qquad \text{Answer}$$

In Fig. 12.8(f) the free-body diagram for the beam to the right of the section at $x = 10$ ft is shown. For vertical equilibrium:

$$\uparrow \Sigma F_y = 0 \qquad V_{10} + 5.917 = 0$$
$$V_{10} = -5.917 \text{ kips} \qquad \qquad \text{(d)}$$
$$-5.917 = -5.917 \qquad \text{OK}$$

Alternative Solution

Reviewing the solution for shearing forces as shown in Eqs. (a)–(c), we see that *the shear at any section is equal to the algebraic sum of the normal forces on the beam to the left of the section.* Forces directed up are positive and down are negative. That is, in equation form

$$V(\text{at section}) = +\uparrow \Sigma F(\text{on beam to left of section}) \qquad (12.1)$$

From Eq. (d) we see that the sign convention must be reversed when the beam to the right of the section is considered. We modify Eq. (12.1) accordingly.

$$V(\text{at section}) = +\downarrow \Sigma \ F(\text{on beam to right of section}) \qquad (12.2)$$

Equations (12.1) and (12.2) are alternative forms of the equation for vertical equilibrium. They permit a direct evaluation of the shear force at a section.

EXAMPLE 12.4

For the beam shown in Fig. 12.8(a), find the bending moment at $x = 2$ ft, 6 ft, and 10 ft.

Solution

In Example 12.3 free-body diagrams were drawn for the various cross sections. We write the equation for *rotational equilibrium about the centroid of the cross section* in each case.

For the section at $x = 2$ ft [Fig. 12.8(c)]:

$$\circlearrowright \Sigma \ M_c = 0 \qquad -6.082(2) + M_2 = 0$$

$$M_2 = 6.082(2) \qquad\qquad\qquad\qquad (\text{e})$$

$$= 12.16 \text{ kip-ft} \qquad\qquad\qquad \text{Answer}$$

For $x = 6$ ft [Fig. 12.8(d)]:

$$\circlearrowright \Sigma \ M_c = 0 \qquad -6.082(6) + 5(3) + M_6 = 0$$

$$M_6 = 6.082(6) - 5(3) \qquad\qquad\qquad (\text{f})$$

$$= 21.49 \text{ kip-ft} \qquad\qquad\qquad \text{Answer}$$

For $x = 10$ ft [Fig. 12.8(e)]:

$$\circlearrowright \Sigma \ M_c = 0 \qquad -6.082(10) + 5(7) + 7(2) + M_{10} = 0$$

$$M_{10} = 6.082(10) - 5(7) - 7(2) \qquad\qquad (\text{g})$$

$$= 11.82 \text{ kip-ft} \qquad\qquad\qquad \text{Answer}$$

In Fig. 12.8(f) the free-body diagram for the beam to the right of the section at $x = 10$ ft is shown. For rotational equilibrium:

Check:

$$\circlearrowright \Sigma \ M_c = -M_{10} + 5.917(2) = 0$$

$$M_{10} = 5.917(2) \qquad\qquad\qquad\qquad (\text{h})$$

$$= 11.83 \text{ kip-ft} \quad \text{OK} \qquad\qquad \text{Answer}$$

Alternative Solution

Reviewing the solution for bending moments as shown in Eqs. (e)–(g), we see that *the bending moment at any section is equal to the algebraic sum of the moments of the forces on the beam to the left of the section about the centroid of the cross section.*

Clockwise moments are positive and counterclockwise moments are negative. That is, in equation form:

$$M(\text{at section}) = (+ \Sigma M_c(\text{on beam to left of section}) \qquad (12.3)$$

From Eq. (h) we see that the sign convention must be reversed when the beam to the right of the section is considered. We modify Eq. (12.3) accordingly.

$$M(\text{at section}) = +) \Sigma M_c(\text{on beam to right of section}) \qquad (12.4)$$

Equations (12.3) and (12.4) are alternative forms of the equation for rotational equilibrium. They permit direct evaluation of the bending moment at a section.

EXAMPLE 12.5

For the beam shown in Fig. 12.9(a), find the shearing force and bending moment at $x = 1$ m, 2.5 m, and 4 m.

Solution

The free-body diagram of the entire beam is shown in Fig. 12.9(a). The resultant of the distributed load on the beam is equal to (40 kN/m) (5 m) = 200 kN and acts at the middle of the beam. The reactions are $A_y = B_y = 100$ kN. For the section at $x = 1$ m [Fig. 12.9(c)] the resultant of the distributed load is (40 kN/m) (1 m) = 40 kN, and it acts at the middle of the 1.0-m length.

From Eq. (12.1),

$$V(\text{at section}) = +\uparrow \Sigma F(\text{on beam left of section})$$
$$V_1 = 100 - 40 = 60 \text{ kN} \qquad \qquad \text{Answer}$$

From Eq. (12.3),

$$M(\text{at section}) = (+ \Sigma M_c(\text{on beam left of section})$$
$$M_1 = 100\,(1) - 40\,(0.5) = 80 \text{ kN·m} \qquad \qquad \text{Answer}$$

At the section $x = 2.5$ m [Fig. 12.9(d)] we apply Eqs. (12.1) and (12.3):

$$V_{2.5} = 100 - 100 = 0 \qquad \qquad \text{Answer}$$
$$M_{2.5} = 100\,(2.5) - 100\,(1.25) = 125 \text{ kN·m} \qquad \qquad \text{Answer}$$

At the section $x = 4$ m [Fig. 12.9(e)] we apply Eqs. (12.1) and (12.3):

$$V_4 = 100 - 160 = -60 \text{ kN} \qquad \qquad \text{Answer}$$
$$M_4 = 100\,(4) - 160\,(2) = 80 \text{ kN·m} \qquad \qquad \text{Answer}$$

We will *check* the results by a free-body diagram of the beam to the right of a section, where $x = 4$ [Fig. 12.9(f)]. From Eq. (12.2),

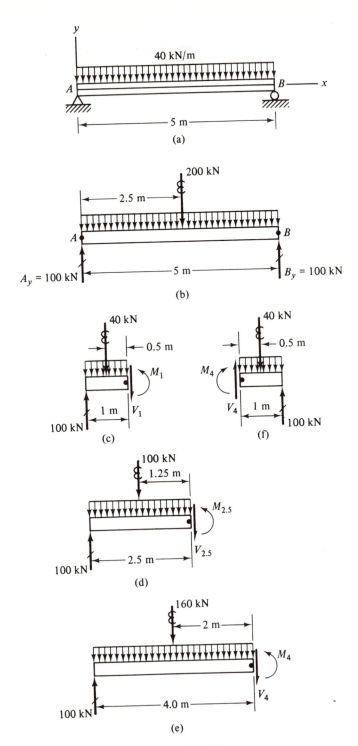

FIGURE 12.9

$$V(\text{at section}) = +\downarrow \Sigma\, F(\text{on beam } \textit{right} \text{ of section})$$

$$V_4 = 40 - 100 = -60 \text{ kN} \qquad \text{OK}$$

From Eq. (12.4),

$$M(\text{at section}) = +\!\!) \Sigma\, M_c(\text{on beam } \textit{right} \text{ of section})$$

$$M_4 = -40(0.5) + 100(1) = 80 \text{ kN·m} \qquad \text{OK}$$

EXAMPLE 12.6

Determine the shear and bending moment for the cantilever beam shown [Fig. 12.10(a)] at $x = 3$ ft and 7 ft.

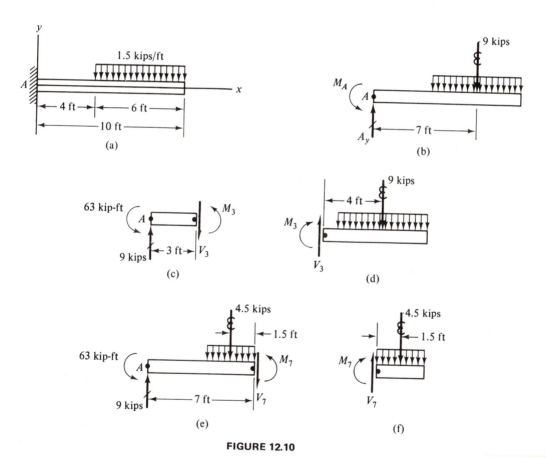

FIGURE 12.10

Solution

The free-body diagram for the entire beam is shown in Fig. 12.10(b). The resultant of the distributed load (1.5 kips/ft) (6 ft) = 9 kips acts at the center of the distributed load $x = 7$ ft. From equilibrium:

$$\uparrow \Sigma \, F_y = 0 \qquad A_y - 9 = 0$$

$$A_y = 9 \text{ kips}$$

$$\circlearrowleft \Sigma \, M_A = 0 \qquad M_A - 9\,(7) = 0$$

$$M_A = 63 \text{ kip-ft}$$

The free-body diagrams of the beam to the left and right of section at $x = 3$ ft are shown in Fig. 12.10(c) and (d). From Eqs. (12.1) and (12.3),

$$V_3 = 9 \text{ kips} \hspace{4cm} \text{Answer}$$

$$M_3 = -63 + 9\,(3) = -36 \text{ kip-ft} \hspace{2cm} \text{Answer}$$

We *check* the results with Eqs. (12.2) and (12.4).

$$V_3 = 9 \text{ kips} \qquad \text{OK}$$

$$M_3 = -9\,(4) = -36 \text{ kip-ft} \qquad \text{OK}$$

The free-body diagrams of the beam to the left and right of a section at $x = 7$ ft are shown in Fig. 12.10(e) and (f). From Eqs. (12.1) and (12.3),

$$V_7 = 9 - 4.5 = 4.5 \text{ kips} \hspace{3cm} \text{Answer}$$

$$M_7 = -63 + 9\,(7) - 4.5\,(1.5) = -6.75 \text{ kip-ft} \hspace{1cm} \text{Answer}$$

We *check* the results with Eqs. (12.2) and (12.4).

$$V_7 = 4.5 \text{ kips} \qquad \text{OK}$$

$$M_7 = -4.5\,(1.5) = -6.75 \text{ kip-ft} \qquad \text{OK}$$

EXAMPLE 12.7

For the beam with overhang shown in Fig. 12.11(a), find the axial forces, shearing forces, and bending moments at sections to the left and right of B ($x = 2$ m), to the left and right of C ($x = 6$ m), halfway between C and D ($x = 7.5$ m), and at the free end D ($x = 9$ m).

Solution

The free-body diagram for the entire beam is shown in Fig. 12.11(b). The resultant of the distributed load has been calculated and displayed on the figure. Equilibrium requires that

$$\rightarrow \Sigma \, F_x = 0 \qquad -A_x + 200 - 150 = 0$$

$$A_x = 50 \text{ kN}$$

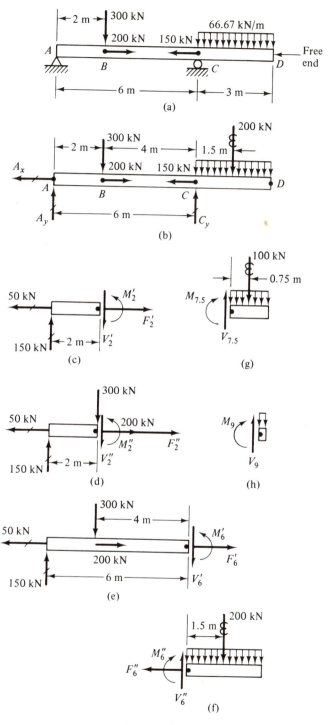

FIGURE 12.11

$$\textstyle\rightthreetimes \Sigma\, M_A = 0 \quad -300\,(2) + C_y 6 - 200\,(7.5) = 0$$

$$C_y = 350 \text{ kN}$$

$$\textstyle\rightthreetimes \Sigma\, M_C = 0 \quad -A_y 6 + 300\,(4) - 200\,(1.5) = 0$$

$$A_y = 150 \text{ kN}$$

Check:

$$\uparrow \Sigma\, F_y = 0 \quad A_y - 300 + 150 - 200 = 0 \quad \text{OK}$$

In Fig. 12.11(c) and (d) free-body diagrams are shown for sections to the left and right of point B. Point B must be excluded from the free-body diagram since the 300-kN load acts at point B. The primed value of axial force, shear, and bending moment (F', V', and M') are reserved for a section to the left of a point and the double-primed values (F'', V'', and M'') for a section to the right. From the equilibrium equation,

$$\rightarrow \Sigma\, F_x = 0 \quad -50 + F_2' = 0$$

$$F_2' = 50 \text{ kN} \qquad\qquad \text{Answer}$$

$$\rightarrow \Sigma\, F_x = 0 \quad -50 + 200 + F_2'' = 0$$

$$F_2'' = -150 \text{ kN} \qquad\qquad \text{Answer}$$

From Eqs. (12.1) and (12.3),

$$V_2' = 150 \text{ kN} \qquad\qquad \text{Answer}$$

$$M_2' = 150\,(2) = 300 \text{ kN·m} \qquad\qquad \text{Answer}$$

and

$$V_2'' = 150 - 300 = -150 \text{ kN} \qquad\qquad \text{Answer}$$

$$M_2'' = 150\,(2) = 300 \text{ kN·m} \qquad\qquad \text{Answer}$$

The free-body diagrams are shown in Fig. 12.11(e) and (f) for sections o the left and right of point C. From equilibrium:

$$\rightarrow \Sigma\, F_x = 0 \quad -50 + 200 + F_6' = 0$$

$$F_6' = -150 \text{ kN} \qquad\qquad \text{Answer}$$

$$\rightarrow \Sigma\, F_x = 0 \qquad\qquad F_6'' = 0 \qquad\qquad \text{Answer}$$

From Eqs. (12.1) and (12.3),

$$V_6' = 150 - 300 = -150 \text{ kN} \qquad\qquad \text{Answer}$$

$$M_6' = 150\,(6) - 300\,(4) = -300 \text{ kN·m} \qquad\qquad \text{Answer}$$

From Eqs. (12.2) and (12.4),

$$V_6'' = 200 \text{ kN}$$ Answer

$$M_6'' = -200(1.5) = -300 \text{ kN·m}$$ Answer

For the section at $x = 7.5$ [Fig. 12.11(g)], we use Eqs. (12.2) and (12.4) to find the shear and bending moment.

$$V_{7.5} = 100 \text{ kN}$$ Answer

$$M_{7.5} = -100(0.75) = -75 \text{ kN·m}$$ Answer

For the free end of the beam, point D, $x = 9$ m [Fig. 12.11(h)], we use Eqs. (12.2) and (12.4) to find the shear and bending moment.

$$V_9 = 0$$ Answer

$$M_9 = 0$$ Answer

PROBLEMS

12.1 and 12.2 For the simply supported beam shown, determine (a) the shear force for sections to the right of $A, B,$ and C and (b) the bending moment for sections at $A, B, C,$ and D.

12.3 and 12.4 Determine (a) the shear force for sections to the right of A and B and (b) the bending moment for sections at $A, B,$ and C for the cantilever beam shown.

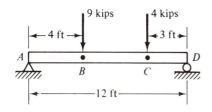

PROB. 12.1

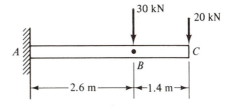

PROB. 12.3

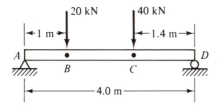

PROB. 12.2

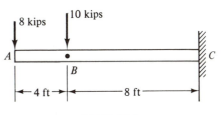

PROB. 12.4

12.5 For the beam with the overhang shown, determine (a) the shear force for sections to the right of A, B, and C and (b) the bending moment for sections at A, C, and D and to the right of B.

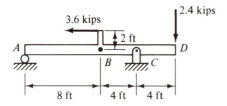

PROB. 12.5

12.6 Determine (a) the shear force for sections to the right of A and B and (b) the bending moment for sections to the right of A and at B for the cantilever beam shown.

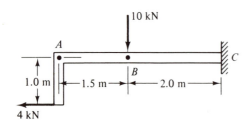

PROB. 12.6

12.7 For the beam with the overhang shown, determine (a) the shear force for sections to the right of A and B and to the left of C and (b) the bending moment for sections at A, B, and C.

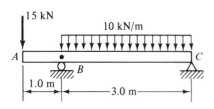

PROB. 12.7

12.8 Determine (a) the shear force for sections to the left of B and to the right of B and C and (b) the bending moments for sections at A, B, C, and D for the beam with the overhang shown.

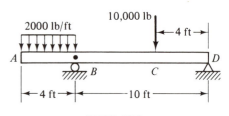

PROB. 12.8

12.9 Determine (a) the shear force for sections to the right of A and B and (b) the bending moment for sections at A, B, and C for the cantilever beam shown.

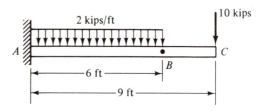

PROB. 12.9

12.10 For the cantilever beam shown, determine (a) the shear force for sections at B and to the right of C and (b) the bending moment for sections at A, B, C, and D.

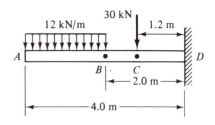

PROB. 12.10

12.5 AXIAL FORCE, SHEARING FORCE, AND BENDING MOMENT DIAGRAMS

In many problems the value of the internal reactions at selected cross sections along the beam may be sufficient. In other problems, however, the values throughout the beam are required. To draw the axial-force, shear-force, and bending-moment diagrams, we need their values at selected cross sections. In Fig. 12.12 we have repeated Fig. 12.11(a) and have drawn the various diagrams from the data obtained in Example 12.7. The

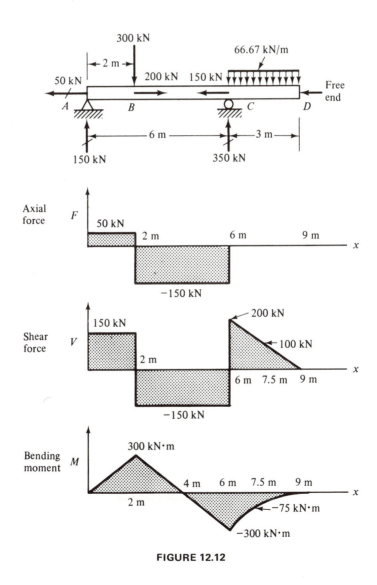

FIGURE 12.12

moment-diagram curve crosses the axis halfway between $x = 2$ m and $x = 6$ m, that is, at $x = 4$ m.

In reviewing the various diagrams, we see that if we are dealing with *concentrated loads* or reactions, the shear diagram consists of straight horizontal lines that are broken only at the load or reaction. The moment diagram consists of straight sloping lines broken at the loads or reactions. When dealing with *uniform loads*, the shear diagram consists of straight sloping lines. The moment diagram consists of curved lines (the curved lines are second-degree curves).

EXAMPLE 12.8

Draw the axial-force, shear-force, and bending-moment diagram for the beam shown in Fig. 12.13(a). The free-body diagram for the entire beam is shown in Fig. 12.13(b).

Solution

From equilibrium, we write

$$\circlearrowright \Sigma M_A = 0 \quad -10(2) - 20(6) + 2(10) + B_y 12 = 0$$

$$B_y = 10 \text{ kips}$$

$$\circlearrowright \Sigma M_B = 0 \quad -A_y 12 + 10(10) + 20(6) - 2(2) = 0$$

$$A_y = 18 \text{ kips}$$

Check:

$$\uparrow \Sigma F_y = 0 \quad A_y - 10 - 20 + 2 + B_y = 0$$

$$18 - 10 - 20 + 2 + 10 = 0 \quad \text{OK}$$

We must calculate the shear to the left and right of $x = 2$ ft, at $x = 4$ ft, at $x = 8$ ft, and to the left and right of $x = 10$ ft. The bending moment must be calculated at $x = 2$ ft, 4 ft, 6 ft, 8 ft, and 10 ft. Free-body diagrams for the various sections are drawn in Fig. 12.13(c)–(k).

Applying Eqs. (12.1) and (12.3) to free-body diagrams shown in Fig. 12.13(c)–(g), we have

$$V_2' = 18 \text{ kips}$$

$$M_2' = 18(2) = 36 \text{ kip-ft}$$

$$V_2'' = 18 - 10 = 8 \text{ kips}$$

$$M_2'' = 18(2) = 36 \text{ kip-ft}$$

$$V_4 = 18 - 10 = 8 \text{ kips}$$

$$M_4 = 18(4) - 10(2) = 52 \text{ kip-ft}$$

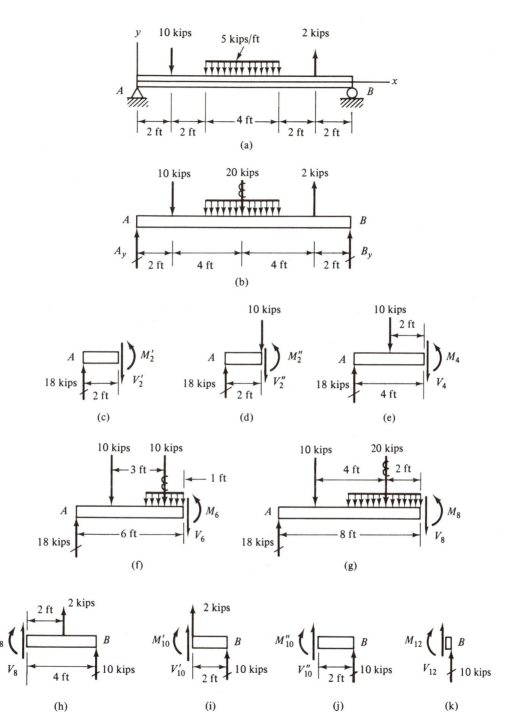

FIGURE 12.13

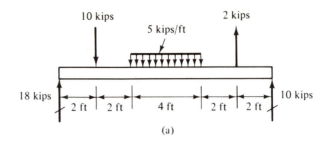

(a)

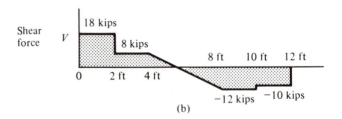

(b)

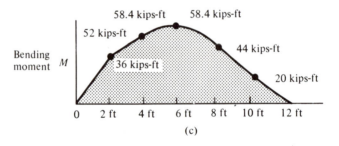

(c)

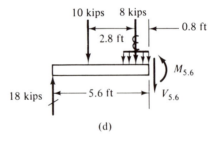

(d)

FIGURE 12.14

$$V_6 = 18 - 10 - 10 = -2 \text{ kips}$$

$$M_6 = 18(6) - 10(4) - 10(1) = 58 \text{ kip-ft}$$

$$V_8 = 18 - 10 - 20 = -12 \text{ kips}$$

$$M_8 = 18(8) - 10(6) - 20(2) = 44 \text{ kip-ft}$$

Applying Eqs. (12.2) and (12.4) to free-body diagrams shown in Fig. 12.13(h)-(k), we have

$$V_8 = -2 - 10 = -12 \text{ kips}$$

$$M_8 = 2(2) + 10(4) = 44 \text{ kip-ft}$$

$$V'_{10} = -2 - 10 = -12 \text{ kips}$$

$$M'_{10} = 10(2) = 20 \text{ kip-ft}$$

$$V''_{10} = -10 \text{ kips}$$

$$M''_{10} = 10(2) = 20 \text{ kip-ft}$$

$$V_{12} = -10 \text{ kips}$$

$$M_{12} = 0$$

From the data we now draw the shear and moment diagrams in Figs. 12.14(b) and (c). The maximum moment occurs when the shear diagram crosses the axis, that is, when the shear is zero. The shear is zero at a distance x_1 along the distributed load such that $8 = 5x_1$ or $x_1 = 1.6$ ft. Therefore, the shear is zero at $x = 4 + 1.6 = 5.6$ ft. A free-body diagram of the beam is shown for $x = 5.6$ ft in Fig. 12.14(d). From rotational equilibrium:

$$\curvearrowright \Sigma M_c = 0 \quad -18(5.6) + 10(3.6) + 8(0.8) + M_{5.6} = 0$$

$$M_{5.6} = 58.4 \text{ kip-ft}$$

12.6 RELATIONS BETWEEN LOADS, SHEARING FORCES, AND BENDING MOMENTS

We shall consider the free-body diagram of a small element of a beam of length Δx [Fig. 12.15(b)] which lies between sections A–A and B–B of

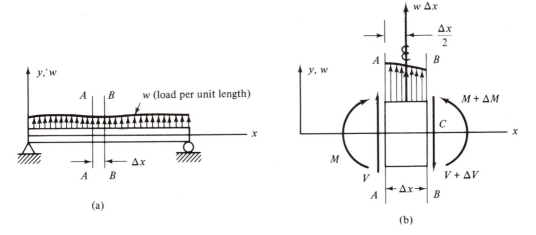

(a)

(b)

FIGURE 12.15

the beam in Fig. 12.15(a). The shear and bending moment on section A–A will be taken as V and M, respectively, and are shown in a positive sense in the figure. Let ΔV represent the change in shear from sections A–A to B–B and ΔM represent the change in bending moment from sections A–A to B–B. Then the shear at section B–B is equal to $V + \Delta V$ and the bending moment at section B–B is equal to $M + \Delta M$. They are shown in a positive sense in the figure. The resultant of the distributed load is the area of the distributed load $w(\Delta x)$, and it acts at the middle of the element as shown. Loads directed up are considered positive.

From equilibrium

$$\uparrow \Sigma\, F_y = 0 \qquad V + w(\Delta x) - (V + \Delta V) = 0$$

$$\Delta V = w(\Delta x) \tag{a}$$

$$\circlearrowright \Sigma\, M_c = 0 \qquad -M - w(\Delta x)\,\frac{(\Delta x)}{2} + (M + \Delta M) - V(\Delta x) = 0$$

$$-w\,\frac{(\Delta x)^2}{2} + \Delta M - V(\Delta x) = 0$$

Now the element Δx is small and the term $w(\Delta x)^2/2$ is even smaller and can be dropped from the equation. We have, therefore,

$$\Delta M = V(\Delta x) \tag{b}$$

Equations (a) and (b) have important physical interpretations. Equation (a) indicates that the change in the shear from sections A–A to B–B is equal to the external load on the beam between those two sections. If we add up all the elements between any sections 1 and 2, we see that the change in shear between 1 and 2 is equal to the external load on the beam between sections 1 and 2 (area under load curve from 1 to 2). In equation form,

$$V_2 - V_1 = \text{external load between 1 and 2}$$

or

$$\Delta V_{1 \to 2} = \text{load}_{1 \to 2} \tag{12.5}$$

Rewriting Eq. (a), we have

$$\frac{\Delta V}{\Delta x} = w \tag{12.6}$$

Therefore, the slope of the shear diagram $\Delta V/\Delta x$ at any point is equal to the value of the load at that point.

Equation (b) can be interpreted in a similar way. The change in bending moment between sections 1 and 2 is equal to the area under the shear curve between 1 and 2. In equation form,

$$M_2 - M_1 = \text{area of shear curve between 1 and 2}$$

or

$$\Delta M_{1 \to 2} = \text{area shear curve}_{1 \to 2} \qquad (12.7)$$

Rewriting Eq. (b), we have

$$\frac{\Delta M}{\Delta x} = V \qquad (12.8)$$

Therefore, the slope of the bending moment diagram $\Delta M/\Delta x$ at any point is equal to the value of the shear force V at that point.

EXAMPLE 12.9

Use the relation between loads, shears, and bending moments to sketch the shear and moment diagrams for the beam shown in Fig. 12.16(a).

Solution

The free-body diagram is shown in Fig. 12.16(b). From the usual equations of equilibrium:

$$\circlearrowright \Sigma M_B = 0 \qquad 30(1) - 40(1.5) - 75(2.25) + D_y 3 = 0$$

$$D_y = 66.25 \text{ kN}$$

$$\circlearrowright \Sigma M_D = 0 \qquad 30(4) - B_y 3 + 40(1.5) + 75(0.75) = 0$$

$$B_y = 78.75 \text{ kN}$$

Check:

$$\uparrow \Sigma F_y = 0 \qquad -30 + B_y - 40 - 75 + D_y = 0$$

$$-30 + 78.75 - 40 - 75 + 66.25 = 0 \qquad \text{OK}$$

From Eq. (12.5) and the free-body diagram, we calculated the shears. The value of the shear to the left of point A is zero. That is,

$$V(\text{left of point } A) = 0$$

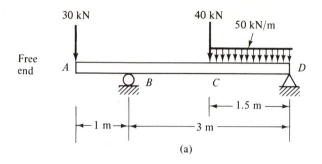

Free
end

(a)

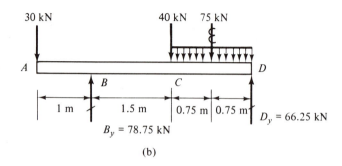

(b)

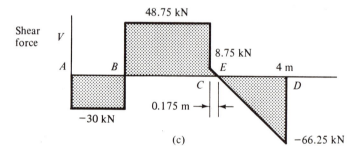

Shear
force

(c)

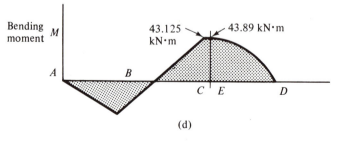

Bending
moment

(d)

FIGURE 12.16

$$\Delta V(\text{left of } A \text{ to right of } A) = -30 \text{ kN}$$

$$V(\text{right of } A) = -30 \text{ kN}$$

$$\Delta V(\text{from } A \text{ to } B) = 0$$

$$V(\text{left of } B) = -30 \text{ kN}$$

$$\Delta V(\text{left of } B \text{ to right of } B) = +78.75 \text{ kN}$$

$$V(\text{right of } B) = 78.75 - 30 = 48.75 \text{ kN}$$

$$\Delta V(\text{from } B \text{ to } C) = 0$$

$$V(\text{left of } C) = 48.75 \text{ kN}$$

$$\Delta V(\text{left of } C \text{ to right of } C) = -40 \text{ kN}$$

$$V(\text{right of } C) = -40 + 48.75 = 8.75 \text{ kN}$$

$$\Delta V(\text{right of } C \text{ to left of } D) = -75 \text{ kN}$$

$$V(\text{left of } D) = 8.75 - 75 = -66.25 \text{ kN}$$

$$\Delta V(\text{left of } D \text{ to right of } D) = +66.25$$

$$V(\text{right of } D) = -66.25 + 66.25 = 0$$

The point where the shear is zero can be found by solving the equation $8.75 = 50x'$. We see that $x' = 0.175$. The section where the shear is zero is at $x = 2.5 + 0.175 = 2.675$ m. The shear diagram is drawn in Fig. 12.16(c). From Eq. (12.6) and the shear diagram, we calculate the following values for bending moments:

$$M_A = 0$$

$$\Delta M_{A \to B} = -30(1) = -30 \text{ kN} \cdot \text{m}$$

$$M_B = -30 \text{ kN} \cdot \text{m}$$

$$\Delta M_{B \to C} = 48.75(1.5) = 73.125 \text{ kN} \cdot \text{m}$$

$$M_C = -30 + 73.125 = 43.125 \text{ kN} \cdot \text{m}$$

$$\Delta M_{C \to E} = \frac{(0.175)(8.75)}{2} = +0.766 \text{ kN} \cdot \text{m}$$

$$M_E = 43.125 + 0.766 = 43.891 \text{ kN} \cdot \text{m}$$

$$\Delta M_{E \to D} = \frac{(1.5 - 0.175)(-66.25)}{2} = -43.891 \text{ kN} \cdot \text{m}$$

$$M_D = 43.891 - 43.891 = 0 \qquad \text{OK}$$

The moment diagram is drawn in Fig. 12.16(d).

EXAMPLE 12.10

Use the relationship between loads, shears, and bending moments to sketch the shear and moment diagram for the beam shown in Fig. 12.17(a).

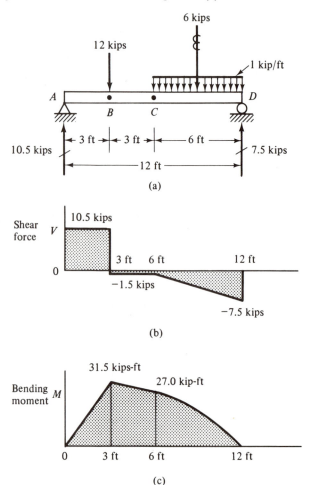

FIGURE 12.17

Solution

The reactions are calculated and displayed in Fig. 12.17(a). From Eq. (12.5) and Fig. 12.17(a) we calculate the shears. The value of shear to the left of A is zero.

$$V_A = 0$$

$$\Delta V(\text{left to right of } A) = 10.5 \text{ kips}$$

$$V(\text{right of } A) = 10.5 \text{ kips}$$

$$\Delta V(\text{left to right of } B) = -12 \text{ kips}$$

$$V(\text{right of } B) = 10.5 - 12 = -1.5 \text{ kips}$$

$$\Delta V(\text{right of } B \text{ to } C) = 0$$

$$V(\text{at } C) = -1.5 \text{ kips}$$

$$\Delta V(C \text{ to left of } D) = 1(-6) = -6.0 \text{ kips}$$

$$V(\text{left of } D) = -1.5 - 6.0 = -7.5 \text{ kips}$$

$$\Delta V(\text{left of } D \text{ to right of } D) = 7.5 \text{ kips}$$

$$V(\text{right of } D) = -7.5 + 7.5 = 0 \quad \text{OK}$$

The shear diagram is drawn in Fig. 12.17(b). The moment diagram will be constructed from the shear diagram and Eq. (12.6) as follows:

$$M_A = 0$$

$$\Delta M_{A \to B} = 10.5(3) = 31.5 \text{ kip-ft}$$

$$M_B = 31.5 \text{ kip-ft}$$

$$\Delta M_{B \to C} = -1.5(3) = -4.5 \text{ kip-ft}$$

$$M_C = 31.5 - 4.5 = 27 \text{ kip-ft}$$

$$\Delta M_{C \to D} = \frac{-(7.5 + 1.5)(6)}{2} = -27 \text{ kip-ft}$$

$$M_D = 27 - 27 = 0 \quad \text{OK}$$

The moment diagram is drawn in Fig. 12.17(c).

PROBLEMS

12.11 through 12.30 Sketch and label the shear-force diagram for the beam shown.

12.31 through 12.50 Sketch and label the bending-moment diagram for the beam shown.

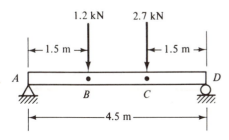

PROB. 12.11 and PROB. 12.31

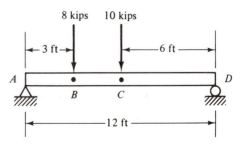

PROB. 12.12 and PROB. 12.32

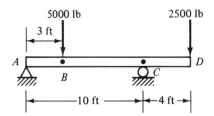

PROB. 12.13 and PROB. 12.33

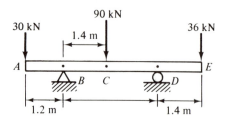

PROB. 12.17 and PROB. 12.37

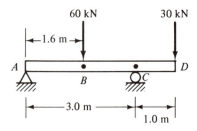

PROB. 12.14 and PROB. 12.34

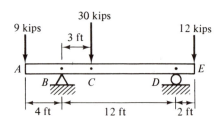

PROB. 12.18 and PROB. 12.38

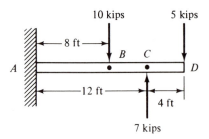

PROB. 12.15 and PROB. 12.35

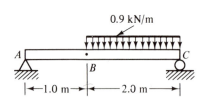

PROB. 12.19 and PROB. 12.39

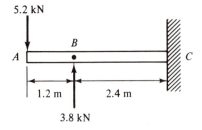

PROB. 12.16 and PROB. 12.36

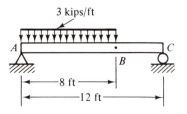

PROB. 12.20 and PROB. 12.40

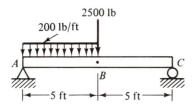

PROB. 12.21 and PROB. 12.41

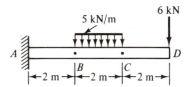

PROB. 12.25 and PROB. 12.45

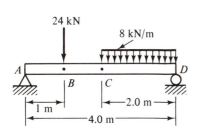

PROB. 12.22 and PROB. 12.42

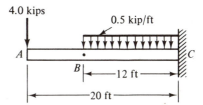

PROB. 12.26 and PROB. 12.46

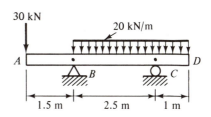

PROB. 12.27 and PROB. 12.47

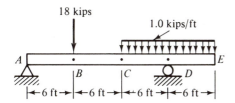

PROB. 12.23 and PROB. 12.43

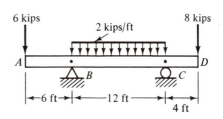

PROB. 12.28 and PROB. 12.48

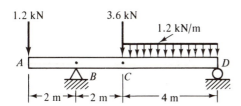

PROB. 12.24 and PROB. 12.44

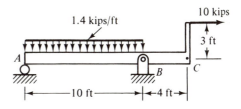

PROB. 12.29 and PROB. 12.49

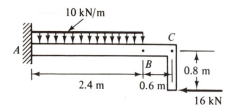

PROB. 12.30 and PROB. 12.50

12.51 Draw, for the beam shown, (a) the shear-force diagram and (b) the bending-moment diagram.

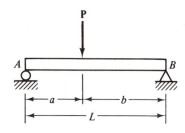

PROB. 12.51

12.52 Draw the diagrams for Prob. 12.51 if $P = 10$ kips, $a = 4$ ft, and $b = 6$ ft.

12.53 Draw the diagrams for Prob. 12.51 if $P = 40$ kN, $a = 2$ m, and $b = 3$ m.

12.54 For the beam shown, draw (a) the shear-force diagram and (b) the bending-moment diagram.

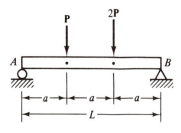

PROB. 12.54

12.55 Draw the diagrams for Prob. 12.54 if $P = 12$ kips and $a = 5$ ft.

12.56 Draw the diagrams for Prob. 12.54 if $P = 48$ kN and $a = 2$ m.

12.57 Draw, for the beam shown, (a) the shear-force diagram and (b) the bending-moment diagram.

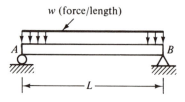

PROB. 12.57

12.58 Draw the diagrams for Prob. 12.57 if $w = 2$ kips/ft and $L = 12$ ft.

12.59 Draw the diagrams for Prob. 12.57 if $w = 12$ kN/m and $L = 4$ m.

12.60 For the beam shown, draw (a) the shear-force diagram and (b) the bending-moment diagram.

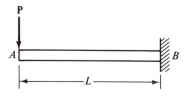

PROB. 12.60

12.61 Draw the diagrams for Prob. 12.60 if $P = 10$ kips and $L = 12$ ft.

12.62 Draw the diagrams for Prob. 12.60 if $P = 25$ kN and $L = 4$ m.

12.63 Draw, for the beam shown, (a) the shear-force diagram and (b) the bending-moment diagram.

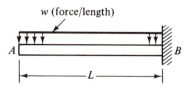

PROB. 12.63

12.64 Draw the diagrams for Prob. 12.63 if $w = 3$ kips/ft and $L = 15$ ft.

12.65 Draw the diagrams for Prob. 12.63 if $w = 18$ kN/m and $L = 5$ m.

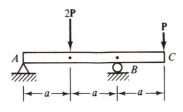

PROB. 12.66

12.66 For the beam shown, draw (a) the shear-force diagram and (b) the bending-moment diagram.

12.67 Draw the diagrams for Prob. 12.66 if $P = 10$ kips and $a = 6$ ft.

12.68 Draw the diagrams for Prob. 12.66 if $P = 40$ kN and $a = 2$ m.

13

Bending and Shearing Stresses in Beams _____

13.1 INTRODUCTION

If no axial force is present in a beam, the bending moment or bending couple at a cross section of the beam is the resultant of internal stresses that act normal to the cross section. The shear force is the resultant of shear stresses that act tangent to the cross section. In this chapter we will be concerned with the magnitude and distribution of these stresses. We begin by considering the nature of the deformations that are assumed to occur when a beam is subject to bending.

13.2 PURE BENDING OF A SYMMETRICAL BEAM

The beam that we consider is subjected to *pure bending:* that is, a bending moment without either shearing or axial forces. In Fig. 13.1(a) the beam

318

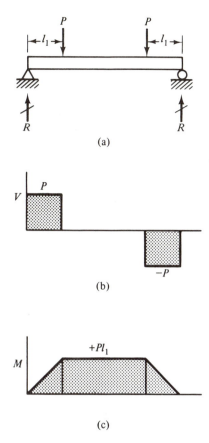

FIGURE 13.1

has been loaded by couples to produce pure bending. Notice the absence of a shearing force in the shear diagram of Fig. 13.1(b) and a constant bending moment in the interval of the beam between the loads in Fig. 13.1(c). Further, the beam has a uniform cross section, the cross section is symmetric about a vertical axis, the loads act in the plane of symmetry, and bending takes place in the same plane.

For purposes of illustration, consider a beam with a rectangular cross section. Horizontal and vertical lines have been drawn on the side of the beam [Fig. 13.2(a)]. In addition, we imagine horizontal and vertical lines drawn on a cross section of the beam [Fig. 13.2(b)]. Two loads are applied to the beam [Fig. 13.3(a) and (b)] to produce pure bending in the middle interval of the beam. As a result of the loads, horizontal lines on the side of the beam form arcs of circles which have a common center at point f a distance of $y = \rho$ (the Greek letter rho) *above* the beam. The vertical lines on the side of the beam remain straight and rotate so they are directed toward the center of the circles at point f. At the same time, imaginary horizontal lines on the cross section form arcs of circles that

319

have a common center at point *g*, a distance $y = -\rho_1$ (rho) *below* the beam. The imaginary vertical lines on the cross section remain straight and rotate so they are directed toward the center of the circle at point *g*. Thus we see that when pure bending produces longitudinal curvature in the *xy* plane, it is accompanied by transverse curvature, also called anticlastic curvature in the *yz* plane. The existence of anticlastic curvature may be observed by bending a rubber eraser between the thumb and forefinger. The deformations of the cross section have negligible effect on most beams and will not be considered in the following discussions.

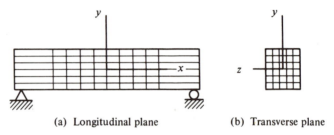

(a) Longitudinal plane   (b) Transverse plane

FIGURE 13.2 Beam before bending.

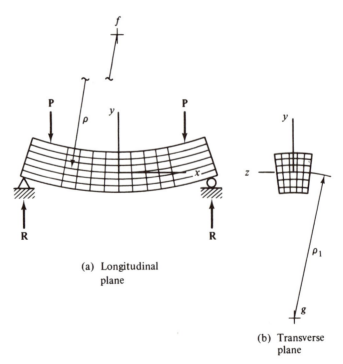

(a) Longitudinal
plane

(b) Transverse
plane

FIGURE 13.3 Beam after bending.

13.3 DEFORMATION GEOMETRY FOR A SYMMETRICAL BEAM IN PURE BENDING

In our discussion of the bending of a rectangular beam, we saw that the vertical straight lines on the side of the beam remained straight after bending. We also saw that they rotated through an angle so they were directed toward the center of the arcs of the circles that were formed by the horizontal lines. Generalizing these observations for a beam of any symmetrical cross section we can say that *cross sections of a beam that are plane and normal to the axis of the beam before bending remain plane and normal to the axis after bending.* Consider part of such a beam, as shown in Fig. 13.4   ?t plane cross section $c'e'$ rotate as a result of bending

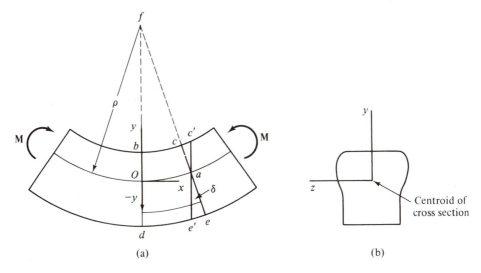

FIGURE 13.4

through an angle to a new position ce. No deformation occurs in the longitudinal fiber oa. Fiber oa is the edge of a surface formed by fibers that do not stretch. The surface extends from the front to the back of the beam and is called the *neutral surface*. Below the neutral surface, longitudinal fibers stretch and, above the neutral surface, longitudinal fibers compress.

The intersection of the neutral surface and the cross section of the beam forms an axis called the *neutral axis*. The stretching of a horizontal fiber a distance $-y$ below the neutral axis at O is given by δ. Therefore, the strain of the fiber is

$$\epsilon = \frac{\delta}{oa} \tag{a}$$

From similar triangles $e'ae$ and ofa,

$$\frac{\delta}{y} = \frac{oa}{\rho} \quad \text{or} \quad \frac{\delta}{oa} = \frac{y}{\rho}$$

Substituting into Eq. (a), we have

$$\epsilon = \frac{-y}{\rho} \tag{13.1}$$

where the negative sign accounts for the fact that tensile strain occurs when y is negative. We see from Eq. (13.1) that the strains of the longitudinal fibers are proportional to the distance from the neutral axis and inversely proportional to the radius of curvature.

13.4 HOOKE'S LAW—DISTRIBUTION OF BENDING STRESS

The deformation geometry that we have considered up to this point depended only on the fact that the beam had a constant cross section and that the cross section and bending were symmetrical with respect to the xy plane. We shall now assume that the material follows Hooke's law for longitudinal stresses and strains; that is, $\sigma = E\epsilon$. Combining Hooke's law and Eq. (13.1), we have

$$\sigma = \frac{-Ey}{\rho} \tag{13.2}$$

Thus we see that the bending stress is proportional to the distance y from the neutral axis and inversely proportional to the radius of curvature ρ.

The distribution for bending stress is shown in Fig. 13.5. The maximum stress occurs at a point that is the farthest distance c from the

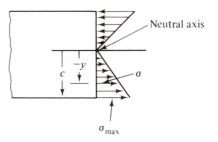

FIGURE 13.5

neutral axis. The stress at any distance y is related to the maximum stress by the expression obtained from similar triangles

$$\frac{\sigma}{-y} = \frac{\sigma_{max}}{c} \quad \text{or} \quad \sigma = \frac{-\sigma_{max}}{c} y \tag{13.3}$$

13.5 BENDING STRESS FORMULA—FLEXURE FORMULA

The position of the neutral axis can be found from the condition that the resultant axial force for the cross section of the beam must be equal to zero.

The element of force ΔF acting on an element of area ΔA of the cross section is equal to the product of the stress σ and the area ΔA, as shown in Fig. 13.6. With the value of σ given by Eq. (13.3), we have

$$\Delta F = \sigma(\Delta A) = \frac{-\sigma_{max}}{c} y(\Delta A) \tag{a}$$

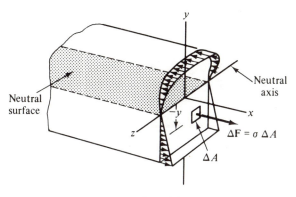

FIGURE 13.6

From equilibrium in the axial direction, the sum of the elements of force ΔF must vanish, that is,

$$\Sigma \frac{-\sigma_{max}}{c} y(\Delta A) = \frac{-\sigma_{max}}{c} \Sigma y(\Delta A) = 0 \tag{b}$$

Since σ_{max} and c are nonzero constants for a given cross section, $\Sigma\, y\Delta A$ must be equal to zero. The summation $\Sigma\, y\Delta A$ represents the moment of the elements of area ΔA of the cross section with respect to the neutral axis. Therefore,

$$\Sigma \, y\Delta A = \bar{y}A = 0$$

where $\bar{y}$ is the coordinate of the centroid measured from the neutral axis and A is the area of the cross section. The centroidal distance $\bar{y} = 0$ because the area A is not zero. Therefore, *the neutral axis passes through the centroid of the cross section of the beam.* Thus the z axis is the neutral axis.

From rotational equilibrium, the bending moment must be equal to the sum of the moments about the neutral axis of the elements of force ΔF, that is,

$$M = \Sigma \, y(\Delta F) \tag{c}$$

Substituting for ΔF from Eq. (a), we have

$$M = \Sigma \, \frac{-\sigma_{max}}{c} \, y^2 \, \Delta A = \frac{-\sigma_{max}}{c} \, \Sigma \, y^2 \, \Delta A \tag{d}$$

Recall that

$$I = \Sigma \, y^2 \, \Delta A \tag{e}$$

is the moment of inertia of the cross-sectional area with respect to the neutral axis (z axis). It has a definite value for a given cross section. (See Sec. 8.7 for a discussion of moments of inertia.) Substituting Eq. (e) in Eq. (d), we have

$$M = \frac{-\sigma_{max}}{c} \, I \quad \text{or} \quad \sigma_{max} = \frac{-Mc}{I} \tag{f}$$

The negative sign in this equation is usually dropped and the sign of the stress determined by inspection. The stress distribution on the cross section must produce a resultant couple that has the same direction as the bending moment. Thus a positive bending moment produces compression on the top of the beam and tension on the bottom of the beam. The reverse is true for a negative bending moment.

Rewriting Eq. (f) without the negative sign, we have

$$\sigma_{max} = \frac{Mc}{I} \tag{13.4}$$

This equation is known as the *bending stress formula* or the *flexure formula*. It gives the *maximum* normal stress on a cross section which has a bending moment M. By eliminating σ_{max} between Eqs. (13.3) and (13.4), we have

$$\sigma = \frac{My}{I} \tag{13.5}$$

This formula gives the normal stress at any point on a cross section a distance y from the neutral axis in terms of the bending moment and the dimensions of the member. The sign of the stress is determined by inspection. Bending moment M will be expressed in inch-pound or newton·meters (N·m), y in inches or meters, and I in in.^{4} or m^4. The normal stress will have units of

$$\frac{(\text{in.-lb})(\text{in.})}{(\text{in.}^4)} = \frac{\text{lb}}{\text{in.}^2} = \text{psi}$$

or, in SI units,

$$\frac{(\text{N·m})(\text{m})}{(\text{m}^4)} = \frac{\text{N}}{\text{m}^2} \text{ or Pa (pascals)}$$

The bending stress formula was developed for a beam subject to pure bending only. However, it can be shown that the formula can also be used when shear as well as bending exists on the cross section.

13.6 FURTHER COMMENTS ON THE BENDING STRESS FORMULA

In many problems involving the bending of a beam, the maximum normal stress is the quantity desired. Both I and c in Eq. (13.4) are constants for a given cross section. Therefore, we define a new constant, $I/c = S$, which is called the *elastic section modulus*. Equation (13.4) can be rewritten in terms of the elastic section modulus as follows:

$$\sigma_{max} = \frac{M}{I/c} \quad \text{or} \quad \sigma_{max} = \frac{M}{S} \tag{13.6}$$

The bending moment will be expressed in inch-pounds or newton·meters (N·m) and the elastic section modulus in in.^{3} or m^3. As before, the normal stress will be expressed in lb/in.^{2} (psi) or N/m^2 (Pa).

The moment of inertia or the elastic section modulus of the cross section with respect to the neutral axis may be found by the methods of Sec. 8.7 or for rolled steel beams from Tables A.5 through A.10 of the Appendix.

13.7 PROBLEMS INVOLVING THE BENDING STRESS FORMULA

In this section we solve several problems involving the normal stress due to the bending of a beam.

EXAMPLE 13.1

A simply supported rectangular wooden beam with a cross section 8 in. by 10 in. and a span of 14 ft supports a 1.5-kip load [Fig. 13.7(a)]. Determine the *maximum* bending stress (a) at a section 2 ft from the left end and (b) for any section. Neglect the weight of the beam.

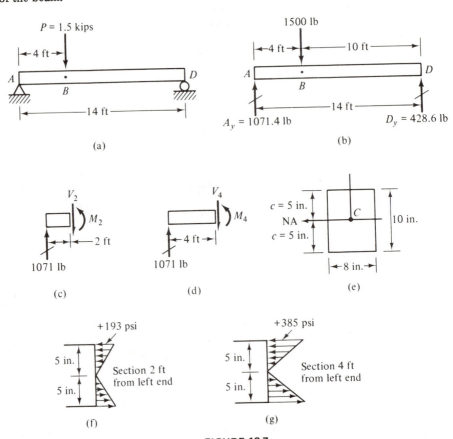

FIGURE 13.7

Solution

The bending moment and the moment of inertia with respect to the neutral axis are required to find the bending stress.

Bending Moment

A free-body diagram for the beam is drawn in Fig. 13.7(b). Equilibrium requires that

$$\curvearrowright \Sigma M_A = 0 \quad -1500(4) + D_y(14) = 0$$
$$D_y = 429 \text{ lb}$$

$$\curvearrowright \Sigma M_D = 0 \quad -A_y(14) + 1500(10) = 0$$
$$A_y = 1071 \text{ lb}$$

Check:

$$\uparrow \Sigma F_y = 0 \quad A_y - 1500 + D_y = 0$$
$$1071 - 1500 + 429 = 0 \quad \text{OK}$$

The bending moment for $x = 2$ ft is determined from the free-body diagram in Fig. 13.7(c). From equilibrium

$$\curvearrowright \Sigma M_c = 0 \quad -1071(2) + M_2 = 0$$
$$M_2 = 2142 \text{ lb-ft} = 25,700 \text{ lb-in.}$$

The maximum bending moment will occur under the concentrated load at $x = 4$ ft. From the free-body diagram shown in Fig. 13.7(d), we have

$$\curvearrowright \Sigma M_c = 0 \quad -1071(4) + M_4 = 0$$
$$M_4 = 4284 \text{ lb-ft} = 51,400 \text{ lb-in.}$$

Moment of Inertia

From Table A.4 of the Appendix, the formula for the moment of inertia with respect to the centroidal or neutral axis is $I = bh^3/12$. For the cross section shown in Fig. 13.7(e),

$$I = \frac{bh^3}{12} = \frac{8(10)^3}{12} = 667 \text{ in.}^4$$

   (a) The maximum stress at a section 2 ft from the left end of the beam is given by the bending stress formula, Eq. (13.4).

$$\sigma_{max} = \frac{Mc}{I} = \frac{25,700(5)}{667} = 192.7 \text{ psi} \qquad \text{Answer}$$

   (b) The maximum stress for any section occurs 4 ft from the left end of the beam, where the bending moment is maximum. The maximum stress is given by the bending stress formula.

$$\sigma_{max} = \frac{Mc}{I} = \frac{51,400(5)}{667} = 385 \text{ psi} \qquad\qquad \text{Answer}$$

The stresses are shown in Fig. 13.7(f) and (g).

EXAMPLE 13.2

A cantilever beam 3.5 m long supports a load of 20 kN on its free end. The beam and cross section are shown in Fig. 13.8(a) and (b). Find the maximum bending stress in the beam.

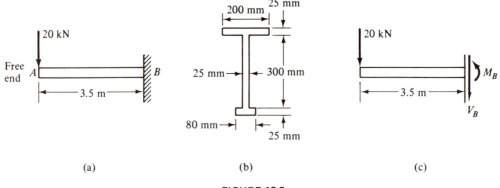

(a)   (b)   (c)

FIGURE 13.8

Solution

The maximum bending moment and moment of inertia of the cross section with respect to the neutral axis (centroidal axis) are required to find the bending stress.

Bending Moment

The maximum bending moment occurs at the fixed end of the cantilever. The free-body diagram of the beam is drawn in Fig. 13.8(c). From equilibrium:

$$\circlearrowleft \Sigma \, M_B = 0 \qquad 20(3.5) + M_B = 0$$

$$M_B = -70 \text{ kN·m}$$

Moment of Inertia

The moment of inertia will be calculated by the method of Sec. 8.7.

    The I with unequal flanges can be divided into three rectangular areas as in Fig. 13.9. The solution for the centroid is tabulated as shown. From the sums in the table,

$$\bar{y} = \frac{\Sigma \, Ay}{\Sigma \, A} = \frac{3.025 \times 10^6}{14.5 \times 10^3} = 208.6 \text{ mm}$$

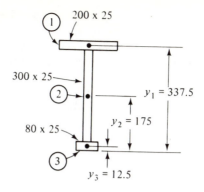

(All dimensions in mm)

FIGURE 13.9

	A	y	Ay
①	5000	337.5	1.688×10^6
②	7500	175	1.312×10^6
③	2000	12.5	0.025×10^6
	$\Sigma A = 14\ 500\ \text{mm}^2$		$\Sigma Ay = 3.025 \times 10^6\ \text{mm}^3$

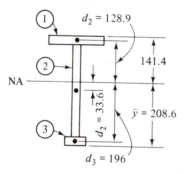

(All dimensions in mm)

FIGURE 13.10

The transfer distances for each area are shown in Fig. 13.10. The moment of inertia for each area about its *own* centroidal axis $I_c = bh^3/12$. For areas ①, ②, and ③,

$$(I_c)_1 = 200\,(25)^3/12 = 260.4 \times 10^3\ \text{mm}^4$$

$$(I_c)_2 = 25\,(300)^3/12 = 56.25 \times 10^6\ \text{mm}^4$$

$$(I_c)_3 = 80\,(25)^3/12 = 104.2 \times 10^3\ \text{mm}^4$$

Thus I_c for areas ① and ③ are negligible compared with area ②. The solution for the moment of inertia about the neutral axis is tabulated as shown. From the sums in the table,

$$I_{NA} = \Sigma I_c + \Sigma (Ad^2) = 56.2 \times 10^6 + 168.5 \times 10^6$$
$$= 224 \times 10^6 \text{ mm}^4 = 224 \times 10^{-6} \text{ m}^4$$

	A	d	I_c	Ad^2
①	5000	128.9	—	83.08×10^6
②	7500	33.6	56.25×10^6	8.47×10^6
③	2000	196.1	—	76.91×10^6
			$\Sigma I_c = 56.2 \times 10^6 \text{ mm}^4$	$\Sigma (Ad^2) = 168.5 \times 10^6 \text{ mm}^4$

The maximum bending stress will occur at the bottom of the beam at a maximum distance $c = 208.6 \times 10^{-3}$ m from the neutral axis. The stress will be compression because the bending moment is negative. From Eq. (13.4),

$$\sigma_{max} = \frac{Mc}{I} = \frac{70 (208.6 \times 10^{-3})}{224 \times 10^{-6}} = 65 \times 10^3 \ \frac{\text{kN}}{\text{m}^2}$$

$$= 65 \text{ MPa C} \qquad\qquad\qquad \text{Answer}$$

EXAMPLE 13.3

A W16 × 64 steel beam with an overhang supports a uniform load, including its own weight of 4 kips/ft, as shown in Fig. 13.11(a). Find the maximum bending stress.

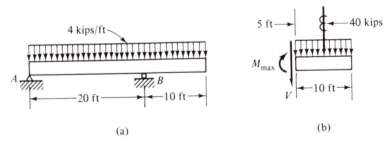

(a)   (b)

FIGURE 13.11

Solution

The maximum bending moment will occur over the support at point B. Isolating the beam to the right of a section at B as shown in Fig. 13.11(b), we calculate the maximum bending moment from equilibrium.

$$\text{)}\Sigma\, M_c = 0 \qquad -M_{max} - 40(5) = 0$$

$$M_{max} = -200 \text{ kip-ft} = 2.4 \times 10^6 \text{ lb-in.}$$

Section properties for the W16 × 64 beam are given in Table A.5 of the Appendix. $S = I/c = 104$ in.^{3}. The maximum stress from Eq. (13.6) is given by

$$\sigma_{max} = \frac{M}{S} = \frac{2.4 \times 10^6}{104} = 23{,}100 \text{ psi} \qquad\qquad \text{Answer}$$

EXAMPLE 13.4

A simply supported uniform beam with a span of 4 m supports a uniform load. The maximum bending stress is 70 000 kN/m^2. The beam and cross section of the beam are shown in Fig. 13.12(a) and (b). Find the uniform load, including the weight of the beam in units of kN/m.

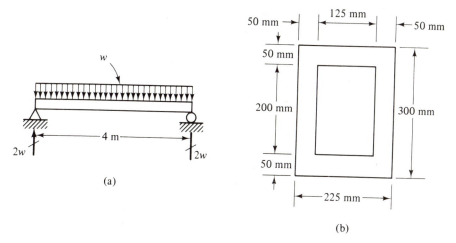

(a)

(b)

FIGURE 13.12

Solution

The reactions for the simply supported beam will be equal to one-half of the *resultant*, $R = 4w$, of the uniformly distributed load w. That is, $R = 4w/2$. The maximum bending moment will occur at the center of the beam. The free-body diagram in Fig. 13.13(a) will be used to find the maximum moment. From rotational equilibrium,

$$\text{)}\Sigma\, M_c = 0 \qquad -2w(2) + 2w(1) + M_2 = 0$$

$$M_2 = 2w \text{ kN·m}$$

The moment of inertia may be found by calculating the moment of inertia of the outside rectangle and subtracting the moment of inertia of the inside

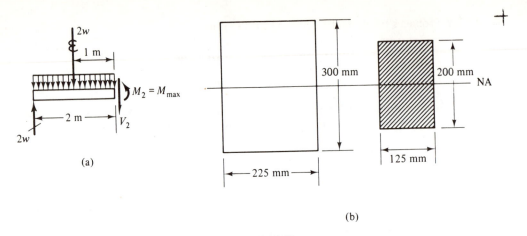

(a)

(b)

FIGURE 13.13

rectangle [Fig. 13.13(b)]. The formula for moment of inertia about a centroidal (neutral) axis is $I = bh^3/12$. Therefore,

$$I = \frac{(225)(300)^3}{12} - \frac{(125)(200)^3}{12} = 422.9 \times 10^6 \text{ mm}^4$$

$$= 422.9 \times 10^{-6} \text{ m}^4$$

In this case, $c = 150$ mm $= 150 \times 10^{-3}$ m.

From the bending stress formula [Eq. (13.6)],

$$\sigma = \frac{Mc}{I} \quad \text{or} \quad M = \sigma \frac{I}{c}$$

$$2w \text{ kN} \cdot \text{m} = 70\ 000\ \frac{\text{kN}}{\text{m}^2} \cdot \frac{422.9 \times 10^{-6} \text{ m}^4}{150 \times 10^{-3} \text{ m}}$$

$$2w = \frac{(70)(10^3)(422.9)(10^{-6})}{(150)(10^{-3})} = 197.4$$

$$w = 98.7 \text{ kN/m} \qquad\qquad \text{Answer}$$

EXAMPLE 13.5

A simply supported W14 × 38 steel beam with a span of 9 ft supports two concentrated loads of 35 kips at its third points, as shown in Fig. 13.14(a). If the maximum allowable bending stress is 24 ksi, would the beam be satisfactory? Neglect the weight of the beam.

Solution

The reactions are each equal to 35 kips. The shearing force and bending moment diagrams are shown in Fig. 13.14(b) and (c). Thus we see that the maximum bending moment is 105 kip-ft.

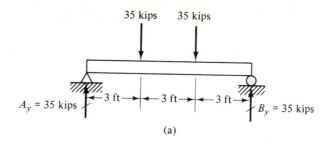

(a)

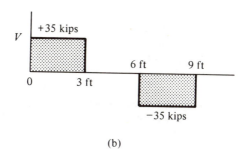

(b)

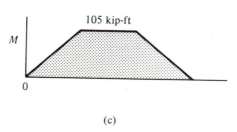

(c)

FIGURE 13.14

Section properties for the W14 × 38 beam are given in Table A.5 of the Appendix. $S = I/c = 54.7$ in.^{3}. The maximum stress from Eq. (13.6) is

$$\sigma_{max} = \frac{M}{S} = \frac{105(12)}{54.7} = 23.0 \text{ ksi}$$

$$= 23.2 \text{ ksi} < \sigma_a = 24 \text{ ksi}$$

Therefore, the beam is satisfactory. Answer

PROBLEMS

13.1 A rectangular beam 5 in. wide and 8 in. deep is acted on by a bending mo- ment of 250,000 lb-in. Determine (a) the maximum tensile and compressive bending stress and (b) the bending stress 1 in. from the top of the beam cross section.

13.2 A beam of rectangular cross section is 125 mm wide and 200 mm deep. If the maximum bending moment is 28.5 kN · m, determine (a) the maximum tensile and compressive bending stress, and (b) the bending stress 25 mm from the top of the section.

13.3 A rectangular beam 50 mm wide and 100 mm deep is subjected to bending. What bending moment will cause a maximum bending stress of 137.9 MN/m² (MPa)?

13.4 Determine the bending moment in a rectangular beam 2 in. wide and 4 in. deep if the maximum bending stress is 20,000 psi.

13.5 A rectangular tube has outside dimensions which are 2 in. wide and 4 in. deep and a wall thickness of 0.25 in. The tube is acted on by a bending moment of 15,000 lb-in. Determine (a) the maximum tensile and compressive bending stress, and (b) the bending stress 0.25 in. from the bottom of the tube.

13.6 A rectangular tube has an outside width of 100 mm, an outside depth of 150 mm, and a wall thickness of 10 mm. If the bending moment is 1.5 kN·m, determine (a) the maximum bending stress in the tube, and (b) the bending stress 10 mm from the top of the tube.

13.7 Determine the bending moment in a rectangular tube with an outside width of 150 mm, an outside depth of 300 mm, and a wall thickness of 20 mm if the maximum bending stress is 15.5 MN/m² (MPa).

13.8 A rectangular tube has outside dimensions which are 8 in. wide and 12 in. deep and a wall thickness of 0.5 in. Determine the bending moment in the

tube if the maximum bending stress is 12 ksi.

13.9 A bending moment of 28,800 lb-ft causes a maximum bending stress of 2400 psi in a rectangular beam. If the width of the beam is one-half the depth of the beam, determine the depth of the beam.

13.10 Determine the depth of a rectangular beam if a bending moment of 38.9 kN·m causes a maximum bending stress of 16.5 MN/m² (MPa). Assume that the width of the beam is one-third the depth of the beam.

13.11 What bending moment will cause a maximum stress of 80.0 MN/m² (MPa) in the beam with the cross section shown?

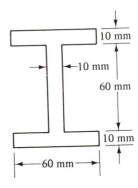

PROB. 13.11

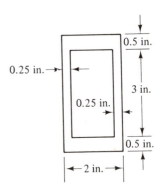

PROB. 13.12

13.12 What bending moment will cause a maximum bending stress of 12,000 psi in the beam with the cross section shown?

13.13 A beam with the cross section shown is acted on by a bending moment of 12,000 lb-in. Determine (a) the maximum tensile and compressive bending stress, and (b) the bending stress 1 in. from the bottom of the beam.

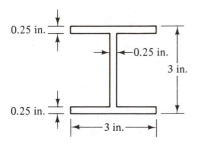

PROB. 13.13

13.14 A bending moment of 13.5 kN·m acts on a beam with the cross section shown. Determine (a) the maximum tensile and compressive bending stress, and (b) the bending stress 50 mm from the bottom of the beam.

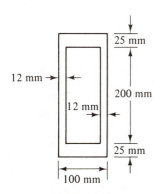

PROB. 13.14

13.15 A simply supported beam has a concentrated load $P = 60$ kN and cross section as shown. Determine (a) the

maximum tensile and compressive bending stress, and (b) the bending stress 1.8 m to the left of point C and 50 mm from the bottom of the beam.

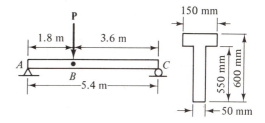

PROB. 13.15

13.16 Determine the maximum tensile and compressive bending stress for the simply supported beam with a concentrated load $P = 14$ kips and cross section as shown.

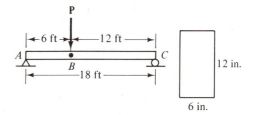

PROB. 13.16

13.17 A simply supported beam has a uniform load $w = 6.5$ kips/ft and cross section as shown. Determine (a) the maximum tensile and compressive bending stress, and (b) the bending stress 3 ft to the right of point A and 3 in. from the bottom of the beam.

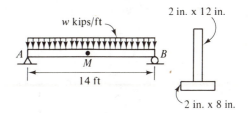

PROB. 13.17

13.18 Compute the maximum tensile and compressive bending stress for the simply supported beam with uniform load $w = 85$ kN/m and cross section as shown.

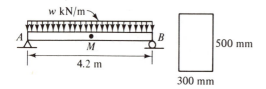

PROB. 13.18

13.19 A cantilever beam has a cross section and concentrated loads $P_1 = 36$ kN and $P_2 = 12$ kN, as shown. Compute (a) the maximum tensile and compressive bending stress, and (b) the bending stress 2 m to the right of point A and 50 mm from the top of the beam.

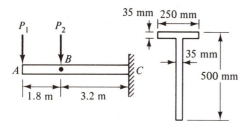

PROB. 13.19

13.20 Determine the maximum tensile and compressive bending stress for the cantilever beam with concentrated loads $P_1 = 3$ kips and $P_2 = 9$ kips and cross section shown.

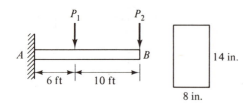

PROB. 13.20

13.21 A cantilever beam has a uniform load $w = 1.10$ kips/ft and cross section as shown. Determine (a) the maximum tensile and compressive bending stress, and (b) the bending stress 10 ft to the left of point B and 2 in. from the top of the beam.

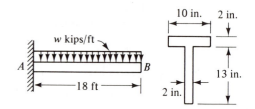

PROB. 13.21

13.22 Compute the maximum tensile and compressive bending stress for the cantilever beam with uniform load $w = 100$ kN/m and cross section shown.

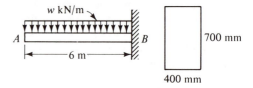

PROB. 13.22

13.23 The allowable tensile and compressive bending stress for the beam shown in the figure for Prob. 13.15 is 8.24 MN/m² (MPa). Determine the allowable load.

13.24 Determine the allowable load P for the beam shown in the figure for Prob. 13.16 if the allowable tensile and compressive bending stress is 10,000 psi.

13.25 The allowable tensile and compressive bending stress for the beam shown in the figure for Prob. 13.17 is 20 ksi. Calculate the allowable uniform load w.

13.26 Compute the allowable uniform load w for the beam shown in the figure for Prob. 13.18 if the allowable tensile bending stress is 46.0 MN/m² (MPa) and the allowable compressive bending stress is 184.0 MN/m² (MPa).

13.27 Determine the allowable load P_2 for the beam shown in the figure for Prob. 13.19 if the allowable tensile and compressive bending stress is 68.7 MN/m² (MPa). Assume that $P_1 = P_2$.

13.28 The allowable tensile bending stress is 6700 psi and the allowable compressive bending stress is 27,000 psi for the beam shown in the figure for Prob. 13.20. Find the allowable load P_1 if we assume that $P_2 = 3P_1$.

13.29 Compute the allowable uniform load w for the beam shown in the figure for Prob. 13.21 if the allowable tensile and compressive bending stress is 12.0 ksi.

13.30 Calculate the allowable uniform load w for the beam shown in the figure for Prob. 13.22 if the allowable tensile and compressive bending stress is 100 MN/m² (MPa).

For Probs. 13.31 through 13.35, see the Appendix for properties of rolled steel sections.

13.31 The W36 × 160 steel cantilever beam is acted on by a concentrated load and uniform load as shown. Determine (a) the maximum tensile and compressive bending stress if $P = 24$ kips and $w = 2$ kips/ft, and (b) the allowable uniform load w if the bending stress is 13.8 ksi and $P = 12w$.

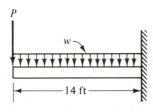

PROB. 13.31

13.32 The W33 × 220 steel simply supported beam is loaded with concentrated loads and uniform load as shown. Determine (a) the maximum tensile and compressive bending stress if $P = 150$ kips and $w = 10$ kips/ft, and (b) the allowable uniform load w if the allowable bending stress is 24 ksi and $P = 15w$.

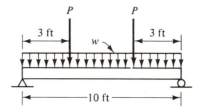

PROB. 13.32

13.33 A W18 × 35 steel beam supports a uniform load with overhanging ends as shown. Determine (a) the maximum tensile and compressive bending stress if $w = 3$ kips/ft, and (b) the allowable uniform load w if the allowable bending stress is 12 ksi.

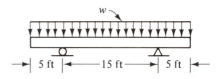

PROB. 13.33

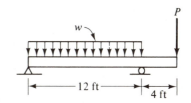

PROB. 13.34

13.34 A W16 × 50 steel beam supports a uniform load and a concentrated load as shown. Determine (a) the maximum

tensile and compressive bending stress
if $w = 1.5$ kips/ft and $P = 15$ kips, and
(b) the allowable load w if the allowable bending stress is 17 ksi and
$P = 10w$.

13.35   A W18 × 55 steel beam supports a
uniform load with an overhanging
end as shown. Determine (a) the maximum bending stress if $w = 2$ kips/ft,
and (b) the allowable uniform load w
if the allowable bending stress is
15.8 ksi.

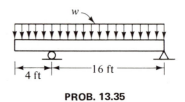

PROB. 13.35

13.8 SHEARING STRESS IN BEAMS

The shear force at a section is the resultant of the shear stresses which act
tangent to that section. We assume that the shear stresses are parallel to
the y axis as shown in Fig. 13.15(a) for a beam that is symmetric with
respect to the xy plane. In addition to the shear stress on the cross section, shear stress of equal magnitude exists on horizontal planes which are
perpendicular to the plane of the cross section. That is, as shown in Fig.
13.15(b), the shear stresses that exist at points on line qq' in the plane

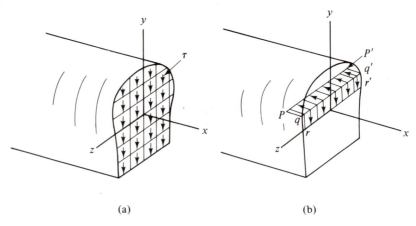

(a)   (b)

FIGURE 13.15

$qq'r'r$ are at every point equal in magnitude to the shear stresses that exist
at points on line qq' in the plane $qq'p'p$. The existence of horizontal shear
stress and its equality to the shear stress on the cross section was to be
expected from the results of Sec. 11.2, where we found that shearing
stresses at a point acting on mutually perpendicular planes are equal.

As an aid in visualizing horizontal shear stress, let us consider

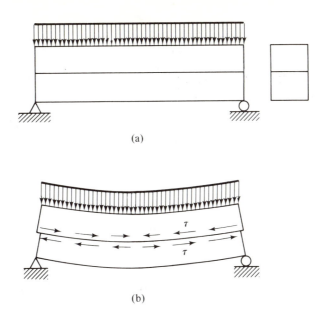

(a)

(b)

FIGURE 13.16

two rectangular beams stacked one on top of the other [Fig. 13.16(a)]. If we neglect friction, the two beams would bend independently, with one beam sliding over the other [Fig. 13.16(b)]. If the beam were solid, this sliding action would be resisted by shearing stresses. We may visualize a solid beam as being made up of many horizontal layers. To bend as a solid beam, shear stresses would be set up between each horizontal layer or on each horizontal plane.

We have seen that the shearing stress τ at a point on a vertical plane (cross section) is equal to the shearing stress at the same point on a horizontal plane. Therefore, the stress is usually referred to as horizontal shearing stress.

13.9 HORIZONTAL SHEARING STRESS FORMULA

In a beam subject to pure bending, no shearing force exists and therefore no horizontal shearing stress. The derivation of the bending stress formula was based on the deformation geometry of pure bending. The shearing stress formula is not based on deformation geometry. In fact, the formula does not satisfy compatible deformation geometry and its use is more restricted than the bending stress formula. However, it gives satisfactory results for most engineering problems.

The shearing stress formula will be derived from the requirements of equilibrium. Consider an element of a beam of length Δx between two sections A-A and B-B as shown in Fig. 13.17(a). A free-body

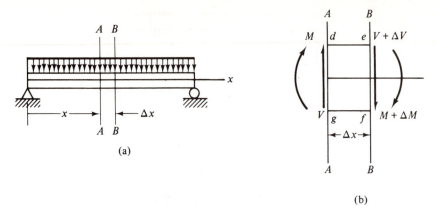

(a)

(b)

FIGURE 13.17

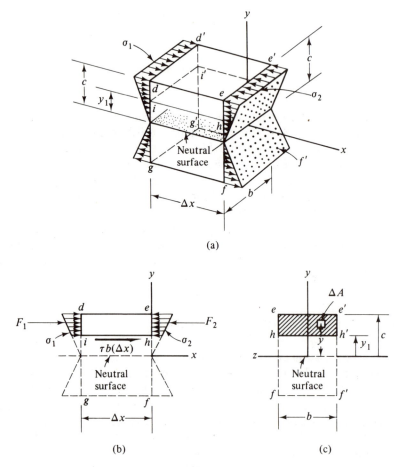

(a)

(b)

(c)

FIGURE 13.18

diagram for the element is drawn in Fig. 13.17(b), where the shearing force and bending moment on the sections are shown. In Fig. 13.18(a) the bending moments have been replaced by bending stresses. The bending stresses at any point on section A-A (plane $dd'g'g$) are given by

$$\sigma_1 = \frac{My}{I} \tag{a}$$

while the bending stresses at any point on section B-B (plane $ee'f'f$) are given by

$$\sigma_2 = \frac{(M + \Delta M)y}{I} \tag{b}$$

To find the shearing stress on a horizontal plane, we cut the element along a horizontal plane a distance y_1 above the neutral surface and draw a free-body diagram of the part of the element above the cut [Fig. 13.18(b)]. The force F_1 is the sum of the stresses σ_1 multiplied by the elements of area ΔA summed over the plane area $dd'i'i$. Thus $F_1 = \Sigma \, \sigma_1 \Delta A$. The stress σ_1 is given by Eq. (a). Therefore,

$$F_1 = \Sigma \frac{My}{I} \Delta A = \frac{M}{I} \Sigma \, y \Delta A \tag{c}$$

where the summation must be carried out for the area above y_1 of the cross section [Fig. 13.18(c)]. Similarly, the force F_2 is the sum of the stress σ_2 multiplied by the elements of area ΔA summed over the plane area $ee'h'h$. Thus $F_2 = \Sigma \sigma_2 \Delta A$. The stress is given by Eq. (b); therefore,

$$F_2 = \Sigma \frac{M + \Delta M}{I} y \Delta A = \frac{M + \Delta M}{I} \Sigma \, y \Delta A$$

or

$$F_2 = \frac{M}{I} \Sigma \, y \Delta A + \frac{\Delta M}{I} \Sigma \, y \Delta A \tag{d}$$

where the summation must be carried out for the area above y_1 of the cross section [Fig. 13.18(c)]. The shear force along the horizontal plane $i'ihh'$ is given by the product of the average shear stress τ multiplied by the shear area $b(\Delta x)$, that is, $\tau b(\Delta x)$. From the equilibrium equation

$\Sigma\, F_x = 0$, we write

$$F_1 - F_2 + \tau b(\Delta x) = 0$$

Substituting Eqs. (c) and (d) for F_1 and F_2, we have

$$\frac{M}{I}\, \Sigma\, y\Delta A - \frac{M}{I}\, \Sigma\, y\Delta A - \frac{\Delta M}{I}\, \Sigma\, y\Delta A + \tau b(\Delta x) = 0$$

or

$$\tau b(\Delta x) = \frac{\Delta M}{I}\, \Sigma\, y\Delta A$$

Dividing both sides of the equation by $b(\Delta x)$, this becomes

$$\tau = \frac{\Delta M\, \Sigma\, y\Delta A}{(\Delta x)\, Ib}$$

In Sec. 12.6 we found that $\Delta M/\Delta x = V$; therefore,

$$\tau = \frac{V\, \Sigma\, y\Delta A}{Ib} \tag{13.7}$$

The summation, $\Sigma\, y\Delta A$, represents the first moment of the area of the cross section of the beam above y_1 with respect to the neutral or z axis. (The first moment of the area below y_1 gives the negative of the first moment of the area above y_1.) Let $\Sigma\, y\Delta A = Q$; then Eq. (13.7) may be written as

$$\tau = \frac{VQ}{Ib} \tag{13.8}$$

If the shearing force V is in pounds (lb) or newtons (N), the first moment of area Q is in in.^{3} or m^3, the moment of inertia in in.^{4} or m^4, and the width b in in. or m, then the average shear stress τ will be in lb/in.^{2} (psi) or newtons/meter^2 (N/m^2).

EXAMPLE 13.6

A rectangular beam 400 mm by 800 mm supports a shear force of 5 kN. Find the horizontal shearing stress at the various levels shown in Fig. 13.19(a) and draw a graph showing the distribution of stress across the depth of the beam.

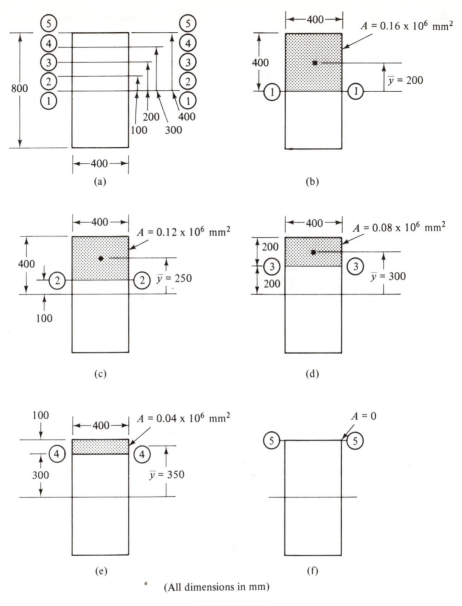

(All dimensions in mm)

FIGURE 13.19

Solution

The moment of inertia of the rectangular cross section with respect to the neutral axis is given by the formula

$$I_{NA} = \frac{bh^3}{12} = \frac{400(800)^3}{12} = 17.067 \times 10^9 \text{ mm}^4$$

The shear force $V = 5$ kN and the width of the section in each case is $b = 400$ mm.

The value of Q for each part of the example will be calculated from the diagrams shown in Fig. 13.19.

Level ①-①

From Fig. 13.19(b):

$$Q = \bar{y}A = 200\,(0.16)\,(10^6) = 32 \times 10^6 \text{ mm}^3$$

From the horizontal shearing stress formula [Eq. (13.8)],

$$\tau = \frac{VQ}{Ib} = \frac{5\,(32)\,(10^6)}{(17.067)\,(10^9)\,(400)} = 0.023\ 44 \times 10^{-3} \ \frac{\text{kN}}{\text{mm}^2}$$

$$= 23.44 \ \frac{\text{kN}}{\text{m}^2} \ (\text{kPa}) \qquad\qquad \text{Answer}$$

Level ②-②

From Fig. 13.19(c):

$$Q = \bar{y}A = 250\,(0.12)\,(10^6) = 30 \times 10^6 \text{ mm}^3$$

and

$$\tau = \frac{VQ}{Ib} = \frac{5\,(30)\,(10^6)}{(17.067)\,(10^9)\,(400)} = 0.02197 \times 10^{-3} \ \frac{\text{kN}}{\text{mm}^2}$$

$$= 21.97 \ \frac{\text{kN}}{\text{m}^2} \ (\text{kPa}) \qquad\qquad \text{Answer}$$

Level ③-③

From Fig. 13.19(d):

$$Q = \bar{y}A = 300\,(0.08)\,(10^6) = 24 \times 10^6 \text{ mm}^3$$

and

$$\tau = \frac{VQ}{Ib} = \frac{5\,(24)\,(10^6)}{(17.067)\,(10^9)\,(400)} = 0.01758 \times 10^{-3} \ \frac{\text{kN}}{\text{mm}^2}$$

$$= 17.58 \ \frac{\text{kN}}{\text{m}^2} \ (\text{kPa}) \qquad\qquad \text{Answer}$$

Level ④-④

From Fig. 13.19(e):

$$Q = \bar{y}A = 350\,(0.04)\,(10^6) = 14 \times 10^6 \text{ mm}^3$$

and

$$\tau = \frac{VQ}{Ib} = \frac{5(14)(10^6)}{(17.067)(10^9)(400)} = 0.01025 \times 10^{-3} \ \frac{kN}{mm^2}$$

$$= 10.25 \ \frac{kN}{m^2} \ (kPa) \qquad\qquad \text{Answer}$$

Level ⑤ - ⑤

From Fig. 13.19(f): $Q = 0$; therefore,

$$\tau = 0 \qquad\qquad \text{Answer}$$

Stress Distribution

The distribution of horizontal shear stress is shown from top to bottom of the beam in Fig. 13.20. The distribution below the neutral axis was calculated in the same way as the distribution above except that for convenience the area used to calculate Q was taken below the shear plane instead of above it. In fact, the shearing stress distribution along the depth of a rectangular beam is parabolic. The maximum value occurs at the neutral axis and it is zero at the top and bottom. This is *not* true for all beam shapes. For example, it is not true for a beam with a triangular cross section.

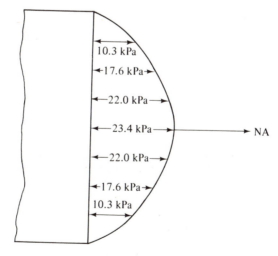

FIGURE 13.20

EXAMPLE 13.7

Derive a formula for the maximum horizontal shearing stress in a beam of rectangular cross section.

Solution

The moment of inertia with respect to the neutral axis $I_{NA} = bh^3/12$. The moment of the area above the neutral axis (plane of maximum shearing stress) $Q = \bar{y}A = bh^2/8$, and the width of the shear plane $b = b$ as shown in Fig. 13.21. Therefore,

$$\tau = \frac{VQ}{Ib} = \frac{V(bh^2/8)}{(bh^3/12)b} = \frac{3}{2}\frac{V}{bh}$$

or

$$\tau_{max} = \frac{3}{2}\frac{V}{A} \tag{13.9}$$

where A is the area of the cross section. Thus the maximum horizontal shearing stress is $1\frac{1}{2}$ times greater than the average value ($\tau_{av} = V/A$).

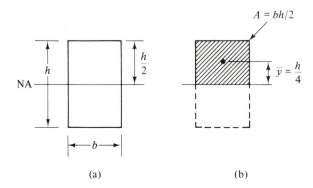

FIGURE 13.21

EXAMPLE 13.8

A simply supported wooden beam with a span of 10 ft supports a load at the middle of its span [Fig. 13.22(a)]. The cross section is 10 in. wide and 14 in. deep. The allowable shearing stress for the wood is 1200 psi. Find P.

Solution

The shearing force diagram is shown in Fig. 13.22(b). The maximum shear force $V = P/2$, the cross-sectional area $A = 10(14) = 140$ in.^{2}, and the allowable shearing stress is 1200 psi. From Eq. (13.9),

$$\tau_{max} = \frac{3V}{2A} \quad \text{or} \quad V = \frac{\tau_{max}\,2A}{3}$$

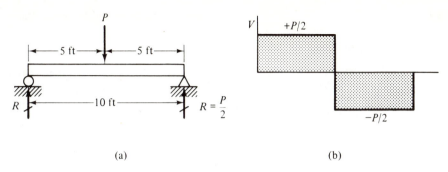

(a)   (b)

FIGURE 13.22

Therefore,

$$\frac{P}{2} = \frac{1200(2)(140)}{3} = 112,000 \text{ lb}$$

or

$$P = 224,000 \text{ lb} \qquad\qquad \text{Answer}$$

EXAMPLE 13.9

Find the maximum shearing stress for the beam and cross section in Fig. 13.23(a) and (b). (The maximum shearing stress occurs at the neutral axis.)

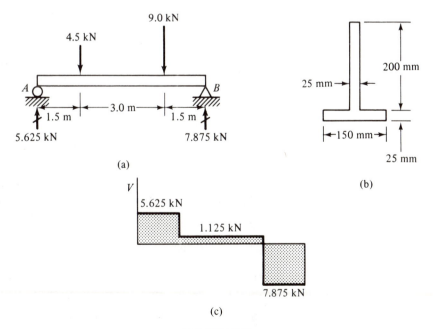

FIGURE 13.23

347

Solution

The reactions were calculated from the equations of equilibrium and the shear diagram drawn in Fig. 13.23(c). The maximum shear force at any section is 7.875 kN. The T-shaped cross section is divided into two rectangular areas, as in Fig. 13.24. The solution for the centroid is tabulated as shown. From the sums in the table,

$$\bar{y} = \frac{\Sigma\, Ay}{\Sigma\, A} = \frac{671.8 \times 10^3}{8.75 \times 10^3} = 76.8 \text{ mm}$$

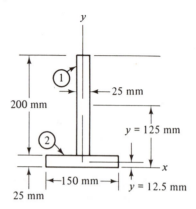

FIGURE 13.24

	A	y	Ay
①	5000	125	625.00×10^3
②	3750	12.5	46.88×10^3
	$\Sigma A = 8750$ mm^2		$\Sigma Ay = 671.88 \times 10^3$ mm^3

The transfer distances for each area are given in Fig. 13.25(a). The solution for the moment of inertia about the neutral axis is tabulated as shown. From the sums in the table,

$$I_{NA} = \Sigma\, I_c + \Sigma\, (Ad^2) = 16.86 \times 10^6 + 27.12 \times 10^6$$
$$= 43.98 \times 10^6 \text{ mm}^4 = 43.98 \times 10^{-6} \text{ m}^4$$

The moment of the cross-sectional area above the neutral axis is determined from Fig. 13.25(b).

$$Q = \bar{y}A = 74.1(25)(148.2) = 0.2745 \times 10^6 \text{ mm}^3$$

The width of the beam at the shear plane (neutral axis) is $b = 25$ mm.

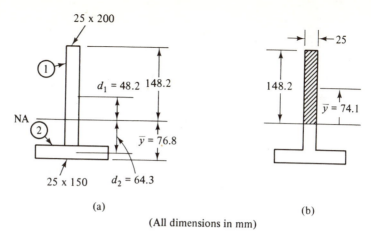

25 x 200

$d_1 = 48.2$ 148.2

25

148.2

NA

$\bar{y} = 74.1$

25 x 150 $d_2 = 64.3$

$\bar{y} = 76.8$

(a)

(b)

(All dimensions in mm)

FIGURE 13.25

	A	d	I_c	Ad^2
①	5000	48.2	ⓐ16.66×10^6	11.62×10^6
②	3750	64.3	ⓑ0.20×10^6	15.50×10^6
			$\Sigma I_c = 16.86 \times 10^6$ mm^4	$\Sigma (Ad^2) = 27.12 \times 10^6$ mm^4

ⓐ$I_c = 25(200)^3/12$
ⓑ$I_c = 150(25)^3/12$

The maximum shearing stress that occurs at the neutral axis is found from the horizontal shearing stress formula. Therefore,

$$\tau_{max} = \frac{VQ}{Ib} = \frac{7.875(0.2745)(10^6)}{(43.98)(10^6)(25)} = 1.966 \times 10^{-3} \ \frac{kN}{mm^2}$$

$$= 1.966 \ \frac{MN}{m^2} \ (MPa) \qquad\qquad \text{Answer}$$

EXAMPLE 13.10

An I-beam with cross section as shown in Fig. 13.26(a) supports a shear force of 75 kips. Find the horizontal shearing stress at the various levels shown and draw a graph showing the distribution of stresses across the depth of the beam.

Solution

The moment of inertia may be found by calculating the moment of inertia of a rectangle 8 in. by 14 in. and subtracting the moment of inertia of two rectangles 3.5 in. by 12 in. That is,

$$I = \frac{8(14)^3}{12} - 2\left[\frac{3.5(12)^3}{12}\right] = 821.3 \ in.^4$$

349

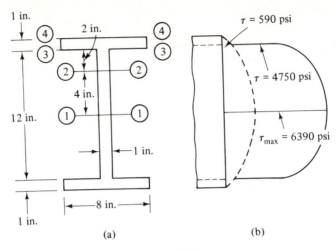

FIGURE 13.26

Therefore,

$$\frac{V}{I} = \frac{75,000}{821.3} = 91.3 \text{ lb/in.}^4$$

Values of Q and b and the corresponding shear stress for the various levels required are tabulated as shown.

Level	A (in.^{2})	$\bar{y}$ (in.)	$Q = A\bar{y}$ (in.^{3})	b (in.)	$\tau = \dfrac{VQ}{Ib}$ (psi)
①-①	8 × 1 = 8 6 × 1 = 6	6.5 3.0	52 ⎱ 18 ⎰ 70	1.0	6390
②-②	8 × 1 = 8 2 × 1 = 2	6.5 5.0	52 ⎱ 10 ⎰ 62	1.0	5660
③-③	8 × 1 = 8	6.5	52	1.0 8.0	4750 590
④-④	0	7.5	0	8.0	0

$\dfrac{V}{I} = 91.3 \text{ lb/in.}^4$

The distribution of shear stress is shown in Fig. 13.26(b). The stress is symmetric with respect to the neutral axis and the maximum stress occurs at the neutral axis. At level ③-③, values of $b = 8$ in. and $b = 1$ in. are used to find the end points of the two curves. The two curves are parts of parabolas.

The maximum shear stress is supported by the web of the beam; therefore, the maximum shear stress is often approximated by dividing the total shear force V by the area of the web to find the average stress.

$$\tau_{av} = \frac{V}{A_{web}}$$

$$= \frac{75,000}{12(1)} = 6250 \text{ psi} \qquad (13.10)$$

We see that the average value is lower than the maximum value by approximately 2 percent. Therefore, it would be a reasonable approximation of the value of the maximum shear stress in the web of the beam.

PROBLEMS

13.36 Determine the maximum shear stress in a rectangular beam 6 in. wide and 8 in. deep produced by a shear force of 38,400 lb.

13.37 A rectangular beam has a cross section 150 mm wide and 200 mm deep. Determine the maximum shear stress caused by a shear force of 165 kN.

13.38 What shear force will cause a maximum shear stress of 8.24 MPa in a rectangular beam 125 mm wide and 225 mm deep?

13.39 A rectangular beam has a cross section 5 in. wide and 9 in. deep. Determine the shear force that will cause a maximum shear stress of 1200 psi.

13.40 The beam shown has a rectangular cross section 4 in. wide and 8 in. deep. Determine the maximum shear stress.

13.41 Calculate the maximum shear stress in the beam shown. The rectangular cross section is 500 mm wide and 900 mm deep.

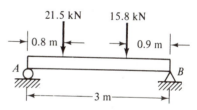

PROB. 13.41

13.42 The rectangular cross section of the beam shown is 300 mm wide and 450 mm deep. If the allowable shear stress is 6.85 MPa, determine the value of the concentrated load **P**.

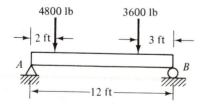

PROB. 13.40

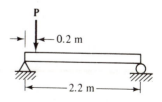

PROB. 13.42

13.43 Determine the value of the concentrated load **P** if the rectangular cross section of the beam shown is 12 in. wide and 18 in. deep. The maximum allowable shear stress is 1200 psi.

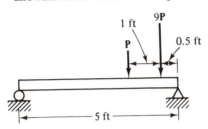

PROB. 13.43

13.44 Determine the maximum shear stress in a rectangular tube 6 in. wide, 8 in. deep and 1 in. thick caused by a shear force of 25,300 lb.

13.45 A rectangular tube has a cross section 125 mm wide, 200 mm deep, and 25 mm thick. Find the maximum shear stress caused by a shear force of 130 kN.

13.46 For the simply supported beam with a uniform load and cross section as shown, calculate (a) the maximum shear stress, (b) the average shear stress [Eq. (13.10)], and (c) the per-

cent difference between the maximum shear stress and average shear stress. [Percent difference = $100\,(\tau_{max} - \tau_{av})/\tau_{max}$.]

13.47 A simply supported beam has a uniform load and cross section as shown. Determine (a) the maximum shear stress, (b) the average shear stress [Eq. (13.10)], and (c) the percent difference between the maximum shear stress and the average shear stress. [Percent difference = $100\,(\tau_{max} - \tau_{av})/\tau_{max}$.]

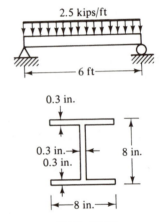

PROB. 13.47

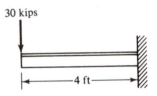

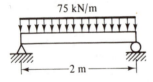

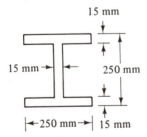

PROB. 13.46

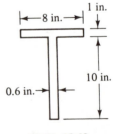

PROB. 13.48

13.48 Calculate the maximum shear stress for the cantilever beam with the cross section shown.

13.49 Find the maximum shear stress for the cantilever beam with cross section shown.

150 kN

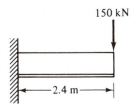

|◄——2.4 m——►|

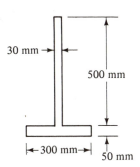

30 mm →|←—

500 mm

|←—300 mm—►| 50 mm

PROB. 13.49

13.50 A beam has a cross section as shown. At a section where the shear force is 10 kN, (a) determine the shear stress

at 25-mm intervals along the depth, and (b) plot the shear stress distribution.

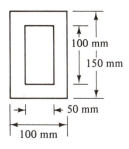

100 mm

150 mm

|←50 mm

100 mm

PROB. 13.50

13.51 A beam has a cross section as shown. If the shear force on the section is 10 kips, (a) calculate the shear stress at 1-in. intervals along the depth, and (b) plot the shear stress distribution.

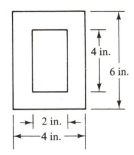

4 in.

6 in.

2 in.

4 in.

PROB. 13.51

13.10 SHEAR FLOW FORMULA

Shear flow q is defined as the force per unit of length along the shear plane. As such, it is equal to the product of the shearing stress and the width of the beam at the shear plane, that is,

$$q = \tau b$$

Substituting for the value of shearing stress from Eq. (13.8) in this equation gives

$$q = \frac{VQ}{I} \qquad (13.11)$$

Shearing force V is in lb. or newtons (N), the first moment of the area Q is in in.^{3} or m^3, and the moment of inertia is in in.^{4} or m^4. Then shear flow q will be lb/in. or newtons/meter (N/m).

The shear flow formula can be used to find the number or spacing of connectors used to fasten members together to form a beam cross section.

EXAMPLE 13.11

The I-beam shown in Fig. 13.27(a) and (b) is made by bolting together three planks 14 ft long. What should be the minimum spacing of the bolts to resist the shearing stress? The bolts can safely resist a shearing force of 1100 lb. Neglect the weight of the beam.

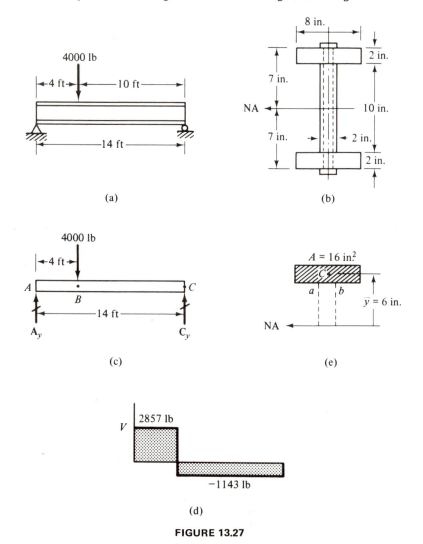

FIGURE 13.27

Solution

The free-body diagram for the beam is drawn in Fig. 13.27(c). From equilibrium:

$$\circlearrowright \Sigma M_A = 0 \quad -4000(4) + Cy\,14 = 0$$

$$Cy = 1143 \text{ lb.}$$

$$\circlearrowright \Sigma M_C = 0 \quad -Ay\,14 + 4000(10) = 0$$

$$Ay = 2857 \text{ lb}$$

Check:

$$\uparrow \Sigma F_y = 0 \quad Ay - 4000 + Cy = 0$$

$$2857 - 4000 + 1143 = 0 \quad \text{OK}$$

The shear diagram is constructed in Fig. 13.27(d). We see that the beam must transmit a maximum shear force $V = 2857$ lb. The cross section of the beam [Fig. 13.27(b)] is symmetric about a horizontal and vertical axis. Therefore, the neutral axis can be located by inspection. The neutral axis is shown in the figure.

The moment of inertia about the neutral axis may be determined by calculating the moment of inertia of a rectangle 8 in. by 14 in. and subtracting the moment of inertia of two rectangles 3 in. by 10 in. That is,

$$I = \frac{8(14^3)}{12} - (2)\,\frac{3(10^3)}{12} = 1329 \text{ in.}^4$$

The shear plane ab where the bolts must resist the shear force is shown in Fig. 13.27(e). The value of Q is equal to the moment of the area above plane ab (shaded area) about the neutral axis. The centroid for this area $\bar{y} = 6$ in. and the shaded area $A = 2(8) = 16$ in.^{2}. Therefore, $Q = \bar{y}A = 6(16) = 96$ in.^{3}. From Eq. (13.10), the shear flow

$$q = \frac{VQ}{I} = \frac{2857(96)}{1329} = 206.4 \text{ lb/in.}$$

Each bolt resists a shearing force $F_V = 1100$ lb and must resist the force developed by the shear flow q over a length s. Therefore,

$$F_V = qs \quad \text{or} \quad s = \frac{F_V}{q} = \frac{1100 \text{ lb}}{206.4 \text{ lb/in.}} = 5.329 \text{ in.}$$

Use a spacing $s = 5.25$ in.

Answer

PROBLEMS

13.52 A box beam is made by nailing together four full-sized planks to form a cross section as shown. What should be the minimum spacing of the nails

to resist a shear force of 1100 N? Each
nail can resist 270 N in shear.

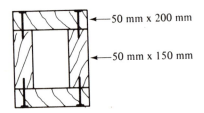

PROB. 13.52

13.53 Four full-sized planks are nailed together to form a box beam with cross
section as shown. Each nail can resist a
shear force of 60 lb. What should be
the minimum spacing of the nails to
resist a shear force of 300 lb?.

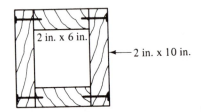

PROB. 13.53

13.54 A wooden beam is made up of three
full-sized planks that are joined together by screws to form a cross section as shown. Each screw can resist
a force of 120 lb. What should be the
minimum spacing of the screws to
resist a shear force of 1200 lb? The

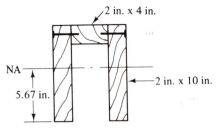

PROB. 13.54

moment of inertia of the cross section
$I_{NA} = 776$ in^4.

13.55 Two planks are joined together by
screws to form a T-shaped cross section as shown. Each screw has a shear
strength of 525 N. What should be the
minimum spacing of the screws to
resist a shear force of 720 N? The
moment of inertia of the cross section
$I_{NA} = 113.5 \times 10^6$ mm^4.

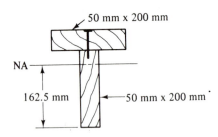

PROB. 13.55

13.56 Two W8 × 15 rolled steel shapes are
joined by bolts to form a cantilever
beam with loading and cross section as
shown. What should be the minimum
spacing of the $\frac{1}{2}$-in.-diameter bolts if
the allowable shear stress for the
bolts is 22 ksi?

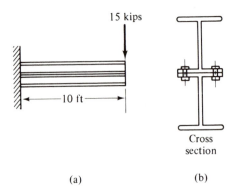

(a)   (b)

PROB. 13.56

13.57 The built-up shape shown in the
figure is used as a beam. The cover
plates are joined to the W10 × 45 by

$\frac{3}{4}$-in.-diameter rivets. What should be the minimum spacing (pitch) of the rivets to resist a shear force of 43.5 kips? The allowable shear stress in the rivets is 20 ksi. The moment of inertia of the cross section $I_{NA} = 644$ in^4.

13.58 For the beam cross section shown, determine the required spacing of the 1.0-in.-diameter rivets. The shear force on the section is 200 kips and the allowable shear stress in the rivets is 20 ksi. The moment of inertia of the cross section $I_{NA} = 8466$ in^4.

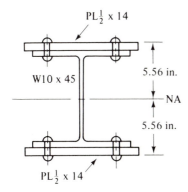

PL$\frac{1}{2}$ x 14

W10 x 45

5.56 in.

NA

5.56 in.

PL$\frac{1}{2}$ x 14

PROB. 13.57

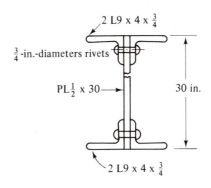

2 L9 x 4 x $\frac{3}{4}$

$\frac{3}{4}$-in.-diameters rivets

PL$\frac{1}{2}$ x 30

30 in.

2 L9 x 4 x $\frac{3}{4}$

PROB. 13.58

Deflection of Beams Due to Bending ____

14.1 INTRODUCTION

The relationship between applied loads and bending and shear stresses in beams was developed in Chapter 13. In addition to limitations on the stresses in the design of beams, deflections must in many cases also be limited. In this chapter two methods for calculating beam deflections will be studied: the *moment-area* method and the *superposition* method. We begin the moment-area method by additional study of the bending-moment diagram.

14.2 BENDING-MOMENT DIAGRAM BY PARTS

When calculating the deflection of a beam by the moment-area method, it is necessary to know the area and location of the centroid of the moment

TABLE 14.1

	Type of load	Moment diagram	Geometric properties
(a)	M L	L $-M$ $-M$ $M_x = -M$ Rectangle	$\bar{x}$ C h b $A = bh \quad \bar{x} = \dfrac{b}{2}$
(b)	P L	L d $-Pd$ $-PL$ $M_x = -Pd$ Triangle	$\bar{x}$ C h b $A = \dfrac{bh}{2} \quad \bar{x} = \dfrac{b}{3}$
(c)	w L	L d $-\dfrac{wd^2}{2}$ $-\dfrac{wL^2}{2}$ $M_x = -\dfrac{wd^2}{2}$ Parabola	$\bar{x}$ C h b $A = \dfrac{bh}{3} \quad \bar{x} = \dfrac{b}{4}$
(d)	w_{max} L	L d $\dfrac{-w_{max}d^3}{6L}$ $\dfrac{-w_{max}L^2}{6}$ $M_x = \dfrac{-w_{max}d^3}{6L}$ Cubic parabola	$\bar{x}$ C h b $A = \dfrac{bh}{4} \quad \bar{x} = \dfrac{b}{5}$

diagram. The calculations are simplified if we imagine that each load and reaction on the beam produces a separate moment diagram. The resulting diagrams are said to be drawn by parts. The sum of all the moment diagrams drawn by parts is equivalent to the usual bending moment diagram for the beam.

Four types of loads will be considered: the couple, the concentrated load, the uniform load, and the triangular load. Table 14.1 shows the types of loads and the shape, area, and location of the centroid of the moment diagrams. The table will be used to draw moment diagrams by parts.

EXAMPLE 14.1

Draw the moment diagram by parts for the cantilever beam shown in Fig. 14.1(a).

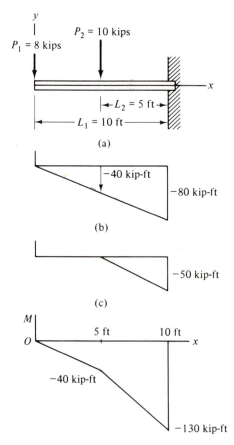

Combined moment diagram

(d)

FIGURE 14.1

Solution

Two concentrated loads act on the cantilever beam. Each concentrated load has a moment diagram that is triangular in shape, as shown in Table 14.1(b). The bending moment is given by

$$M_x = -Pd$$

where x is the location measured from the left end of the beam, P the value of the concentrated load, and d the distance from the load.

Values of the bending moment from $P_1 = 8$ kips are shown in Fig. 14.1(b). At a distance $d = 5$ ft from the load at $x = 5$ ft,

$$M_5 = -(8)(5) = -40 \text{ kip-ft}$$

and at a distance $d = 10$ ft from the load at $x = 10$ ft,

$$M_{10} = -(8)(10) = -80 \text{ kip-ft}$$

Values of the bending moment for $P_2 = 10$ kips are shown in Fig. 14.1(c). At a distance $d = 5$ ft from the load at $x = 10$ ft,

$$M_{10} = -(10)(5) = -50 \text{ kip-ft}$$

The combined moment diagram is shown in Fig. 14.1(d).

EXAMPLE 14.2

Draw the moment diagram by parts for the cantilever beam shown in Fig. 14.2(a).

Solution

A concentrated load and a distributed load act on the cantilever beam. The concentrated load has a moment diagram as given in Table 14.1(b). Values of the bending moment for $P = 36$ kN are shown in Fig. 14.2(b). At a distance $d = 1.5$ m from the load at $x = 1.5$ m,

$$M_{1.5} = -36(1.5) = -54 \text{ kN} \cdot \text{m}$$

and at a distance $d = 4.5$ m from the load at $x = 4.5$ m,

$$M_{4.5} = -36(4.5) = -162 \text{ kN} \cdot \text{m}$$

The distributed load has a moment diagram that is parabolic in shape, as given in Table 14.1(c). The bending moment is given by

$$M_x = \frac{-wd^2}{2}$$

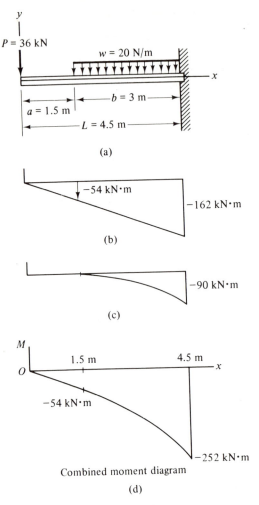

FIGURE 14.2

where x is the location measured from the left end of the beam, w the value of the distributed load, and d the distance along the distributed load. Values of the bending moment for $w = 20$ N/m are shown in Fig. 14.2(c). At a distance $d = 3$ m along the uniform load at $x = 4.5$ m,

$$M_{4.5} = \frac{-20(3)^2}{2} = -90 \text{ kN} \cdot \text{m}$$

The combined moment diagram is shown in Fig. 14.2(d).

EXAMPLE 14.3

Draw the moment diagram by parts for the simply supported beam shown in Fig. 14.3(a).

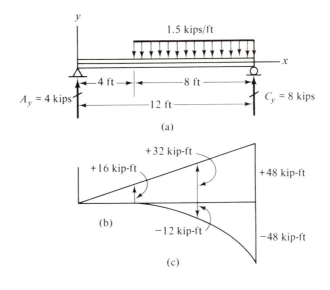

(a)

(b)

(c)

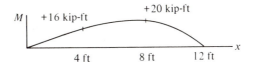

Combined moment diagram

(d)

FIGURE 14.3

Solution

The reactions were determined from the equations of equilibrium and are shown in Fig. 14.3(a). The moment diagram for the concentrated upward reaction of 4 kips gives a positive triangular moment diagram [Fig. 14.3(b)] because the reaction is directed upward. The bending moment at a section x at a distance d from the concentrated load P is given by $M_x = Pd$. Therefore,

$$M_4 = 4(4) = 16 \text{ kip-ft}$$
$$M_8 = 4(8) = 32 \text{ kip-ft}$$

and

$$M_{12} = 4(12) = 48 \text{ kip-ft}$$

The moment diagram for the uniformly distributed downward load of 1.5 kips/ft gives a negative parabolic moment diagram [Fig. 14.3(c)]. The bending moment at a section x at a distance d along the distributed load is given by

$$M_x = -\frac{wd^2}{2}$$

Therefore,

$$M_8 = - \frac{1.5\,(4)^2}{2} = -12 \text{ kip-ft}$$

and

$$M_{12} = - \frac{1.5\,(8)^2}{2} = -48 \text{ kip-ft}$$

The combined moment diagram is shown in Fig. 14.3(d).

EXAMPLE 14.4

Draw the moment diagram by parts for the simply supported beam shown in Fig. 14.4(a).

Solution

The reactions were determined. The partial uniform downward load of 15 N/m on the beam from $x = 0$ to $x = 2$ m was replaced by a uniform downward load of 15 N/m from $x = 0$ to $x = 5$ m and a uniform upward load of 15 N/m from $x = 2$ m to $x = 5$ m. The replacement loads when added together give the original load. The replacement loads and reactions are shown in Fig. 14.4(b).

The moment diagram for the concentrated upward reaction or load of 24 N gives a positive triangular moment diagram [Fig. 14.4(d)]. The bending moment at a section x at a distance d from the concentrated load P is given by

$$M_x = Pd$$

Therefore,

$$M_1 = 24\,(1) = 24 \text{ N} \cdot \text{m}$$

$$M_2 = 24\,(2) = 48 \text{ N} \cdot \text{m}$$

$$M_{2.5} = 24\,(2.5) = 60 \text{ N} \cdot \text{m}$$

and

$$M_5 = 24\,(5) = 120 \text{ N} \cdot \text{m}$$

The moment diagram for the distributed upward uniform load of 15 N/m gives a positive parabolic moment diagram [Fig. 14.4(c)]. The bending moment at a section x at a distance d along the distributed load is given by

$$M_x = \frac{wd^2}{2}$$

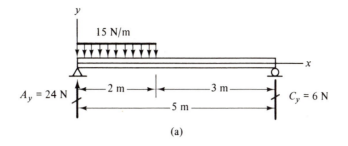

(a)

(b)

1.875 N·m   67.5 N·m

(c)

(d)

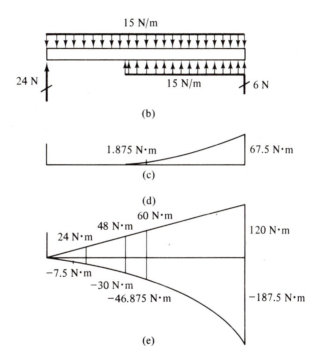

24 N·m
48 N·m
60 N·m
120 N·m
−7.5 N·m
−30 N·m
−46.875 N·m
−187.5 N·m

(e)

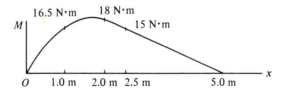

16.5 N·m   18 N·m

15 N·m

M

O 1.0 m   2.0 m 2.5 m   5.0 m x

Combined moment diagram (different scale)

(f)

FIGURE 14.4

365

Therefore,

$$M_{2.5} = \frac{15\,(0.5)^2}{2} = 1.875 \text{ N} \cdot \text{m}$$

and

$$M_5 = \frac{15\,(3)^2}{2} = 67.5 \text{ N} \cdot \text{m}$$

The moment diagram for the distributed downward uniform load of 15 N/m gives a negative parabolic moment diagram [Fig. 14.4(e)]. The formula for the bending moment is

$$M_x = -\frac{wd^2}{2}$$

with x and d as previously defined. Therefore,

$$M_1 = -\frac{15\,(1)^2}{2} = -7.5 \text{ N} \cdot \text{m}$$

$$M_2 = -\frac{15\,(2)^2}{2} = -30 \text{ N} \cdot \text{m}$$

$$M_{2.5} = -\frac{15\,(2.5)^2}{2} = -46.875 \text{ N} \cdot \text{m}$$

and

$$M_5 = -\frac{15\,(5)^2}{2} = -187.5 \text{ N} \cdot \text{m}$$

The combined moment diagram in Fig. 14.4(f) is not drawn to the same scale as the moment diagram by parts.

14.3 THE MOMENT-AREA METHOD

A beam is bent by loads as shown in Fig. 14.5(a). Before the loads were applied, the neutral surface, line AB, was straight. After the loads are applied, the line is curved. The curved line is called the *deflection curve*. We will now develop relationships between the changes in the shape of the deflection curve and the moments that produce these changes.

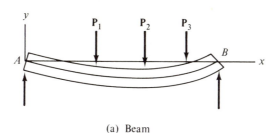

(a) Beam

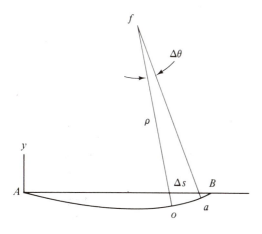

(b) Deflection curve

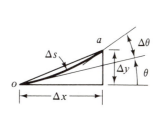

(c) Detail of deflection curve

FIGURE 14.5

Consider the deflection curve shown in Fig. 14.5(b). The radius of curvature of the arc Δs is ρ (rho). The angle $\Delta\theta$ in radians is equal to $\Delta s/\rho$. Therefore,

$$\frac{1}{\rho} = \frac{\Delta\theta}{\Delta s} \tag{a}$$

The slope θ of the deflective curve is assumed at every point to be small [Fig. 14.5(c)]. Consequently, the arc length Δs is approximately equal to Δx. With this substitution in Eq. (a), we have

$$\frac{1}{\rho} = \frac{\Delta\theta}{\Delta x} \tag{14.1}$$

Eliminating the normal bending stress σ between Eqs. (13.2) and (13.4), that is, $\sigma = Ey/\rho$ and $\sigma = My/I$, another equation for $1/\rho$ is obtained. That is,

$$\frac{1}{\rho} = \frac{M}{EI} \tag{14.2}$$

Combining Eqs. (14.1) and (14.2), we write

$$\frac{\Delta\theta}{\Delta x} = \frac{M}{EI}$$

or

$$\Delta\theta = \frac{M}{EI} \Delta x \tag{14.3}$$

The M/EI Diagram

To draw the *M/EI* diagram for a beam, the value of the bending moment *M* at each cross section of the beam is divided by the value of the flexural stiffness *EI* at that cross section. For a beam of constant cross section made of one material, *EI* is constant and the *shape* of the *M/EI* diagram and the moment diagram is the same.

In Fig. 14.6 we show a loaded beam, the *M/EI* diagram, and the deflection curve. For illustration the deflection curve is exaggerated [Fig. 14.6(c)]. The change in slope of the deflection curve from *o* to *a* is equal to $\Delta\theta$. From Eq. (14.3), $\Delta\theta$ is equal to the shaded area from *o* to *a* of the *M/EI* diagram. The total change in slope between *A* and *B* is obtained by summing up all the changes in slope $\Delta\theta$ from *A* to *B*. That is,

$$\theta_B - \theta_A = \Sigma \ \Delta\theta \ (\text{from } A \text{ to } B) = \Sigma \ \frac{M}{EI} \ \Delta x \ (\text{from } A \text{ to } B) \tag{14.4}$$

Therefore, *the change in slope between A and B on a deflection curve is equal to the area of the M/EI diagram between A and B.* Equation (14.4) can be restated for numerical applications in a simpler form:

$$\theta_B - \theta_A = A^* \ (\text{from } A \text{ to } B) \tag{14.5}$$

where A^* represents the area of the *M/EI* diagram. This is the *first moment-area theorem*.

The quantity Δt in Fig. 14.6 is due to the bending of an element of the beam and is equal to $x_B \ \Delta\theta$. By summing up this effect from *A* to *B*, we obtain the vertical distance *B'B*, which is called the tangent deviation $t_{B/A}$. That is,

$$t_{B/A} = \Sigma \ \Delta t \ (\text{from } A \text{ to } B) = \Sigma \ x_B \ \Delta\theta \ (\text{from } A \text{ to } B)$$

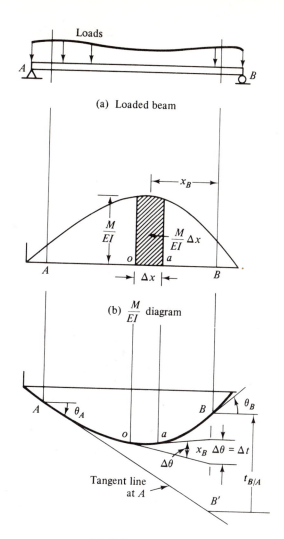

(a) Loaded beam

(b) $\dfrac{M}{EI}$ diagram

(c) Deflection curve

FIGURE 14.6

Substituting for $\Delta\theta$ from Eq. (14.3), we obtain

$$t_{B/A} = \Sigma\, x_B\ \frac{M}{EI}\ \Delta x \text{ (from } A \text{ to } B)\tag{14.6}$$

Therefore, *the deviation of B from a tangent line at A is equal to the first moment about a vertical line through B of the area of the M/EI diagram between A and B.* If we use the definition of the centroid, Eq. (14.6) can be restated for numerical application in simpler form:

$$t_{B/A} = \bar{x} \text{ (from } B) \, A^* \text{ (from } A \text{ to } B)\tag{14.7}$$

where A^* represents the area of the M/EI diagram. This is the *second moment-area theorem*.

	1	2	3	4	5
	A^* (from A to B)	$\bar{x}_A$ (in.)	$\bar{x}_A A^*$ (A to B)	$\bar{x}_B$ (in.)	$\bar{x}_B A^*$ (A to B)
①	$\dfrac{120}{2}(3 \times 10^{-4}) = 0.18$	80	1.440	40	0.720
②	$-\dfrac{60}{2}(1.5 \times 10^{-4}) = -0.0045$	100	-0.450	20	-0.090
			$t_{A/B} = 0.990$ in.		$t_{B/A} = 0.630$ in.

Consider the deflection curve shown in Fig. 14.7(a) and the M/EI diagram that has been drawn by parts in Fig. 14.7(b). The deviation of A from a tangent line at B, $t_{A/B}$, is shown in Fig. 14.7(a). The areas A^* of the M/EI diagram from A to B and the centroids measured from A, $\bar{x}_A$, are shown in column 1 and 2 of the table. In column 3 the product of areas and centroidal distances are tabulated. The sum of column 3 gives

$$t_{A/B} = 0.990 \text{ in.}$$

Similarly, the deviation of B from a tangent line at A is shown in Fig. 14.7(a). The centroids measured from B, $\bar{x}_B$, are shown in column 4 of

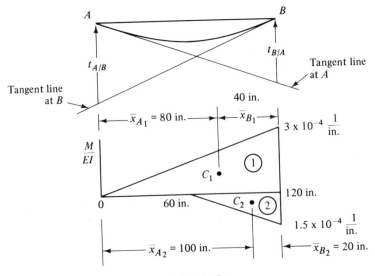

FIGURE 14.7

the table. In column 5 the products of the areas and the centroidal distances are tabulated. The sum of column 5 gives

$$t_{B/A} = 0.630 \text{ in.}$$

To find the deflection of a simply supported beam requires both the first and second moment-area theorems. However, the deflection and tangent deviation can be made equal in a cantilever beam problem. Therefore, the deflection can be found from the second moment-area theorem only.

14.4 DEFLECTION OF A CANTILEVER BEAM BY THE MOMENT-AREA METHOD

A loaded cantilever beam and the deflection curve for the beam is shown in Fig. 14.8(a) and (b). Since the beam is clamped at B, the slope of the deflection curve is zero at B. Thus the tangent deviation of any point on

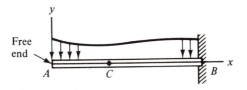

(a) Cantilever beam

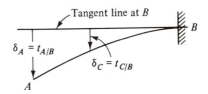

(b) Deflection curve

FIGURE 14.8

the beam from the tangent line at B is directly equal to the deflection of the beam at that point. That is,

$$t_{A/B} = \delta_A \text{ (the deflection at } A)$$

and

$$t_{C/B} = \delta_C \quad \text{(the deflection at } C\text{)}$$

EXAMPLE 14.5

Find the maximum deflection for the cantilever beam shown in Fig. 14.9(a).

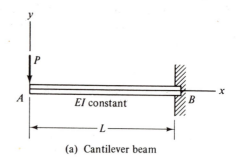

(a) Cantilever beam

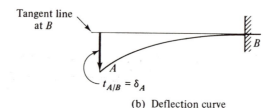

(b) Deflection curve

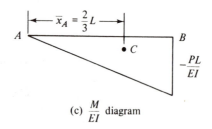

(c) $\dfrac{M}{EI}$ diagram

FIGURE 14.9

Solution

The maximum deflection will occur at the free end A. The deviation of A from a tangent line at B, $t_{A/B}$, is equal to the maximum deflection δ_A [Fig. 14.9(b)]. The M/EI diagram between A and B and the centroidal distance measured from a vertical line through A are shown in Fig. 14.9(c). The required calculations are shown in the table. From the second moment-area theorem, the tangent deviation $t_{A/B}$ is equal to the first moment about a vertical line through A of the M/EI diagram between A and B. That is,

$$t_{A/B} = \bar{x}_A A^* \text{ (from } A \text{ to } B)$$

From the table,

$$t_{A/B} = -\frac{PL^3}{3EI}$$

The maximum deflection $\delta_{max} = t_{A/B}$; therefore,

$$\delta_{max} = -\frac{PL^3}{3EI}$$

The negative sign indicates that the deflection is down.

Solution Table for Examples 14.5 and 14.6

A^* (from A to B)	$\bar{x}_A$	$\bar{x}_A A^*$ (A to B)
$\dfrac{L}{2}\left(-\dfrac{PL}{EI}\right) = -\dfrac{PL^2}{2EI}$	$\dfrac{2L}{3}$	$-\dfrac{PL^3}{3EI}$
$\theta_B - \theta_A$		$t_{A/B}$

EXAMPLE 14.6

Find the slope at the free end of the cantilever beam in Example 14.5.

Solution

From the first moment-area theorem, the change in slope between A and B is equal to the area of the M/EI diagram between A and B. That is,

$$\theta_B - \theta_A = A^* \text{ (from } A \text{ to } B)$$

From the table,

$$\theta_B - \theta_A = -\frac{PL^2}{2EI}$$

The beam is clamped or fixed at B; therefore, $\theta_B = 0$ and

$$\theta_A = \frac{PL^2}{2EI} \measuredangle \theta_A \qquad\qquad \text{Answer}$$

EXAMPLE 14.7

Find the slope at the free end and the maximum deflection for the cantilever beam shown in Fig. 14.10(a). $E = 30 \times 10^6$ psi and $I = 204$ in.^{4}.

Solution

The deflection curve and M/EI diagram are shown in Fig. 14.10. The calculations are tabulated as shown.

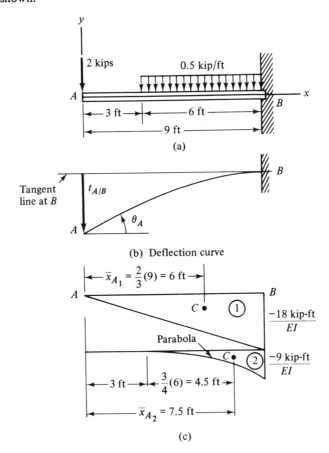

FIGURE 14.10

From the first moment-area theorem,

$$\theta_B - \theta_A = A^* \text{ (from } A \text{ to } B)$$

Tabulated in column 1 we have

$$\theta_B - \theta_A = -\frac{99 \text{ kip-ft}^2}{EI} = -\frac{1.426(10^7) \text{ lb-in.}^2}{EI}$$

	1	2	3
	A^* (from A to B) $(kip\text{-}ft^2)/EI$	$\bar{x}_A$ (ft)	$\bar{x}_A A^*$ (A to B) $(kip\text{-}ft^3)/EI$
①	$\dfrac{9(-18)}{2EI} = -\dfrac{81}{EI}$	6	$-\dfrac{486}{EI}$
②	$\dfrac{6(-9)}{3EI} = -\dfrac{18}{EI}$	7.5	$-\dfrac{135}{EI}$
	$\Sigma -\dfrac{99}{EI}$ $\theta_B - \theta_A$		$\Sigma -\dfrac{621}{EI}$ $t_{A/B}$

The flexural stiffness $EI = 30(10^6)(204) = 6.12(10^9)$ lb-in.^{2} and the slope $\theta_B = 0$. Therefore, the slope at the free end of the cantilever beam

$$\theta_A = \frac{1.426(10^7)}{6.12(10^9)} = 0.00233 \text{ rad}$$

or

$$\theta_A = (0.00233 \text{ rad}) \frac{180°}{\pi \text{ rad}} = 0.1335° \measuredangle \theta_A \qquad \text{Answer}$$

From the second moment-area theorem,

$$t_{AB} = \bar{x}_B A^* \text{ (from A to B)}$$

Tabulated in column 3 we have

$$t_{AB} = -\frac{621 \text{ kip-ft}^3}{EI} = -\frac{1.073(10^9) \text{ lb-in.}^3}{EI}$$

The flexural stiffness $EI = 6.12(10^9)$ lb-in.^{2} and $\delta_{max} = t_{A/B}$; therefore,

$$\delta_{max} = -\frac{1.073(10^9) \text{ lb-in.}^3}{6.12(10^9) \text{ lb-in.}^2}$$

or

$$\delta_{max} = -0.1753 \text{ in.} \qquad \text{Answer}$$

The negative sign indicates that the deflection is down.

PROBLEMS

14.1 through 14.10 Use the moment-area method to find (a) the maximum deflection, and (b) the slope at the free end of the cantilever beam shown in the figure. Draw M/EI diagram by parts, assume that EI is constant, and neglect the weight of the beam.

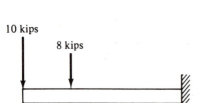

PROB. 14.1

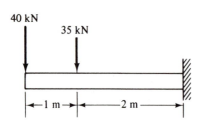

PROB. 14.2

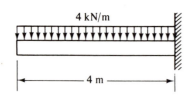

PROB. 14.3

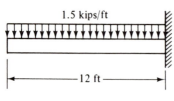

PROB. 14.4

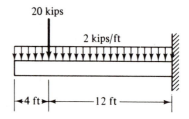

PROB. 14.5

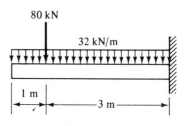

PROB. 14.6

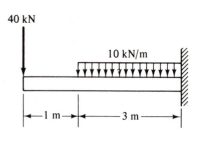

PROB. 14.7

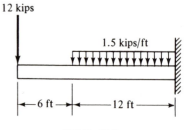

PROB. 14.8

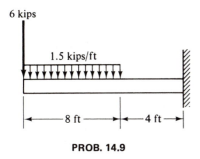

PROB. 14.9

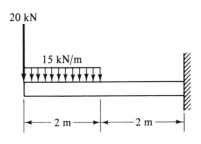

PROB. 14.10

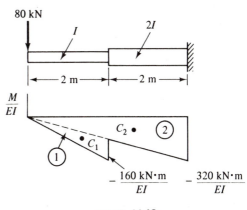

PROB. 14.12

14.11 through 14.14 Use the moment-area method to find (a) the maximum deflection, and (b) the slope at the free end of the cantilever beam shown in the figure. Neglect the weight of the beam. (*Hint:* Draw the M/EI diagram by parts as shown.)

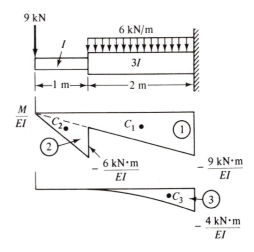

PROB. 14.13

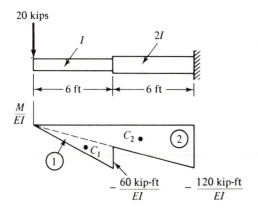

PROB. 14.11

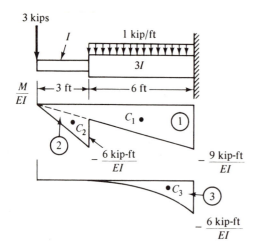

PROB. 14.14

14.15 Determine by the moment-area method
(a) the maximum deflection, and (b)

the slope at the free end of the cantilever beam shown. Assume that $E =$
30×10^6 psi and neglect the weight
of the beam.

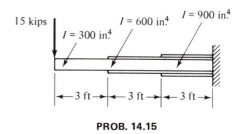

PROB. 14.15

14.16 Determine by the moment-area method
(a) the maximum deflection, and (b)
the slope at the free end of the cantilever beam shown. Assume that $E =$
200×10^6 kN/m^2 (kPa) and neglect
the weight of the beam.

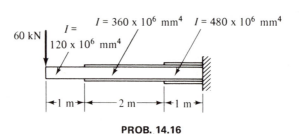

PROB. 14.16

14.5 DEFLECTION OF THE SIMPLY SUPPORTED BEAM
BY THE MOMENT-AREA METHOD

For a simply supported beam with the loads all directed either down or
up, we will calculate the deflection at the middle of the beam rather than
the maximum value of the deflection. In most cases the deflection at the
middle is not substantially different from the maximum deflection and is
much easier to calculate.

Consider the loaded simply supported beam [Fig. 14.11(a)] and
its deflection curve [Fig. 14.11(b)]. The deviation of M from a tangent
line drawn at A, $t_{M/A}$, and the deviation of B from a tangent line drawn at
A, $t_{B/A}$, are shown in the figure. They can both be calculated from the
second moment-area theorem. The *magnitude* of the deflection at the
middle of the beam

$$|\delta_M| = ST - t_{M/A}$$

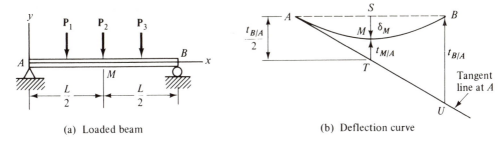

(a) Loaded beam   (b) Deflection curve

FIGURE 14.11

From similar triangles ABU and AST, $ST = t_{B/A}/2$; therefore,

$$\delta_M = -\left(\frac{t_{B/A}}{2} - t_{M/A}\right) \tag{14.8}$$

where the negative sign indicates that the deflection is down.

EXAMPLE 14.8

Calculate the deflection at the middle of the 12-ft span of a simply supported beam with a concentrated load of 10 kips, 9 ft from one of the supports, as shown in Fig. 14.12(a).

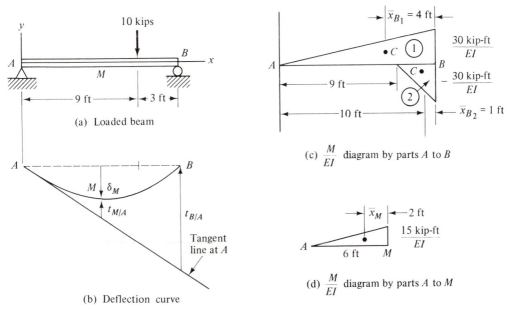

(a) Loaded beam

(b) Deflection curve

(c) $\dfrac{M}{EI}$ diagram by parts A to B

(d) $\dfrac{M}{EI}$ diagram by parts A to M

FIGURE 14.12

Solution

The deflection curve is shown in Fig. 14.12(b). To find the value of $t_{B/A}$, we draw the
M/EI diagram by parts in Fig. 14.12(c). The calculations are shown in the table. From
the sum in column 3,

$$t_{B/A} = \frac{675 \text{ kip-ft}^3}{EI}$$

	1 A* (from A to B) (kip-ft²)/EI	2 $\bar{x}_B$ (ft)	3 $\bar{x}_B A^*$ (A to B) (kip-ft³)/EI
①	$\dfrac{12(30)}{2EI} = \dfrac{180}{EI}$	4	$\dfrac{720}{EI}$
②	$\dfrac{3(-30)}{2EI} = -\dfrac{45}{EI}$	1	$-\dfrac{45}{EI}$
			$\Sigma \dfrac{675}{EI}$ $t_{B/A}$

To determine the value of $t_{B/M}$, we draw the M/EI diagram by parts in Fig. 14.12(d).
Calculating the value of $t_{M/A}$, we have

$$t_{M/A} = \bar{x}_M A^* \text{ (from } A \text{ to } M) = 2\left(\frac{6}{2}\right)\frac{15}{EI}$$

$$= \frac{90 \text{ kip-ft}^3}{EI}$$

Therefore, from Eq. (14.8),

$$\delta_M = -\left(\frac{t_{B/A}}{2} - t_{M/A}\right) = -\left(\frac{675}{2EI} - \frac{90}{EI}\right)$$

$$= -\frac{247.5 \text{ kip-ft}^3}{EI} = -\frac{4.277(10^8) \text{ lb-in.}^3}{EI} \qquad \text{Answer}$$

The formula for the maximum deflection [Table 14.2(e) of Sec. 14.6] is

$$\delta_{max} = -\frac{Pb(L^2 - b^2)^{3/2}}{9\sqrt{3}\,LEI}$$

For $P = 10,000$ lb, $L = 144$ in., and $b = L/4 = 36$ in., the maximum deflection

$$\delta_{max} = - \frac{4.348(10^8)\ \text{lb-in.}^3}{EI}$$

The deflection at the middle of the beam and the maximum deflection differ by approximately 1.6 percent.

EXAMPLE 14.9

A simply supported beam with a span of 12 ft has a concentrated load of 10 kips and a distributed load of 1.5 kips/ft, as shown in Fig. 14.13(a). The reactions have been calculated and the moment diagram drawn by parts [Fig. 14.13(b)]. Find the deflection at the middle of the beam. $E = 30 \times 10^3$ ksi, $I = 240$ in.^{4}.

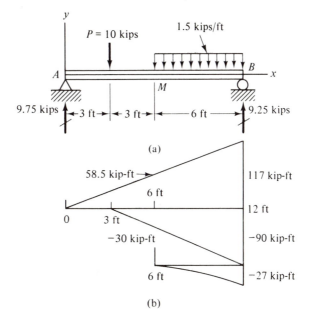

FIGURE 14.13

Solution

To find the tangent deviation of B from A, we construct the M/EI diagram [Fig. 14.14(a)] and tabulate the calculations as shown. The calculations for the tangent deviation of M from A are shown in Fig. 14.14(b) and tabulated as shown. The deflection at the middle of the beam is given by Eq. (14.7):

$$\delta_M = - \left(\frac{t_{B/A}}{2} - t_{M/A} \right)$$

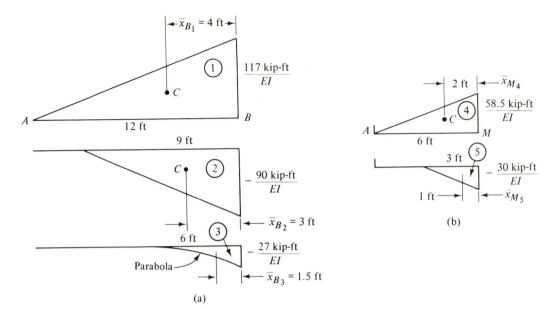

(a)

FIGURE 14.14

Substituting for $t_{B/A}$ and $t_{M/A}$ the sums shown in the tables, we have

$$\delta_M = -\left(\frac{1512}{2EI} - \frac{306}{EI}\right) = -\frac{450 \text{ kip-ft}^3}{EI}$$

or

$$\delta_M = -\frac{7.78(10^5) \text{ kip-in.}^3}{EI}$$

	1	2	3
	A^* *(from A to B)* $(kip\text{-}ft^2)/EI$	$\bar{x}_B$ (ft)	$\bar{x}_B A^*$ *(A to B)* $(kip\text{-}ft^3)/EI$
①	$\dfrac{12(117)}{2EI} = \dfrac{702}{EI}$	4	$\dfrac{2808}{EI}$
②	$\dfrac{9(-90)}{2EI} = -\dfrac{405}{EI}$	3	$-\dfrac{1215}{EI}$
③	$\dfrac{6(-27)}{3EI} = -\dfrac{54}{EI}$	1.5	$-\dfrac{81}{EI}$
		Σ	$\dfrac{1512}{EI}$ $t_{B/A}$

	1	2	3
	A^* *(from A to M)* $(kip\text{-}ft^2)/EI$	$\bar{x}_M$ (ft)	$\bar{x}_M A^*$ *(A to M)* $(kip\text{-}ft^3)/EI$
④	$\dfrac{6(58.5)}{2EI} = \dfrac{175.5}{EI}$	2	$\dfrac{351}{EI}$
⑤	$\dfrac{3(-30)}{2EI} = -\dfrac{45}{EI}$	1	$-\dfrac{45}{EI}$
		Σ	$\dfrac{306}{EI}$ $t_{M/A}$

The flexural stiffness $EI = 30(10^3)(240) = 7.2(10^6)$ kip-in.^{2}; therefore,

$$\delta_M = -\frac{7.78(10^5) \text{ kip-in.}^3}{7.2(10^6) \text{ kip-in.}^2} = 0.108 \text{ in.} \qquad \text{Answer}$$

PROBLEMS

14.17 through 14.26 Use the moment-area method to find the deflection at the middle of the simply supported beam shown in the figure. Draw M/EI diagram by parts, assume that EI is constant, and neglect the weight of the beam.

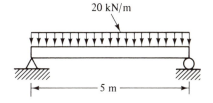

PROB. 14.20

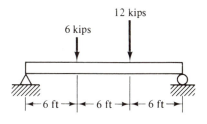

PROB. 14.17

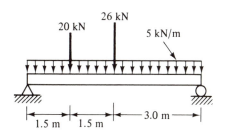

PROB. 14.21

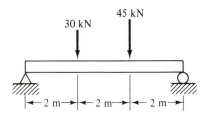

PROB. 14.18

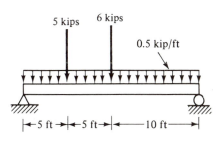

PROB. 14.22

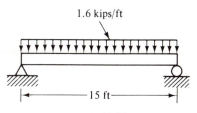

PROB. 14.19

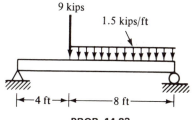

PROB. 14.23

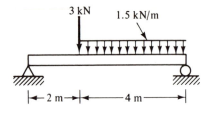

PROB. 14.24

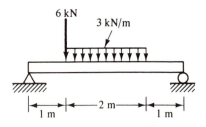

PROB. 14.25

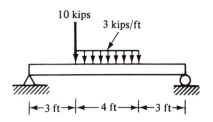

PROB. 14.26

14.27 through 14.29 Use the moment-area method to find the deflection at the middle of the simply supported beam shown in the figure. Neglect the weight of the beam. (*Hint:* Draw the *M/EI* diagram by parts as shown.)

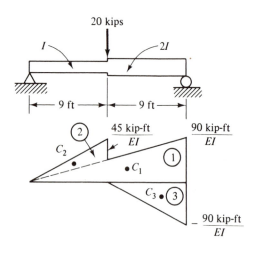

PROB. 14.28

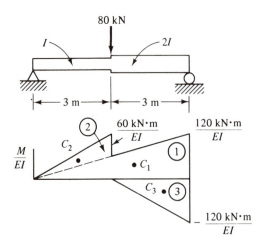

PROB. 14.27

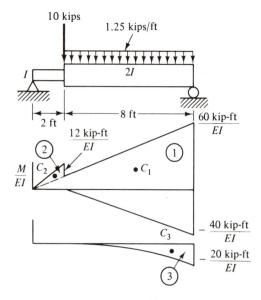

PROB. 14.29

14.30 Determine by the moment-area method
the deflection at the middle of the
simply supported beam shown in the
figure. Assume that $E = 30 \times 10^6$ psi
and neglect the weight of the beam.

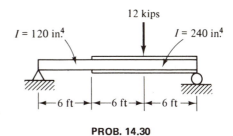

PROB. 14.30

14.6 THE SUPERPOSITION METHOD

Deflections, slopes, and deflection equations for various cantilever and
simply supported beams are shown in Table 14.2 All deflections and
slopes used in this section were taken from the table.

Consider the two identical cantilever beams shown in Fig.
14.15(a) and (b). One of the beams supports a concentrated load on the
free end and the other beam supports a uniformly distributed load over
the entire length. The maximum deflections came from diagrams (b) and
(c) of Table 14.2. In the superposition method we superimpose both
loads on a third identical beam. The reactions and deflections are equal to
the sum of the reactions and deflections of the individual beams [Fig.
14.15(c)]. The method is illustrated in the examples that follow.

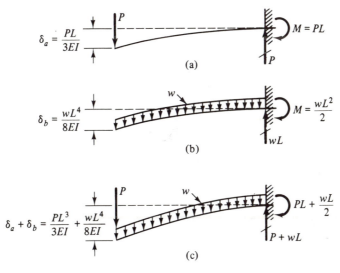

FIGURE 14.15

TABLE 14.2

(a) Cantilever beam – concentrated load P at any point.

$$\delta = \frac{Px^2}{6EI}(3a - x)$$
$$\text{for } x < a$$

$$\delta = \frac{Pa^2}{6EI}(3x - a)$$
$$\text{for } x > a$$

$$\delta_{max} = \frac{Pa^2(3L - a)}{6EI}$$

$$\theta_{max} = \frac{Pa^2}{2EI}$$

(b) Cantilever beam – concentrated load P at the free end.

$$\delta = \frac{Px^2}{6EI}(3L - x)$$

$$\delta_{max} = \frac{PL^3}{3EI}$$

$$\theta_{max} = \frac{PL^2}{2EI}$$

(c) Cantilever beam – uniformly distributed load of w force per unit length.

$$\delta = \frac{wx^2}{24EI}(x^2 + 6L^2 - 4Lx)$$

$$\delta_{max} = \frac{wL^4}{8EI}$$

$$\theta_{max} = \frac{wL^3}{6EI}$$

(d) Cantilever beam – couple M applied at the free end.

$$\delta = \frac{Mx^2}{2EI}$$

$$\delta_{max} = \frac{ML^2}{2EI}$$

$$\theta_{max} = \frac{ML}{EI}$$

δ is positive downward.

TABLE 14.2 (cont.)

(e) Beam freely supported at the ends – concentrated load at any point.

$$\delta = \frac{Pbx}{6LEI}(L^2 - x^2 - b^2) \quad \text{for } x < a$$

$$\delta = \frac{Pb}{6LEI}\left[\frac{L}{b}(x-a)^3 - x^3 + (L^2 - b^2)x\right]$$

$$\delta_M^* = \frac{Pa}{48EI}(3L^2 - 4a^2) \quad \text{for } a > b$$

$$\delta_M^* = \frac{Pb}{48EI}(3L^2 - 4b^2) \quad \text{for } b > a$$

$$\delta_{\max} = \frac{Pb(L^2 - b^2)^{3/2}}{9\sqrt{3}\,LEI}$$

$$\text{at } x = \sqrt{\frac{L^2 - b^2}{3}}$$

$$\theta_L = \frac{Pab(2L - a)}{6LEI}$$

$$\theta_R = \frac{Pab(2L - b)}{6LEI}$$

$R_L = \frac{Pb}{L}$ $R_R = \frac{Pa}{L}$ *Middle of beam

(f) Beam freely supported at ends – concentrated load P at the center.

$$\delta = \frac{Px}{48EI}(-4x^2 + 3L^2)$$
$$\text{for } x < \frac{L}{2}$$

$$\delta = \frac{Px}{48EI}\left[\frac{8}{x}\left(x - \frac{L}{2}\right)^3 - 4x^2 + 3L^2\right]$$
$$\text{for } x > \frac{L}{2}$$

$R_L = \frac{P}{2}$ $R_R = \frac{P}{2}$

$$\delta_{\max} = \frac{PL^3}{48EI}$$
$$\text{at } x = \frac{L}{2}$$

$$\theta_{\max} = \frac{PL^2}{16EI}$$

(g) Beam freely supported at the ends – uniformly distributed load of w force per unit length.

$$\delta = \frac{wx}{24EI}(L^3 - 2Lx^2 + x^3)$$

$R_L = \frac{wL}{2}$ $R_R = \frac{wL}{2}$

$$\delta_{\max} = \frac{5wL^4}{384EI}$$
$$\text{at } x = \frac{L}{2}$$

$$\theta_{\max} = \frac{wL^3}{24EI}$$

(h) Beam freely supported at the ends – couple M at the right end.

$$\delta = \frac{Mx}{6EIL}(L^2 - x^2)$$

$$\theta_L = \frac{ML}{6EI} \qquad \theta_R = \frac{ML}{3EI}$$

$R_L = \frac{M}{L}$ $R_R = \frac{M}{L}$

$$\delta_{\max} = \frac{ML^2}{9\sqrt{3}\,EI}$$
$$\text{at } x = \frac{L}{\sqrt{3}}$$

δ is positive downward.

EXAMPLE 14.10

Use the methods of superposition to find the deflection at the free end of the cantilever beam shown in Fig. 14.16.

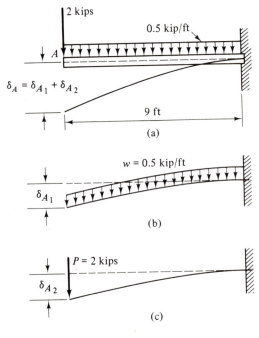

FIGURE 14.16

Solution

Two basic loadings are used as shown in diagrams (b) and (c) of Table 14.2. The deflection at A is equal to the sum of the deflections shown in the table. That is,

$$\delta_A = \frac{PL^3}{3EI} + \frac{wL^4}{8EI}$$

For $P = 2$ kips, $w = 0.5$ kip/ft, and $L = 9$ ft, the deflection

$$\delta_A = \frac{2(9)^3}{3EI} + \frac{0.5(9)^4}{8EI}$$

or

$$\delta_A = \frac{896 \text{ kip-ft}^3}{EI} \qquad\qquad \text{Answer}$$

EXAMPLE 14.11

Use the method of superposition to find the deflection at the middle of the simply supported beam shown in Fig. 14.17.

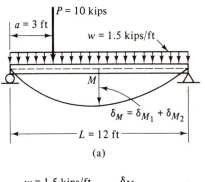

$P = 10$ kips

$a = 3$ ft

$w = 1.5$ kips/ft

M

$\delta_M = \delta_{M_1} + \delta_{M_2}$

$L = 12$ ft

(a)

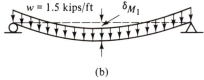

$w = 1.5$ kips/ft δ_{M_1}

(b)

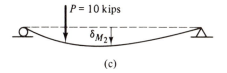

$P = 10$ kips

δ_{M_2}

(c)

FIGURE 14.17

Solution

Two basic loadings are used. See diagrams (e) and (g) of Table 14.2. The deflection at M is equal to the sum of the deflections shown in the table. That is,

$$\delta_M = \frac{Pa}{48EI} (3L^2 - 4a^2) + \frac{5wL^4}{384EI}$$

For $P = 10$ kips, $w = 1.5$ kips/ft, $a = 3$ ft, and $L = 12$ ft, the deflection

$$\delta_M = \frac{10(3)}{48EI} [3(12)^2 - 4(3)^2] + \frac{5(1.5)(12)^4}{384EI}$$

or

$$\delta_M = \frac{652 \text{ kip-ft}^3}{EI} \qquad \text{Answer}$$

EXAMPLE 14.12

Use the method of superposition to find the deflection at the middle of the simply supported beam shown in Fig. 14.18.

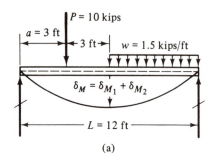

(a)

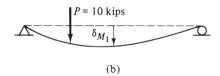

(b)

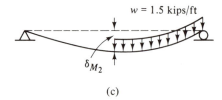

(c)

FIGURE 14.18

Solution

Two basic loadings are used. The concentrated load shown in diagram (e) of Table 14.2 can be used directly. However, the uniform load over half the beam will produce half the deflection of the uniform load over the entire beam shown in diagram (g) of Table 14.2. Therefore,

$$\delta_M = \frac{Pa}{48EI}(3L^2 - 4a^2) + \frac{5wL^4}{2(384)EI}$$

For $P = 10$ kips, $w = 1.5$ kips/ft, $a = 3$ ft, and $L = 12$ ft, the deflection

$$\delta_M = \frac{10(3)}{48EI}[3(12)^2 - 4(3)^2] + \frac{5(1.5)(12)^4}{2(384)EI}$$

or

$$\delta_M = \frac{450 \text{ kip-ft}^3}{EI}$$ Answer

This agrees with the results obtained in Example 14.9 by the moment-area method.

EXAMPLE 14.13

Use the method of superposition to find the deflection at the free end of the cantilever beam shown in Fig. 14.19(a).

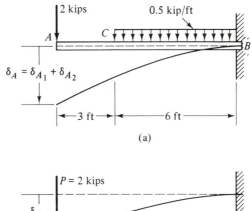

(a)

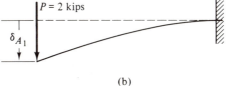

(b)

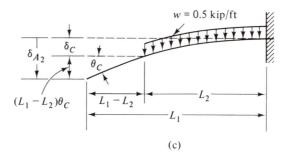

(c)

FIGURE 14.19

Solution

The concentrated load at the free end of the cantilever beam produces a deflection at A as shown in Fig. 14.19(b). The uniform loading produces a deflection at C and because the beam is continuous and has a slope at C an additional deflection occurs at A equal to $(L_1 - L_2)\theta_C$ as shown in Fig. 14.19(c). Therefore, the deflection

$$\delta_A = \delta_{A_1} + \delta_C + (L_1 - L_2)\theta_C \tag{a}$$

Deflections and slopes are given in Table 14.2. The deflection δ_A is shown in diagram (b) and the deflection δ_C and slope θ_C are shown in diagram (c). Substituting values from the diagrams into Eq. (a), we have

$$\delta_A = \frac{PL_1^3}{3EI} + \frac{wL_2^4}{8EI} + (L_1 - L_2)\frac{wL_2^3}{6EI}$$

For $P = 2$ kips, $w = 0.5$ kip/ft, $L_2 = 6$ ft, $L_1 = 9$ ft, and $(L_1 - L_2) = 3$ ft, the deflection

$$\delta_A = \frac{2(9)^3}{3EI} + \frac{0.5(6)^4}{8EI} + \frac{3(0.5)(6)^3}{6EI}$$

or

$$\delta_A = \frac{621 \text{ kip-ft}^3}{EI} \qquad\qquad \text{Answer}$$

This answer agrees with the results obtained in Example 14.7 by the moment-area method.

PROBLEMS

In Probs. 14.31 through 14.40, using Table 14.2 and the method of superposition, determine (a) the maximum deflection, and (b) the slope at the free end of the cantilever beam shown in the figure. Assume that EI is constant and neglect the weight of the beam.

14.31   Use the figure for Prob. 14.1.

14.32   Use the figure for Prob. 14.2.

14.33   Use the figure for Prob. 14.3.

14.34   Use the figure for Prob. 14.4.

14.35   Use the figure for Prob. 14.5.

14.36   Use the figure for Prob. 14.6.

14.37   Use the figure for Prob. 14.7.

14.38   Use the figure for Prob. 14.8.

14.39   Use the figure for Prob. 14.9.

14.40   Use the figure for Prob. 14.10.

In Probs. 14.41 through 14.46, using Table 14.2 and the method of superposition, determine the deflection at the middle of the simply supported beam shown in the figure. Assume that EI is constant and neglect the weight of the beam.

14.41   Use the figure for Prob. 14.17.

14.42   Use the figure for Prob. 14.18.

14.43   Use the figure for Prob. 14.19.

14.44   Use the figure for Prob. 14.20.

14.45   Use the figure for Prob. 14.21.

14.46   Use the figure for Prob. 14.22.

14.7 STATICALLY INDETERMINATE BEAMS BY THE SUPERPOSITION METHOD

Statically indeterminate beams, beams for which there are more reactions than can be determined by the equations of equilibrium, can be solved by the superposition method. The following examples will illustrate the method.

EXAMPLE 14.14

Determine the reactions for the statically indeterminate beam shown in Fig. 14.20(a).

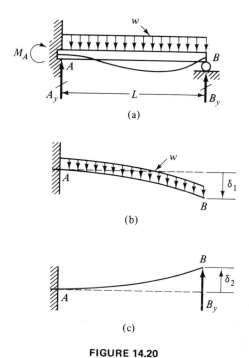

FIGURE 14.20

Solution

The loads on the statically indeterminate beam can be thought of as involving the uniform load w on a cantilever beam and the concentrated reaction or load B_y on the end of an identical cantilever beam [Fig. 14.20(b) and (c)]. The downward deflection δ_1 at B due to the uniform load must be equal to the upward deflection δ_2 at B due to the concentrated load. That is,

$$\delta_1 = \delta_2$$

393

From diagrams (b) and (c) of Table 14.2,

$$\frac{wL^4}{8EI} = \frac{B_y L^3}{3EI}$$

or

$$B_y = \frac{3}{8} wL \qquad\qquad \text{Answer}$$

With one reaction known, the other two can be found from the equilibrium equations as follows:

$$\circlearrowright \Sigma\, M_A = 0 \qquad -M_A - wL\left(\frac{L}{2}\right) + B_y L = 0$$

$$M_A = -\frac{wL^2}{2} + \frac{3wL^2}{8} = -\frac{wL^2}{8} \qquad\qquad \text{Answer}$$

$$\uparrow \Sigma\, F_y = 0 \qquad A_y - wL + B_y = 0$$

$$A_y = \frac{5}{8} wL \qquad\qquad \text{Answer}$$

Check:

$$\circlearrowright \Sigma\, M_B = 0 \qquad -M_A + wL\,\frac{L}{2} - A_y L = 0$$

$$\frac{wL^2}{8} + \frac{wL^2}{2} - \frac{5wL^2}{8} = 0 \quad \text{OK}$$

EXAMPLE 14.15

Determine the reaction for the statically indeterminate beam shown in Fig. 14.21(a).

Solution

The loads can be thought of as made up of a concentrated load at the middle of a simply supported beam and a couple on the left end of an identical simply supported beam, as shown in Fig. 14.21(b) and (c). The slope must be zero at A; therefore,

$$\theta_{A_1} + \theta_{A_2} = 0$$

From diagrams (f) and (h) of Table 14.2,

$$\frac{PL^2}{16EI} + \frac{M_A L}{3EI} = 0$$

$$M_A = -\frac{3}{16} PL \qquad\qquad \text{Answer}$$

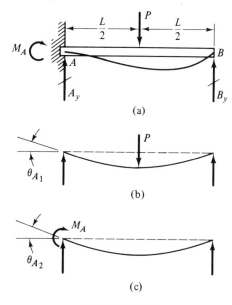

FIGURE 14.21

With one of the reactions known, the other two can be found from the equilibrium equations as follows:

$$\circlearrowright \Sigma \, M_A = 0 \qquad -M_A - P\frac{L}{2} + B_y L = 0$$

$$B_y = -\frac{3}{16}P + \frac{P}{2} = \frac{5P}{16} \qquad \text{Answer}$$

$$\circlearrowright \Sigma \, M_B = 0 \qquad -M_A - A_y L + P\frac{L}{2} = 0$$

$$A_y = \frac{3}{16}P + \frac{P}{2} = \frac{11P}{16} \qquad \text{Answer}$$

Check:

$$\uparrow \Sigma \, F_y = 0 \qquad A_y - P + B_y = 0$$

$$\frac{5}{16}P - P + \frac{11}{16}P = 0 \qquad \text{OK}$$

An alternative solution, consisting of two cantilever beams that are fixed at A with the deflection at B made equal to zero, as in Example 14.14, is also possible.

EXAMPLE 14.16

Determine the reaction for the statically indeterminate beam shown in Fig. 14.22(a).

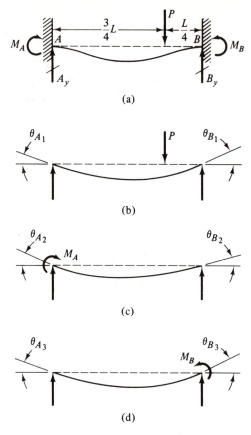

FIGURE 14.22

Solution

The beam fixed on both ends has a zero slope at A and B. By adding the three simple beams shown in Fig. 14.22(b)-(d) and requiring that

$$\theta_{A_1} + \theta_{A_2} + \theta_{A_3} = 0 \qquad \text{(a)}$$

and

$$\theta_{B_1} + \theta_{B_2} + \theta_{B_3} = 0 \qquad \text{(b)}$$

we can determine the value of the reactions M_A and M_B. From diagram (e) of Table 14.2 with $a = 3L/4$ and $b = L/4$,

$$\theta_{A_1} = \frac{5PL^2}{128EI} \quad \text{and} \quad \theta_{B_1} = \frac{7PL^2}{128EI}$$

The other angles can be found directly from diagram (h) of Table 14.2. Substituting values of θ_A into Eq. (a),

$$\frac{5PL^2}{128EI} + \frac{M_A L}{3EI} + \frac{M_B L}{6EI} = 0 \qquad \text{(c)}$$

and the values of θ_B into Eq. (b), we obtain

$$\frac{7PL^2}{128EI} + \frac{M_A L}{6EI} + \frac{M_B L}{3EI} = 0 \qquad \text{(d)}$$

Multiplying Eqs. (c) and (d) by $384\,EI/L$, they reduce to

$$15PL + 128M_A + 64M_B = 0$$
$$21PL + 64M_A + 128M_B = 0$$

Solving for M_B and M_A, we have

$$M_B = -\frac{9}{64}\,PL \quad \text{and} \quad M_A = -\frac{3}{64}\,PL \qquad \text{Answer}$$

From equilibrium:

$$\circlearrowleft \Sigma\, M_A = 0 \qquad -M_A - \frac{3PL}{4} + B_y L + M_B = 0$$

$$B_y = \frac{M_A}{L} + \frac{3PL}{4L} - \frac{M_B}{L}$$

Substituting for M_A and M_B, we obtain

$$B_y = -\frac{3P}{64} + \frac{3P}{4} + \frac{9P}{64} = \frac{54}{64}\,P \qquad \text{Answer}$$

From equilibrium:

$$\circlearrowleft \Sigma\, M_B = 0 \qquad -M_A - A_y L + \frac{PL}{4} + M_B = 0$$

$$A_y = -\frac{M_A}{L} + \frac{P}{4} + \frac{M_B}{L}$$

Substituting for M_A and M_B, we obtain

$$A_y = \frac{3P}{64} + \frac{P}{4} - \frac{9P}{64} = \frac{10}{64}\,P \qquad \text{Answer}$$

Check:

$$\uparrow \Sigma \, F_y = 0 \quad A_y - P + B_y = 0$$

$$\frac{10}{64} P - P + \frac{54}{64} P = 0 \quad \text{OK}$$

PROBLEMS

In Probs. 14.47 through 14.64, use the equations of equilibrium and Table 14.2 together with superposition to find the reactions for the statically indeterminate beam shown in the figure. Assume EI constant and neglect the weight of the beam.

14.47 Find the reactions for a uniform load w and span L as shown. Use (a) Table 14.2, diagrams (b) and (c); and (b) Table 14.2, diagrams (g) and (h).

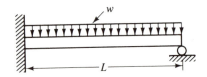

PROB. 14.47

14.48 Find the reactions for Prob. 14.47 if $w = 1.5$ kips/ft and $L = 12$ ft.

14.49 Find the reactions for Prob. 14.47 if $w = 20$ kN/m and $L = 4$ m.

14.50 Find the reactions for the beam shown if $a = 2L/3$ and $b = L/3$. Use (a) Table 14.2, diagrams (a) and (b); and (b) Table 14.2, diagrams (e) and (h).

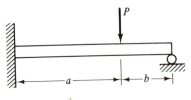

PROB. 14.50

14.51 Find the reactions for Prob. 14.50 if $P = 80$ kN, $a = 2$ m, and $b = 1$ m.

14.52 Find the reactions for Prob. 14.50 if $P = 20$ kips, $a = 12$ ft, and $b = 6$ ft.

14.53 Find the reaction for the beam shown if $a = 3L/4$ and $b = L/4$. Use (a) Table 14.2, diagrams (a), (b), and (c); and (b) Table 14.2, diagrams (e), (g), and (h).

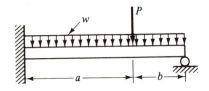

PROB. 14.53

14.54 Find the reactions for Prob. 14.53 if $P = 20$ kips, $w = 2$ kips/ft, $a = 12$ ft, and $b = 4$ ft.

14.55 Find the reactions for Prob. 14.53 if $P = 80$ N, $w = 32$ kN/m, $a = 3$ m, and $b = 1$ m.

14.56 Find the reactions for the beam shown if $a = 3L/5$ and $b = 2L/5$. Use (a) Table 14.2, diagrams (e) and (h); and (b) Table 14.2, diagrams (a), (b), and (d).

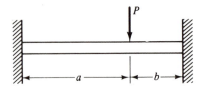

PROB. 14.56

14.57 Find the reactions for Prob. 14.56 if $P = 60$ kN, $a = 1.5$ m, and $b = 1$ m.

14.58 Find the reactions for Prob. 14.56 if $P = 25$ kips, $a = 6$ ft, and $b = 4$ ft.

14.59 Find the reactions for the beam shown. Use (a) Table 14.2, diagrams (b), (c), and (d); and (b) Table 14.2, diagrams (g) and (h).

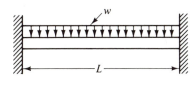

PROB. 14.59

14.60 Find the reactions for Prob. 14.59 if $w = 4$ kips/ft and $L = 12$ ft.

14.61 Find the reactions for Prob. 14.59 if $w = 30$ kN/m and $L = 4$ m.

14.62 Find the reactions for the beam shown if $a = L/4$ and $b = 3L/4$. Use (a) Table 14.2, diagrams (e), (g), and (h); and (b) Table 14.2, diagrams (a), (b), (c), and (d).

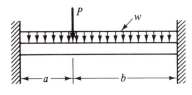

PROB. 14.62

14.63 Find the reactions for Prob. 14.62 if $P = 30$ kips, $w = 2$ kips/ft, $a = 3$ ft, and $b = 9$ ft.

14.64 Find the reactions for Prob. 14.62 if $P = 50$ kN, $w = 15$ kN/m, $a = 1$ m, and $b = 3$ m.

15

Combined Stresses—
Mohr's Circle _____

15.1 INTRODUCTION

The basic formulas for calculating the stress in a member were derived in
previous chapters. In deriving the formulas, it was assumed that a single
internal reaction was acting at a cross section of the member and that the
member was made from material for which stress was proportional to
strain. The basic formulas are listed for reference in the table on page 401:

In this chapter we consider problems in which two or more
internal reactions act at a cross section of the member. The stress from
each internal reaction will be superimposed or added together to find the
resulting stresses. The use of the superposition method is valid if the pres-
ence of one internal reaction does not affect the stresses due to another.
Only problems where superposition is valid will be considered in the
following sections.

Internal Reactions	Stress	Equation No.
1. Axial force (P)	$\sigma = \dfrac{P}{A}$	(9.3)
2. Bending moment (M)	$\sigma = \dfrac{My}{I}$	(13.5)
	$\sigma_{max} = \dfrac{Mc}{I}$	(13.4)
3. Torsion (T)		
(For circular bar)	$\tau = \dfrac{Tr}{J}$	(11.8)
(For circular bar)	$\tau_{max} = \dfrac{Tc}{J}$	(11.9)
4. Shear force (V)	$\tau = \dfrac{VQ}{Ib}$	(13.8)
(For rectangular bar)	$\tau_{max} = \dfrac{3V}{2A}$	(13.9)

15.2 AXIAL FORCES AND BENDING MOMENTS

Axial forces and bending moments produce normal stresses on the transverse cross section of a bar. These stresses are

$$\sigma_1 = \pm \frac{P}{A} \quad \text{and} \quad \sigma_2 = \pm \frac{My}{I}$$

The stresses are both perpendicular to the cross section and can be added algebraically. Therefore, we write

$$\sigma = \sigma_1 + \sigma_2 = \pm \frac{P}{A} \pm \frac{My}{I} \tag{15.1}$$

The stresses produced by a tensile force and a positive moment are shown in Fig. 15.1(b). The stresses at the top and bottom of the beam are given by

$$\sigma_T = \frac{P}{A} - \frac{Mc_T}{I} \quad \text{and} \quad \sigma_B = \frac{P}{A} + \frac{Mc_B}{I}$$

The positive sign indicates tension and the negative sign indicates compression. The combined neutral axis has been shifted to point N above the centroid of the cross section. The maximum stress in this case occurs at the bottom of the bar.

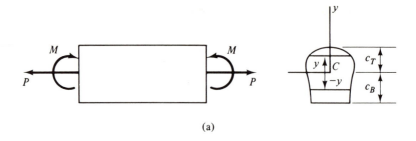

(a)

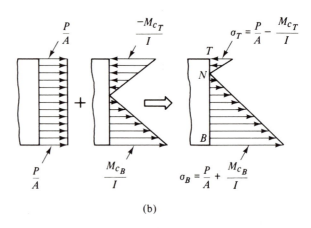

(b)

FIGURE 15.1

EXAMPLE 15.1

A rectangular bar with a cross section 2 in. by 8 in. is subject to an axial tensile force of 20 kips and a bending moment of 60 kip-in., as shown in Fig. 15.2(a). Determine the stress distribution across the bar.

Solution

The cross-sectional area of the bar $A = bh = 2(8) = 16$ in.^{2} and the moment of inertia with respect to the bending neutral axis $I = bh^3/12 = 2(8)^3/12 = 85.33$ in.^{4}. The axial stress $\sigma_1 = P/A = 20,000/16 = 1250$ psi and the maximum bending stress $\sigma_2 = Mc/I = 60,000(4)/85.3 = 2810$ psi. Because the bending moment is positive, the stress at the top of the bar $\sigma = \sigma_1 - \sigma_2 = 1250 - 2810 = -1560$ psi and at the bottom of the bar $\sigma = \sigma_1 + \sigma_2 = 1250 + 2810 = 4060$ psi. The combined neutral axis is located above the centroid of the cross section where the axial stress and bending stress are equal. That is,

$$\frac{P}{A} = \frac{My}{I}$$

$$y = \frac{PI}{AM} = \frac{20,000(85.3)}{16(60,000)} = 1.78 \text{ in.}$$

402

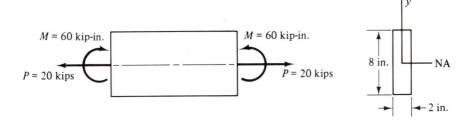

(a)

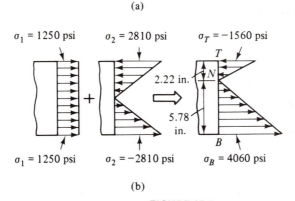

$\sigma_1 = 1250$ psi $\qquad \sigma_2 = 2810$ psi $\qquad \sigma_T = -1560$ psi

2.22 in.

5.78 in.

$\sigma_1 = 1250$ psi $\qquad \sigma_2 = -2810$ psi $\qquad \sigma_B = 4060$ psi

(b)

FIGURE 15.2

Thus the neutral axis is $4.0 - 1.78 = 2.22$ in. from the top of the bar. The distribution of stresses and location of the combined neutral axis are shown in Fig. 15.2(b).

EXAMPLE 15.2

A simply supported I-beam with span, cross section, and loads is shown in Fig. 15.3(a) and (b). Find the stress distribution in the beam at the cross section of maximum tensile and compressive stress.

Solution

The I-shaped cross section can be thought of as made up of a positive rectangle with base $b = 0.250$ m and height $h = 0.250$ m and two negative rectangles with bases $b = 0.120$ m and heights $h = 0.220$ m. Thus the area $A = 0.250 (0.250) - 2 (0.120)(0.220) = 9.7 \times 10^{-3}$ m^2 and the moment of inertia with respect to the bending neutral axis.

$$I_{NA} = \frac{0.250(0.250)^3}{12} - \frac{2(0.120)(0.220)^3}{12} = 112.6 \times 10^{-6} \text{ m}^4$$

The axial stress

$$\sigma_1 = \frac{P}{A} = \frac{300}{9.7 \times 10^{-3}} = 30.9 \times 10^3 \ \frac{\text{kN}}{\text{m}^2}$$

$$= 30.9 \ \frac{\text{MN}}{\text{m}^2} \ \text{(MPa)}$$

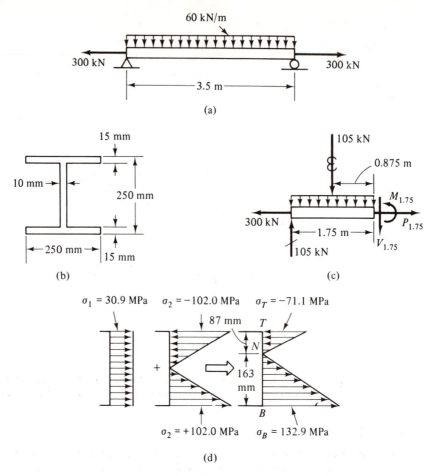

FIGURE 15.3

To find the bending stress, we must determine the maximum bending moment. The maximum bending moment acts on a cross section at the middle of the beam. From the free body diagram shown in Fig. 15.3(c), we write

$$M_{1.75} = 105\,(1.75) - 105\,(0.875)$$

$$= 91.9 \text{ kN} \cdot \text{m}$$

Therefore, the bending stress

$$\sigma_2 = \frac{Mc}{I} = \frac{91.9\,(0.125)}{112.6 \times 10^{-6}} = 102.0 \times 10^3 \frac{\text{kN}}{\text{m}^2}$$

$$= 102.0 \frac{\text{MN}}{\text{m}^2} \text{ (MPa)}$$

Because the bending moment is positive, the stress at the top of the beam $\sigma = \sigma_1 - \sigma_2 = 30.9 - 102.0 = -71.1$ MPa and at the bottom of the beam $\sigma = \sigma_1 + \sigma_2 = 30.9 +$

404

102.0 = 132.9 MPa. The combined neutral axis is located above the centroid of the cross section, where the axial stress and bending stress are equal. That is,

$$\frac{P}{A} = \frac{My}{I}$$

$$y = \frac{PI}{AM} = \frac{300(112 \times 10^{-6})}{(9.7 \times 10^{-3})(91.9)} = 0.0377 \text{ m}$$

Thus the neutral axis is 125 - 38 = 87 mm from the top of the beam. The distribution of stresses and location of the combined neutral axis are shown in Fig. 15.3(d).

EXAMPLE 15.3

Find the normal stress distribution on section *A-A* of the short post with eccentric load *P* = 15 kips and cross section as shown in Fig. 15.4(a) and (b).

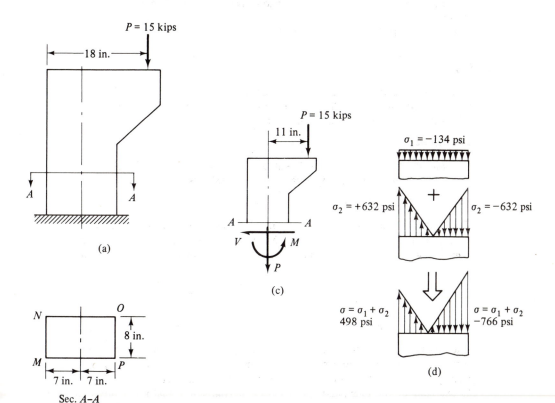

FIGURE 15.4

Solution

The cross-sectional area $A = 8(14) = 112$ in.^{2} and the moment of inertia $I_{NA} = 8(14)^3/12 = 1829$ in.^{4}. The internal reactions at section A–A are found from the free-body diagram [Fig. 15.4(c)] and the equations of equilibrium. The shear force $V = 0$, the axial force $P = -15$ kips, and the bending moment $M = 15(11) = 165$ kip-in. The axial stress

$$\sigma_1 = \frac{P}{A} = \frac{-15,000}{112} = -134 \text{ psi} \quad \text{(compression)}$$

The maximum bending stresses

$$\sigma_2 = \pm\frac{Mc}{I} = \pm\frac{165,000(7)}{1829} = \pm632 \text{ psi} \quad \text{(tension or compression)}$$

The axial and bending stresses are added together [Fig. 15.4(d)] to produce a tensile stress $\sigma = -134 + 632 = 498$ psi on side MN and a compressive stress $\sigma = -134 - 632 = -766$ psi on side OP of the cross section.

EXAMPLE 15.4

The C-clamp shown in Fig. 15.5(a) is tightened until a force $P = 2.5$ kN is exerted on an object in the clamp. Find the stress distribution in the clamp at the cross section of maximum tensile and compressive stress.

Solution

The maximum bending moment will occur on section A–A. To find the maximum bending moment, we begin by locating the neutral axis of the cross section.

Location of Neutral Axis

The T-shaped cross section [Fig. 15.5(b)] can be divided into area ①, the stem of the T, and area ②, the flange of the T, as shown in Fig. 15.5(c). The calculations for the location of the centroid are tabulated on page 408. From the sums in the table,

$$\bar{y} = \frac{\Sigma Ay}{\Sigma A} = \frac{8500}{500} = 17 \text{ mm}$$

Moment of Inertia

The moment of inertia with respect to the neutral axis for each area is found from the formula for the centroidal moment of inertia, $I_c = bh^3/12$, and the parallel-axis theorem $I_{NA} = I_c + Ad^2$. The composite moment of inertia with respect to the neutral axis is the sum of the individual moments of inertia. The transfer distances are shown in Fig. 15.5(d) and are tabulated on page 408. From the sums in the table,

$$I_{NA} = \Sigma I_c + \Sigma (Ad^2) = 24.2 \times 10^3 + 48.0 \times 10^3 = 72.2 \times 10^3 \text{ mm}^4$$

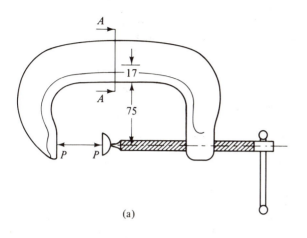

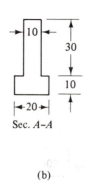

Sec. A–A

(a)   (b)

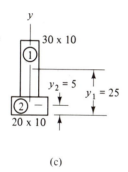

(c)

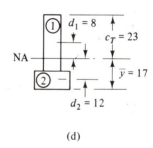

(d)

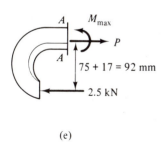

(e)

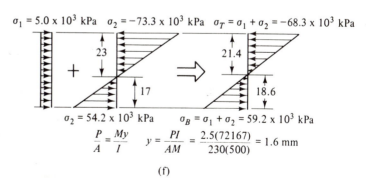

$\sigma_1 = 5.0 \times 10^3$ kPa $\sigma_2 = -73.3 \times 10^3$ kPa $\sigma_T = \sigma_1 + \sigma_2 = -68.3 \times 10^3$ kPa

$\sigma_2 = 54.2 \times 10^3$ kPa $\sigma_B = \sigma_1 + \sigma_2 = 59.2 \times 10^3$ kPa

$$\frac{P}{A} = \frac{My}{I} \qquad y = \frac{PI}{AM} = \frac{2.5(72167)}{230(500)} = 1.6 \text{ mm}$$

(f)

(All dimensions in mm)

FIGURE 15.5

	b	h	A	y	Ay
①	10	30	300	25	7500
②	20	10	200	5	1000
			$\Sigma A = 500$ mm^2		$\Sigma Ay = 8500$ mm^2

	A	d	I_c	Ad^2
①	300	8	22.5×10^3	19.2×10^3
②	200	-12	1.67×10^3	28.8×10^3
			$\Sigma I_c = 24.17 \times 10^3$ mm^4	$\Sigma (Ad^2) = 48.0 \times 10^3$ mm^4

Internal Reactions

From the free-body diagram of Fig. 15-5(e) and the equations of equilibrium, we write $\Sigma F_x = -2.5 + P = 0$ or $P = 2.5$ kN (tension) and

$$\rotatebox{0}{\supset} \Sigma M_c = -2.5(92) + M_{max} = 0 \quad \text{or} \quad M_{max} = 230 \text{ kN} \cdot \text{mm}$$

Axial and Bending Stresses

The axial stress

$$\sigma_1 = \frac{P}{A} = \frac{2.5}{500} = 0.005 \frac{\text{kN}}{\text{mm}^2}$$

or

$$\sigma_1 = 5.0 \times 10^3 \frac{\text{kN}}{\text{m}^2} \text{ (kPa) (tension)}$$

The bending stress at the top of the cross section

$$\sigma_2 = \frac{-Mc_T}{I} = \frac{-230(23)}{72.2 \times 10^3} = -73.3 \times 10^{-3} \frac{\text{kN}}{\text{mm}^2}$$

or

$$\sigma_2 = -73.3 \times 10^3 \frac{\text{kN}}{\text{m}^2} \text{ (kPa) (compression)}$$

The bending stress at the bottom of the cross section

$$\sigma_2 = \frac{Mc_B}{I} = \frac{230(17)}{72.2 \times 10^3} = 54.2 \times 10^{-3} \ \frac{kN}{mm^2}$$

or

$$\sigma_2 = 54.2 \times 10^3 \ \frac{kN}{m^2} \ (kPa) \quad (\text{tension})$$

Stress Distribution

The stress distribution is shown in Fig. 15.5(f), where the stress at the top of the beam

$$\sigma_T = \sigma_1 - \sigma_2 = (5.0 - 73.3) \times 10^3 \ \frac{kN}{m^2}$$

or

$$\sigma_T = -68.3 \times 10^3 \ kPa \quad (\text{compression})$$

and the stress at the bottom of the beam

$$\sigma_B = \sigma_1 + \sigma_2 = (5.0 + 54.2) \times 10^3 \ \frac{kN}{m^2}$$

or

$$\sigma_B = 59.2 \times 10^3 \ kPa \quad (\text{tension})$$

PROBLEMS

15.1 A steel bracket is loaded as shown. (a) Determine the maximum tensile and compressive stress, and (b) plot the distribution of normal stress along *A-B*.

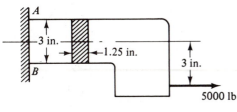

PROB. 15.1

15.2 A C-shaped machine part has dimensions and loads as shown in the figure.

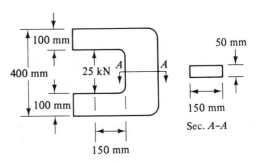

PROB. 15.2

For cross section *A–A*, (a) calculate the maximum tensile and compressive stress, and (b) plot the distribution of normal stress.

15.3 A bracket has dimensions and is loaded as shown. (a) Determine the maximum tensile and compressive stress, and (b) plot the distribution of normal stress, along *A–B*.

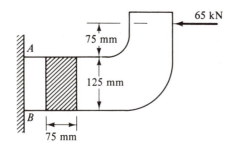

65 kN

75 mm

125 mm

75 mm

A

B

PROB. 15.3

15.4 A horizontal beam consisting of two 7 × 9.8 channels back to back are loaded and supported as shown in the figure. Calculate the maximum tensile and compressive stress on a cross section (a) to the right of *B*, and (b) to the left of *B*. See Table A.7 in the Appendix for properties of the C7 × 9.8.

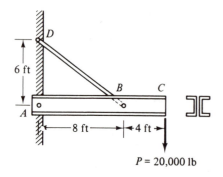

D

6 ft

B   *C*

A

8 ft — 4 ft

P = 20,000 lb

PROB. 15.4

15.5 An L-shaped bracket is supported and loaded as shown. For cross section *B–B*, (a) determine the maximum

tensile and compressive stress, and (b) plot the distribution of normal stress.

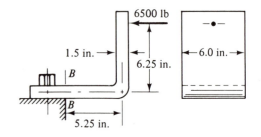

6500 lb

1.5 in.

6.25 in.

6.0 in.

B

B

5.25 in.

PROB. 15.5

15.6 A davit supports a load *P* = 48 kN as shown in the figure. For cross section *A–A*, (a) calculate the maximum tensile and compressive stress, and (b) plot the distribution of normal stress.

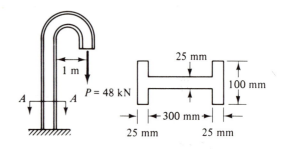

1 m

A   *A*

P = 48 kN

25 mm

100 mm

300 mm

25 mm   25 mm

Sec. *A–A*

PROB. 15.6

15.7 The C-clamp shown in the figure is tightened until the force *P* = 6500 N. For cross section *A–A*, (a) determine the maximum tensile and compressive

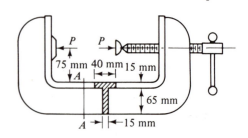

P   *P*

75 mm 40 mm 15 mm

A

65 mm

A — 15 mm

PROB. 15.7

stress, and (b) plot the distribution of normal stress.

15.8 A machine member has dimensions and a load as shown. For cross section A–B, (a) determine the maximum tensile and compressive stress, and (b) plot the distribution of normal stress.

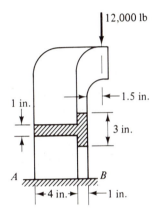

PROB. 15.8

15.9 Two forces act on the open link of a chain as shown in the figure. For cross section A–A, (a) calculate the maximum tensile and compressive stress, and (b) plot the distribution of normal stress.

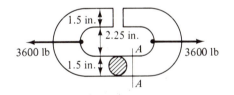

PROB. 15.9

15.10 A 75-mm-diameter round bar is formed into a machine part with a shape as shown. For cross section B–B, (a) determine the maximum tensile and compressive stress, and (b) plot the distribution of normal stress.

15.11 A machined member has an axial load applied as shown in the figure. For

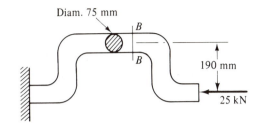

PROB. 15.10

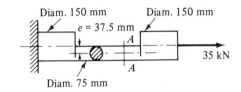

PROB. 15.11

cross section A–A, (a) determine the maximum tensile and compressive stress, and (b) plot the distribution of normal stress.

15.12 An offset link has dimensions and loads as shown. For cross section B–B, (a) find the maximum tensile and compressive stress, and (b) plot the distribution of normal stress.

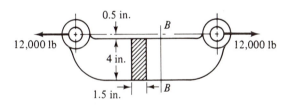

PROB. 15.12

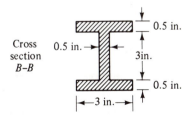

PROB. 15.13

15.13   Solve Prob. 15.12 with cross section *B–B* modified as shown in the figure.

15.14   A slotted offset link has dimensions and loads as shown in the figure. For cross section *A–A*, (a) determine the maximum tensile and compressive stress, and (b) plot the distribution of normal stress.

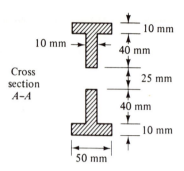

PROB. 15.15

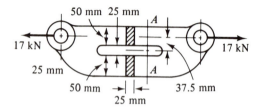

PROB. 15.14

15.15   Solve Prob. 15.14 with cross section *A–A* modified as shown in the figure.

15.16   A 2-in.-diameter bar is bent and loaded as shown. Determine the maximum tensile and compressive stress on a cross section at support *A*.

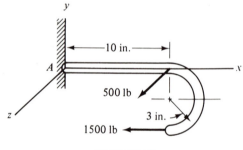

PROB. 15.16

15.3   UNSYMMETRICAL BENDING

Problems arise in which the axis of symmetry of the cross section of a beam is inclined at an angle with the vertical [Fig. 15.6(a)] or the load acts at an angle with the axis of symmetry [Fig. 15.6(b)]. These problems can be solved by general bending stress formulas. However, such formulas are complicated and can be avoided by the method of superposition.

      In the superposition method the load is resolved into components that are parallel and perpendicular to the y or symmetric axis. The component of the load parallel to the y axis produces bending about the z axis, M_z. The bending stresses produced are proportional to the distance from the z axis, y, and inversely proportional to the centroidal moment of inertia of the cross section with respect to the z axis. That is,

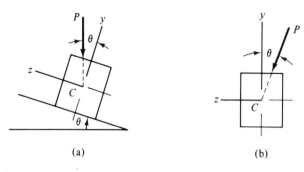

(a)   (b)

FIGURE 15.6

$$\sigma_1 = \pm \frac{M_z y}{I_z}$$

The component of the load perpendicular to the y axis produces bending about the y axis, M_y. The bending stresses produced are proportional to the distance from the y axis, z, and inversely proportional to the centroidal moment of inertia of the cross section with respect to the y axis. That is,

$$\sigma_2 = \pm \frac{M_y z}{I_y}$$

By superposition, the stress at any point of the cross section is equal to the sum of the stresses; that is,

$$\sigma = \sigma_1 + \sigma_2 = \pm \frac{M_z y}{I_z} \pm \frac{M_y z}{I_y} \tag{15.2}$$

The following example will illustrate the method.

EXAMPLE 15.5

A simply supported wooden beam with a cross section 4 in. by 6 in. and a span of 10 ft is used to support a uniformly distributed load of 0.1 kip/ft [Fig. 15.7(a)]. The beam cross section is inclined at an angle of $30°$ with the vertical plane [Fig. 15.7(b)]. Find the maximum bending stress and the neutral axis.

Solution

The maximum moment occurs at the middle of the beam and can be found from the free-body diagram [Fig. 15.7(c)] and the rotational equation of equilibrium. That is,

$$\circlearrowleft \Sigma M_c = -w \left(\frac{L}{2}\right)\left(\frac{L}{2}\right) + w \left(\frac{L}{2}\right)\left(\frac{L}{4}\right) + M_{\text{max}} = 0$$

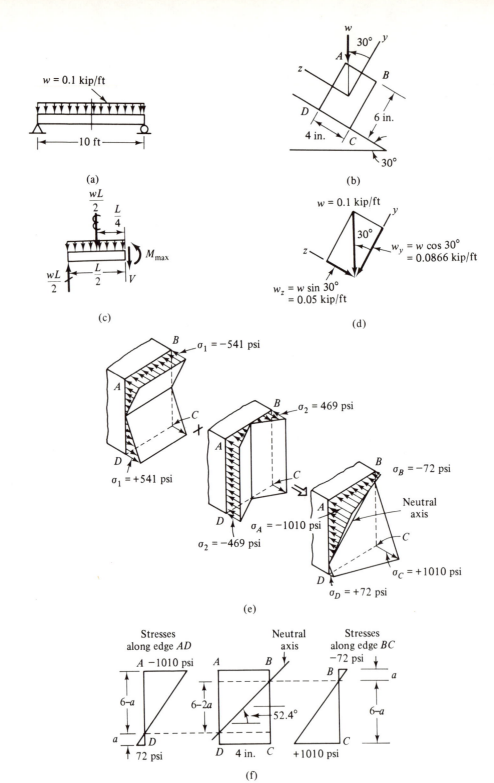

$w = 0.1$ kip/ft

10 ft

(a)

w

30°

y

A

z

B

D

4 in.

C

6 in.

30°

(b)

$\dfrac{wL}{2}$

$\dfrac{L}{4}$

M_{max}

$\dfrac{wL}{2}$

$\dfrac{L}{2}$

V

(c)

$w = 0.1$ kip/ft

30°

y

$w_y = w \cos 30°$
$= 0.0866$ kip/ft

z

$w_z = w \sin 30°$
$= 0.05$ kip/ft

(d)

B $\sigma_1 = -541$ psi

A

C

D

$\sigma_1 = +541$ psi

B $\sigma_2 = 469$ psi

A

C

D

$\sigma_A = -1010$ psi

$\sigma_2 = -469$ psi

B $\sigma_B = -72$ psi

Neutral axis

A

C

D

$\sigma_A = -1010$ psi

$\sigma_C = +1010$ psi

$\sigma_D = +72$ psi

(e)

Stresses along edge AD

A -1010 psi

$6-a$

a

D

72 psi

Neutral axis

A B

$6-2a$

$52.4°$

D 4 in. C

Stresses along edge BC

-72 psi

B

a

$6-a$

C

$+1010$ psi

(f)

FIGURE 15.7

or $M_{max} = wL^2/8$. From Fig. 15.7(d), we see that the components of the load parallel and perpendicular to the y axis are

$$w_y = w \cos \theta = 0.1 \cos 30° = 0.0866 \text{ kips/ft}$$

$$w_z = w \sin \theta = 0.1 \sin 30° = 0.0500 \text{ kips/ft}$$

Bending About the z Axis

The maximum bending moment and moment of inertia about the z axis are given by

$$M_z = \frac{w_y L^2}{8} = \frac{0.0866 (10)^2}{8} = 1.082 \text{ kip-ft}$$

$$I_z = \frac{bh^3}{12} = \frac{4 (6)^3}{12} = 72 \text{ in.}^4$$

Therefore, the maximum bending stress

$$\sigma_1 = \pm \frac{M_z y_{max}}{I_z} = \pm \frac{1082 (12) (3)}{72} = \pm 541 \text{ psi}$$

At A and B the stress is compression and at C and D the stress is tension as shown in Fig. 15.7(e).

Bending Moment About the y Axis

The maximum bending moment and moment of inertia about the y axis are given by

$$M_y = \frac{w_E L^2}{8} = \frac{0.05 (10)^2}{8} \quad \text{or} \quad M_y = 0.625 \text{ kip-ft}$$

$$I_y = \frac{bh^3}{12} = \frac{6 (4)^3}{12} = 32 \text{ in.}^4$$

Therefore, the bending stress

$$\sigma_2 = \pm \frac{M_y z_{max}}{I_y} = \pm \frac{625 (12) (2)}{32} = \pm 469 \text{ psi}$$

At A and D the stress is compression and at B and C the stress is tension, as shown in Fig. 15.7(e).

Combined Stresses

From superposition the stresses at A, B, C, and D are given by

$$\sigma_A = -541 - 469 = -1010 \text{ psi}$$

$$\sigma_B = -541 + 469 = -72 \text{ psi}$$

$$\sigma_C = +541 + 469 = +1010 \text{ psi}$$

and

$$\sigma_D = +541 - 469 = +72 \text{ psi}$$

The combined stresses are shown in Fig. 15.7(e).

Neutral Axis

The neutral axis can be located by finding the points of zero stress on the sides AD and BC of the beam. The stress distribution on the two sides of the beam is shown in Fig. 15.7(f). From similar triangles we write $a/72 = (6 - a)/1010$ or $a = 0.399$ in. The slope of the neutral axis $\theta = \arctan (6 - 2a)/4 = 52.4°$, as shown in Fig. 15.7(f).

PROBLEMS

15.17 A horizontal W18 × 55 rolled steel beam 15 ft long supports a uniform load of 2 kips/ft. The cross section of the beam is inclined at a slope of 15°. Determine the maximum tensile and compressive stress in the beam if the ends are simply supported. See Table A.5 in the Appendix for properties of steel beams.

15.18 A horizontal simply supported W460 × 82 steel beam 4.5 m long supports a uniform load of 25 kN/m. The cross section of the beam is inclined at a slope of 20°. Determine the maximum tensile and compressive stress in the beam. See Table A.5 (SI Units) in the Appendix for properties of steel beams.

15.19 A simply supported W8 × 40 rolled steel beam 15 ft long supports a con-

centrated load at midspan. The load is applied through the centroid of the cross section of the beam. If the cross section of the beam is inclined at a slope of 15° and the allowable bending stress is 24 ksi, what load can the beam support? Neglect the weight of the beam. See Table A.5 in the Appendix for properties of steel beams.

15.20 A W200 × 60 rolled steel cantilever beam 1.5 m long supports a concentrated load which is applied at its free end. If the cross section of the beam is inclined at a slope of 15° and the allowable bending stress is 165 MPa, what load can the beam support? Neglect the weight of the beam. See Table A.5 (SI Units) in the Appendix for properties of steel beams.

15.4 ECCENTRICALLY LOADED MEMBERS

The eccentrically loaded member (Fig. 15.8) is a special case where the load produces both axial and bending stresses. The normal stress at any point on a cross section $ABCD$ of the member is found by superposition. Adding the axial stress and the bending stresses about both the y and z axes, we write

$$\sigma_x = \pm \frac{P}{A} \pm \frac{M_y z}{I_y} \pm \frac{M_z y}{I_z} \tag{a}$$

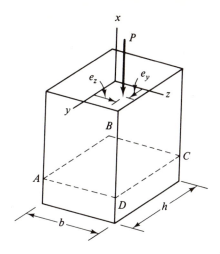

FIGURE 15.8

The moments of the load P about the y and z axes are $M_y = Pe_z$ and $M_z = Pe_y$. Substituting the moments into Eq. (a), we have

$$\sigma_x = \pm \frac{P}{A} \pm \frac{Pe_z z}{I_y} \pm \frac{Pe_y y}{I_z} \qquad \text{(b)}$$

Notice that the distribution of stress is the same for any cross section of the member.

EXAMPLE 15.6

A concrete pier (Fig. 15.8) with cross section $b = 24$ in. by $h = 20$ in. supports an axial compressive load $P = 120$ kips. The load is located at $e_y = 6$ in. and $e_z = 5$ in. on top of the member. Determine the normal stresses at the four corners of the pier.

Solution

Axial Load

The axial stress is compressive and is given by

$$\sigma_1 = -\frac{P}{A} = \frac{-120}{480} = -0.25 \text{ ksi}$$

Bending About the y Axis

The moment of inertia about the y axis is given by $I_y = hb^3/12 = 20(24)^3/12 = 23,040$ in.^{4}. Therefore, the maximum bending stress

$$\sigma_2 = \pm \frac{Pe_z z_{max}}{I_y} = \pm \frac{120(5)12}{23,040}$$

$$= \pm 0.3125 \text{ ksi}$$

The stresses at D and C are compressive and at A and B tensile.

Bending About the z Axis

The moment of inertia about the z axis is given by $I_z = bh^3/12 = 24(20)^3/12 = 16,000$ in.[4]. Therefore, the maximum bending stress

$$\sigma_3 = \pm \frac{Pe_y\, y_{max}}{I_z} = \pm \frac{120(6)(10)}{16,000}$$

$$= \pm 0.45 \text{ ksi}$$

The stresses at A and D are compressive and at B and C tensile.

Combined Stresses

From superposition the stresses at A, B, C, and D are added to give

$$\sigma_A = -0.25 + 0.3125 - 0.45 = -0.3875 \text{ ksi} \qquad \text{Answer}$$

$$\sigma_B = -0.25 + 0.3125 + 0.45 = +0.5125 \text{ ksi} \qquad \text{Answer}$$

$$\sigma_C = -0.25 - 0.3125 + 0.45 = -0.1125 \text{ ksi} \qquad \text{Answer}$$

$$\sigma_D = -0.25 - 0.3125 - 0.45 = -1.0125 \text{ ksi} \qquad \text{Answer}$$

PROBLEMS

15.21 A concrete pier 0.55 m by 0.6 m supports a load of 440 kN at point E as shown. Determine the stresses at the four corners of the pier and locate the neutral axis if point E is located (a) distances $e_y = 100$ mm and $e_z = 0$, and (b) distances $e_y = 100$ mm and $e_z = 125$ mm from the axes of symmetry of the cross section. (*Hint:* The neutral axis can be found by locating points of zero stress on two sides of the pier.)

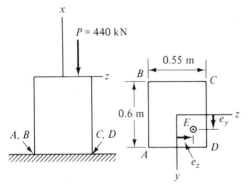

PROB. 15.21

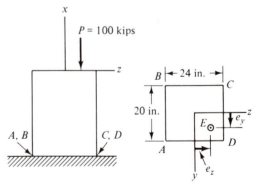

PROB. 15.22

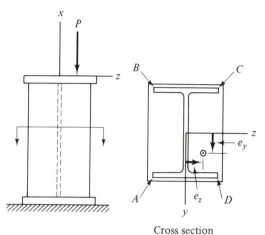

Cross section
PROB. 15.23 and PROB. 15.24

15.22 A concrete pier 20 in. by 24 in. supports a load of 100 kips at point E as shown. Determine the stresses at the four corners of the pier and locate the neutral axis if point E is located at (a) distances $e_y = 0$ and $e_z = 4$ in., and (b)

distances $e_y = 6$ in. and $e_z = 4$ in. from the axes of symmetry for the cross section. (*Hint:* The neutral axis can be found by locating points of zero stress on two sides of the pier.)

15.23 A W10 × 112 rolled steel section is used as a short compression member to support a load of $P = 125$ kips. Plates are welded to each end of the member. The load is applied at distances $e_y = 2$ in. and $e_z = 0.7$ in. in from the axes of symmetry as shown. Determine the stresses at corners A, B, C, and D of the flanges.

15.24 A W250 × 167 rolled steel section is used as a short compression member to support a load $P = 550$ kN. Plates are welded to each end of the member. The load is applied at distances $e_y = 50$ mm and $e_z = 15$ mm from the axes of symmetry as shown. Determine the stresses at corners A, B, C, and D of the flanges.

15.5 PLANE STRESS

An element of volume that has been cut from a body is shown in Fig. 15.9(a). The edges are parallel to the x, y, and z axes. The sides will be

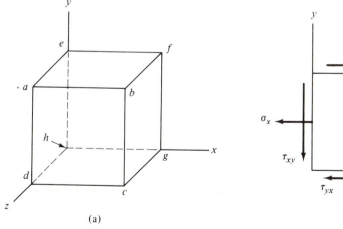

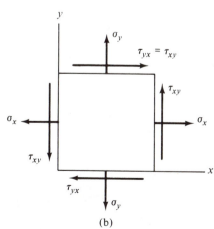

(a)

(b)

FIGURE 15.9

identified in terms of the coordinate axis normal to the side. Thus side *bcgf* is called an *x* plane and side *aefb* is called a *y* plane.

Figure 15.9(b) shows the stress components acting on the same element in a state of plane stress. The normal stresses are indicated by σ_x and σ_y. (The subscript denotes the plane on which the stress acts.) They may be caused by direct tension or compression or bending or any combination of these. The shear stresses are indicated by τ_{xy} and τ_{yx}. (The first subscript denotes the plane on which the stress acts and the second subscript denotes the direction of the stress.) Recall that the shear stress on perpendicular planes must be equal (Sec. 11.2). Therefore, $\tau_{xy} = \tau_{yx}$. The shear stresses may be caused by direct shear or torsion or a combination of the two. The stresses are all positive for the element shown in Fig. 15.9(b). Since many problems involve a state of plane stress, we will consider plane stress in some detail.

15.6 STRESS COMPONENTS ON AN OBLIQUE PLANE

To find the stress components on a plane that is oblique with the *x* and *y* plane and normal to the *z* plane, we cut a section through the element of volume along the desired plane, as shown in Fig. 15.10(a).

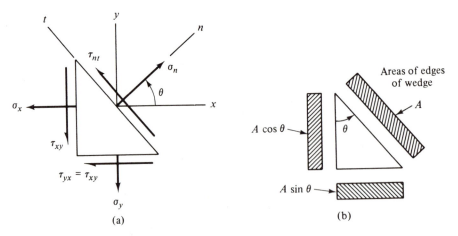

(a)

(b)

FIGURE 15.10

Stresses on the oblique plane are indicated by σ_n and τ_{nt}, where *n* and *t* indicate the normal and tangential axes. The normal axis forms an angle θ with the *x* axis and the tangential axis forms an angle θ with the *y* axis.

To draw the free-body diagram of the wedge, each stress must be multiplied by the area on which the stress acts to change it to force.

Areas of the edges of the wedge are shown in Fig. 15.10(b). The resulting free-body diagram is shown in Fig. 15.11(a). For convenience the forces are drawn outward from a common point in Fig. 15.11(b). For equilibrium in the normal and tangential directions,

$$\measuredangle \, \Sigma \, F_n = 0 \qquad \sigma_n A - \tau_{xy} A \sin \theta \cos \theta - \sigma_x A \cos \theta \cos \theta$$
$$- \tau_{xy} A \cos \theta \sin \theta - \sigma_y A \sin \theta \sin \theta = 0$$
$$\sigma_n = \sigma_x \cos^2 \theta + \sigma_y \sin^2 \theta + 2 \tau_{xy} \sin \theta \cos \theta \quad (15.3)$$

$$\measuredangle \, \Sigma \, F_t = 0 \qquad \tau_{nt} A + \tau_{xy} A \sin \theta \sin \theta + \sigma_x A \cos \theta \sin \theta$$
$$- \tau_{xy} A \cos \theta \cos \theta - \sigma_y A \sin \theta \cos \theta = 0$$
$$\tau_{nt} = (\sigma_y - \sigma_x) \sin \theta \cos \theta + \tau_{xy} (\cos^2 \theta - \sin^2 \theta)$$
$$(15.4)$$

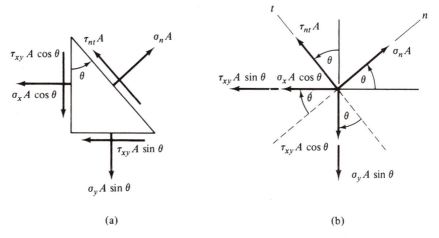

(a)   (b)

FIGURE 15.11

Equations (15.3) and (15.4) can be expressed in a more convenient form. By introducing the trigonometric identities

$$\cos^2 \theta = \frac{1 + \cos 2\theta}{2}$$

$$\sin^2 \theta = \frac{1 - \cos 2\theta}{2}$$

$$\sin \theta \cos \theta = \frac{\sin 2\theta}{2}$$

we obtain

$$\sigma_n = \frac{\sigma_x + \sigma_y}{2} + \frac{\sigma_x - \sigma_y}{2} \cos 2\theta + \tau_{xy} \sin 2\theta \qquad (15.5)$$

$$\tau_{nt} = -\frac{\sigma_x - \sigma_y}{2} \sin 2\theta + \tau_{xy} \cos 2\theta \qquad (15.6)$$

The stresses for any oblique plane could be found from Eqs. (15.5) and (15.6). However, they are rarely used. A simpler method is to use a semi-graphical procedure that is based on the equations.

15.7 MOHR'S CIRCLE OF STRESS

Equations (15.5) and (15.6) can be combined to form the equation of a circle in a plane with σ_n and τ_{nt} as the coordinates. The resulting circle is called Mohr's circle of stress. Normal stresses are plotted along a horizontal axis. Tension is positive and plotted to the right and compression is negative and plotted to the left. Shear stresses are plotted along a vertical axis. A special sign convention for shear stress is required for construction of Mohr's circle. If the shear stresses tend to rotate the element clockwise, they are positive and are plotted up on the vertical axis. If the shear stresses tend to rotate the element counterclockwise, they are negative and are plotted down on the vertical axis. See Fig. 15.12.

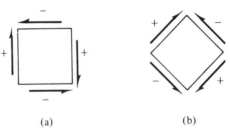

(a)   (b)

FIGURE 15.12

The method for constructing Mohr's circle [Fig. 15.13(b)] for the stresses shown on the element in Fig 15.13(a) is summarized in the following step-by-step list.

1. On σ and τ axes plot point x having coordinates $+\sigma_x$ and $+\tau_{xy}$ and point y having coordinates $+\sigma_y$ and $-\tau_{xy}$. (Shear stresses τ_{xy} tend to rotate element clockwise and are positive, while shear stresses τ_{yx} tend to rotate element counterclockwise and are negative. Remember that they are always numerically equal.)

2. Draw a line between points x and y. Label the intersection of the line with the σ axis point O. The coordinates of point O are $\sigma = (\sigma_x + \sigma_y)/2$ and $\tau = 0$.

3. With O as the center, draw Mohr's circle through points x and y. The radius of the circle is

$$R = \left[\left(\frac{\sigma_x - \sigma_y}{2} \right)^2 + \tau_{xy}^2 \right]^{1/2}$$

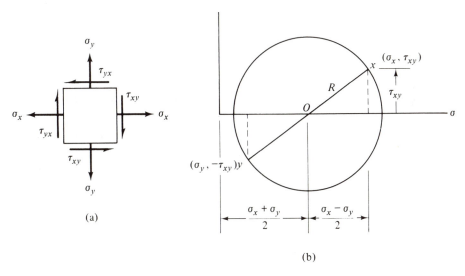

FIGURE 15.13

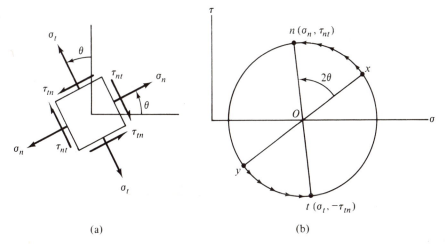

FIGURE 15.14

The method for finding stresses on planes of the element that form angles of θ with respect to the x and y axes as shown in Fig. 15.14(a) are summarized as follows. Rotate the diameter xOy by an angle of 2θ in the *same* direction as for the element [Fig. 15.14(b)]. Coordinate x becomes coordinate n and coordinate y becomes coordinate t. The coordinates of point n give values of the stresses σ_n and σ_{nt} and the coordinates of point t give values of the stresses σ_t and $-\sigma_{tn}$.

EXAMPLE 15.7

The state of stress of an element is shown in Fig. 15.15(a). Use Mohr's circle to find the stresses on planes forming angles of $30°$ with the x and y axes as shown in Fig. 15.15(b).

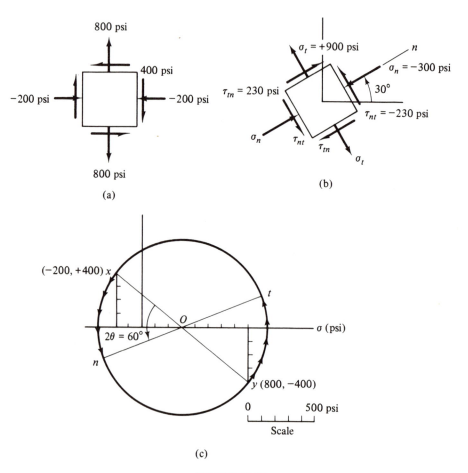

(a)

(b)

(c)

FIGURE 15.15

Solution

The coordinates of point x are $\sigma_x = -200$ psi and $\tau_{xy} = 400$ psi. The coordinates of point y are $\sigma_y = 800$ psi and $\tau_{xy} = -400$ psi. A scale is selected and Mohr's circle is drawn in Fig. 15.15(c). The diameter xOy is rotated counterclockwise by $2\theta = 60°$. Coordinate x becomes coordinate n and coordinate y becomes coordinate t. Scaling values from the circle, the coordinates of n and t are

$$\sigma_n = -300 \text{ psi} \qquad \tau_{nt} = -230 \text{ psi} \qquad\qquad \text{Answer}$$

$$\sigma_t = +900 \text{ psi} \qquad \tau_{tn} = +230 \text{ psi} \qquad\qquad \text{Answer}$$

15.8 PRINCIPAL STRESSES

On planes where the maximum and minimum normal stresses occur there are no shearing stresses. These planes are the *principal planes* of stress and the stresses acting on the principal planes are called the *principal stresses*. Mohr's circle can be used to find the principal planes and the principal stresses. The following example will illustrate the method.

EXAMPLE 15.8

The state of stress of an element is shown in Fig. 15.16(a). Use Mohr's circle to find the principal stresses and the planes of stress. Show your answer on a sketch of the element.

Solution

The coordinates of point x are $\sigma_x = 11$ MPa and $\tau_{xy} = -5$ MPa. The coordinates of point y are $\sigma_y = 3$ MPa and $\tau_{yx} = 5$ MPa. A scale is selected and Mohr's circle is drawn in Fig. 15.16(b). The maximum and minimum normal stress must lie along the σ axis. Therefore, the diameter must be rotated counterclockwise by an angle of $2\theta_1$. Coordinate x becomes coordinate ① and coordinate y becomes coordinate ②. The results can be scaled from the circle or determined from the geometry of the circle. The center O of the circle is at

$$\sigma = \frac{\sigma_x + \sigma_y}{2} = \frac{11 + 3}{2} = 7 \text{ MPa}$$

The radius of the circle

$$R = \left[\left(\frac{\sigma_x - \sigma_y}{2} \right)^2 + \tau_{xy}^2 \right]^{1/2} = [(4)^2 + (5)^2]^{1/2}$$

$$= 6.40 \text{ MPa}$$

Therefore, the maximum and minimum stresses are

$$\sigma_{\max} = \sigma_1 = 7 + 6.40 = 13.40 \text{ MPa} \qquad\qquad \text{Answer}$$

$$\sigma_{\min} = \sigma_2 = 7 - 6.40 = 0.60 \text{ MPa} \qquad\qquad \text{Answer}$$

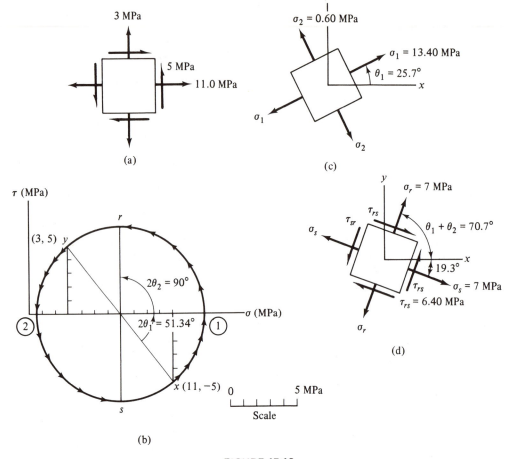

FIGURE 15.16

The principal planes are found from

$$\tan 2\theta_1 = \frac{|\tau_{xy}|}{(\sigma_x - \sigma_y)/2} = \frac{5}{4}$$

$$2\theta_1 = 51.34° \qquad \theta_1 = 25.7° \angle \qquad \text{Answer}$$

The answers are shown in Fig. 15.16(c).

15.9 MAXIMUM SHEAR STRESS

It should be clear from a study of Mohr's circle that the shear stress has a maximum value when the diameter of the circle is in a vertical position. In that position, the magnitude of the shear stress is equal to the radius of the circle, the planes of maximum shear stress form angles of 45° with the planes of principal stress, and the normal stress is equal to the value of σ at the center of the circle.

EXAMPLE 15.9

Determine the maximum shear stresses and their planes and the normal stresses on those planes for Example 15.8. Show your answers on a sketch of the element.

Solution

In Fig. 15.16(b) we must rotate the horizontal diameter counterclockwise by $2\theta_2 = 90°$ or $\theta_2 = 45°$. Coordinate ① becomes coordinate r and coordinate ② becomes s. The results can be scaled from the circle or determined from the geometry of the circle.

$$\sigma_r = \sigma_s = \frac{\sigma_x + \sigma_y}{2} = 7 \text{ MPa} \qquad \text{Answer}$$

$$\tau_{\text{max}_1} = \tau_{rs} = R = 6.40 \text{ MPa} \qquad \text{Answer}$$

$$\tau_{\text{max}_2} = \tau_{sr} = -R = -6.40 \text{ MPa} \qquad \text{Answer}$$

$$\theta = \theta_1 + \theta_2 = 25.7° + 45° = 70.7° \measuredangle \qquad \text{Answer}$$

The answers are shown in Fig. 15.16(d).

PROBLEMS

15.25 Sketch Mohr's circle for the state of stress shown in the figures.

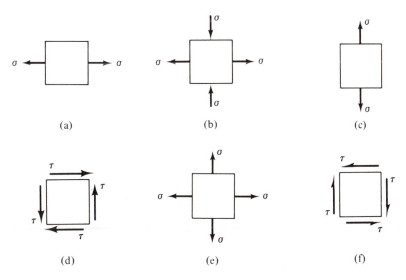

(a) (b) (c)

(d) (e) (f)

PROB. 15.25

15.26 through 15.35 The state of stress is shown in the figure. Use Mohr's circle to find the stresses on planes forming angles of θ with the x and y axes as shown.

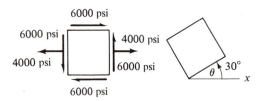

PROB. 15.26

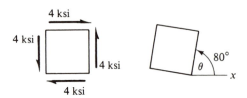

PROB. 15.30

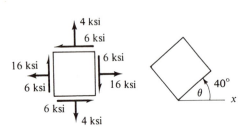

PROB. 15.27

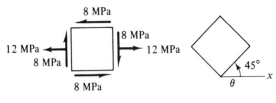

PROB. 15.31

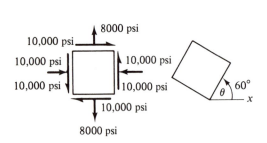

PROB. 15.28

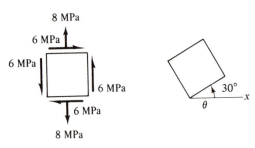

PROB. 15.32

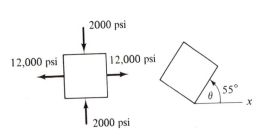

PROB. 15.29

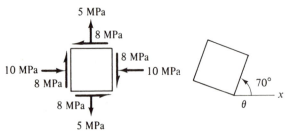

PROB. 15.33

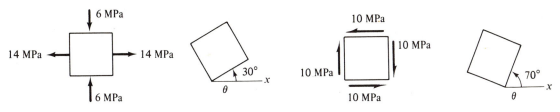

PROB. 15.34   **PROB. 15.35**

In Probs. 15.36 through 15.45, the state of stress is shown in the figure. Use Mohr's circle to find (a) the principal stresses and planes of principal stress, and (b) the maximum shear stress and planes of maximum shear stress. Show your results on a sketch.

15.36   Use the figure for Prob. 15.26.
15.37   Use the figure for Prob. 15.27.

15.38   Use the figure for Prob. 15.28.
15.39   Use the figure for Prob. 15.29.
15.40   Use the figure for Prob. 15.30.
15.41   Use the figure for Prob. 15.31.
15.42   Use the figure for Prob. 15.32.
15.43   Use the figure for Prob. 15.33.
15.44   Use the figure for Prob. 15.34.
15.45   Use the figure for Prob. 15.35.

15.10 AXIAL STRESS

A condition of axial stress exists when normal stress occurs in one direction only. Such is the case for an axially loaded member, as discussed in Chapter 9. From equilibrium we determined the normal and shear stresses for an oblique section. The same results can be obtained by Mohr's circle. The following example will illustrate the method.

EXAMPLE 15.10

The state of stress is shown in Fig. 15.17(a). Use Mohr's circle to find the shear stress and normal stress on a plane forming an angle θ with the x axis.

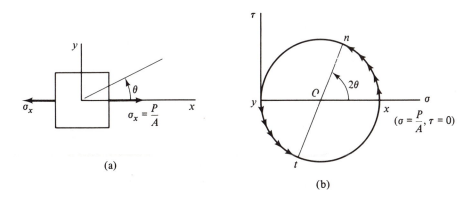

(a)   (b)

FIGURE 15.17

Solution

The coordinates of point x are $\sigma_x = P/A$ and $\tau_{xy} = 0$. The coordinates of point y are $\sigma_y = 0$ and $\tau_{yx} = 0$. Mohr's circle is drawn in Fig. 15.17(b). The center of the circle is at $\sigma = \sigma_x/2 = P/2A$ and $\tau = 0$ and the radius of the circle $R = \sigma_x/2 = P/2A$. Rotate the diameter yOx counterclockwise through an angle of 2θ. Coordinate x becomes coordinate n and coordinate y becomes coordinate t. Therefore,

$$\sigma_n = \frac{P}{2A} + \frac{P}{2A} \cos 2\theta = \frac{P}{2A} (1 + \cos 2\theta)$$

From the trigonometric identity,

$$\frac{1 + \cos 2\theta}{2} = \cos^2 \theta$$

$$\sigma_n = \frac{P}{A} \cos^2 \theta \quad \text{and} \quad \tau_{nt} = \frac{P}{2A} \sin 2\theta$$

The same results were obtained in Sec. 9.8 as Eqs. (9.8) and (9.9).

15.11 BIAXIAL STRESS—THIN-WALLED PRESSURE VESSEL

A condition of biaxial stress exists when normal stresses occur in two mutually perpendicular directions. A good example of biaxial stress occurs on the outside of a pressure vessel.

Let Fig. 15.18(a) represent a closed thin-walled pressure vessel such as a boiler. A pressure vessel can be classified as thin-walled if the

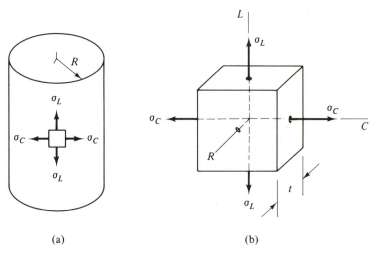

(a)   (b)

FIGURE 15.18

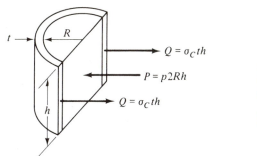
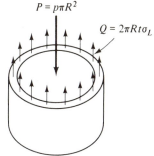

FIGURE 15.19

ratio of the wall thickness t to the radius R of the cylinder is 0.1 or less. Pressure above atmospheric pressure (gauge pressure) on the inside of the curved walls causes the tensile stress σ_C, called the *circumferential stress*. Pressure on the ends of the cylinder causes tensile stress σ_L, called *longitudinal stress*. The stresses are shown on an element of the wall in Fig. 15.18(b). The wall thickness t is measured in the radial direction R perpendicular to the circumferential and longitudinal directions, C and L, respectively.

The stresses may be calculated by free-body diagrams of selected parts of the pressure vessel together with the enclosed fluid under a uniform internal gauge pressure p. The free-body diagram of Fig. 15.19(a) can be used to determine the circumferential stress. Summing forces—stress multiplied by wall area and pressure multiplied by projected inside area—parallel to the stress σ_C, we obtain

$$\rightarrow \Sigma \ F_C = 0 \quad 2Q - P = 0$$

$$2\sigma_C th - p2Rh = 0$$

$$\sigma_C = \frac{pR}{t} \tag{15.7}$$

The free-body diagram of Fig. 15.19(b) can be used to determine the longitudinal stress. Summing forces—stress multiplied by ring shaped wall area and pressure multiplied by projected circular area—parallel to the stress σ_L

$$\uparrow \Sigma \ F_L = 0 \quad Q - P = 0$$

$$2Rt\sigma_L - p\pi R^2 = 0$$

$$\sigma_L = \frac{pR}{2t} \tag{15.8}$$

Notice from Eqs. (15.7) and (15.8) that the longitudinal stress is one-half the circumferential stress.

It can be shown that the normal stress in the walls of the pressure vessel, the radial direction [Fig. 15.18(b)], vary from $\sigma_R = -p$ on the inside of the vessel to $\sigma_R = 0$ on the outside of the vessel.

The tensile stress in a *spherical pressure vessel* with a uniform pressure is given by Eq. (15.8).

EXAMPLE 15.11

A steel boiler 2 m in diameter has a wall thickness of 20 mm. The internal pressure is 1.4 MPa. On the outside surface of the cylindrical shell, determine (a) the maximum shearing stress in a plane parallel to the radial direction, (b) the maximum shearing stress for any plane, and (c) on the inside surface of the cylindrical shell, determine the maximum shearing stress.

Solution

The circumferential stress is given by Eq. (15.7):

$$\sigma_C = \frac{pR}{t} = \frac{1.4(1)}{0.020} = 70 \text{ MPa}$$

and the longitudinal stress is given by Eq. (15.8):

$$\sigma_L = \frac{pR}{2t} = 35 \text{ MPa}$$

The radial stress on the inside surface of the shell is

$$\sigma_R = -p = -1.4 \text{ MPa}$$

and on the outside surface of the shell

$$\sigma_R = 0$$

Outside Surface of Shell

The stresses on the outside surface of the shell are $\sigma_C = 70$ MPa, $\sigma_L = 35$ MPa, and $\sigma_R = 0$. Thus we have a case of biaxial stress. However, a small volume element on the surface of the shell is a three-dimensional body and the maximum shear stress may not occur in a plane perpendicular to the circumferential and longitudinal axes. Therefore, the elements must be viewed along each of the three axes to find the maximum shear stress. Such views are shown in Fig. 15.20(a)–(c). The corresponding Mohr's circles are drawn in Fig. 15.20(d). The maximum shear stress can be determined from the circles.

(a) The maximum shearing stress in a plane parallel to the radial direction is found from the circle for an element viewed along the radial axis $\tau_{max} = 17.5$ MPa. The stresses act on planes parallel to the radial axis which make angles of $45°$ and $135°$ with the circumferential axis.

(b) The maximum shearing stress for any plane is found from the circle for an element viewed along the longitudinal axis $\tau_{max} = 35$ MPa. The stresses act on

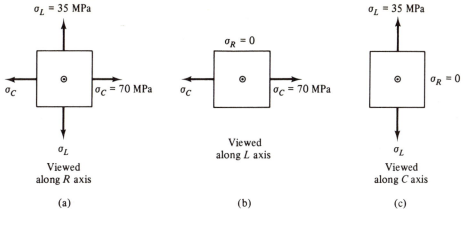

$\sigma_L = 35$ MPa

σ_C $\sigma_C = 70$ MPa

σ_L

Viewed
along R axis

(a)

$\sigma_R = 0$

σ_C $\sigma_C = 70$ MPa

Viewed
along L axis

(b)

$\sigma_L = 35$ MPa

$\sigma_R = 0$

σ_L

Viewed
along C axis

(c)

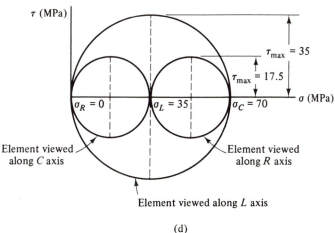

τ (MPa)

$\tau_{max} = 35$

$\tau_{max} = 17.5$

σ (MPa)

$\sigma_R = 0$ $\sigma_L = 35$ $\sigma_C = 70$

Element viewed
along C axis

Element viewed
along R axis

Element viewed along L axis

(d)

FIGURE 15.20

planes parallel to the longitudinal axis which make angles of $45°$ and $135°$ with the circumferential axis.

Inside Surface of Shell

    (c) The state of stress in part (c) is not biaxial. However, the method for parts (a) and (b) can be used here with the addition of $\sigma_R = -1.4$ MPa. Mohr's circles are shown for the three views in Fig. 15.21. The maximum shearing stress for any plane

$$\tau_{max} = \frac{\sigma_C - \sigma_R}{2} = \frac{70 + 1.4}{2} = 35.7 \text{ MPa}$$

The stresses act on planes parallel to the longitudinal axis, which make angles of $45°$ and $135°$ with the circumferential axis. The stresses determined in part (c) differ

433

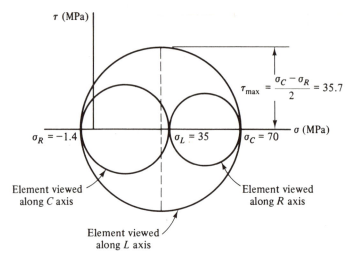

FIGURE 15.21

from those in part (b) by 2 percent. Therefore, the radial stress is usually neglected in a thin-walled pressure vessel.

15.12 PURE SHEAR

A condition of pure shear exists when only shearing stresses exist in two mutually perpendicular directions, as shown in Fig. 15.22(a). Construct Mohr's circle as shown in Fig. 15.22(b). The maximum tensile and compressive stresses (principal stresses) act on planes that form angles of 45°

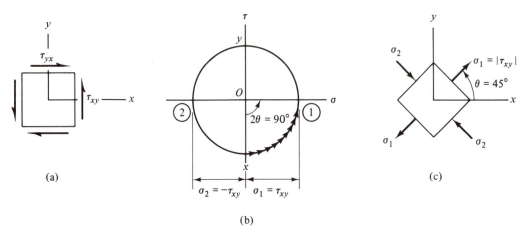

FIGURE 15.22

and $135°$ with the x axis and their values $\sigma_1 = -\sigma_2 = \tau_{xy}$. The principal stresses are shown in Fig. 15.22(c).

EXAMPLE 15.12

A hollow circular shaft has an outside diameter of 6 in. and an inside diameter of 4 in. Determine the maximum torque that can be applied if the tensile and compressive stresses are limited to 10,000 psi and 30,000 psi, respectively, and the shear stress is limited to 12,500 psi.

Solution

An element on the surface of the circular shaft is acted on by shear stresses as shown in Fig. 15.23(a). Therefore, the shaft is in a state of pure shear. In pure shear the principal stresses in tension and compression are equal to the maximum shear stress

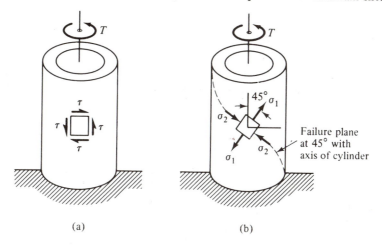

(a)　　　　　　　　　　　　　(b)

FIGURE 15.23

[See Fig. 15.22(c)]. Since the smallest allowable stress is in tension, the tensile stress will control. From the torsion formula, Eq. (11.9), and the condition that the maximum tensile stress is equal to the maximum shear stress,

$$\sigma_{max} = \tau_{max} = \frac{Tc}{J} \tag{a}$$

The polar moment of inertia

$$J = \frac{\pi}{32}(d_o^4 - d_i^4) = \frac{\pi}{32}[(6)^4 - (4)^4] = 102.1 \text{ in.}^4$$

$$c = \frac{d_o}{2} = 3 \text{ in.}$$

From Eq. (a),

$$T = \frac{\sigma_{max} J}{c} = \frac{10{,}000\,(102.1)}{3} = 340 \times 10^3 \text{ lb-in.}$$

$$= 28.4 \times 10^3 \text{ lb-ft} \qquad\qquad\qquad \textbf{Answer}$$

If the torque is increased until failure occurs in tension, the shaft will fail along a helix as shown in Fig. 15.23(b). The helix forms an angle of $45°$ with the axis of the shaft.

15.13 COMBINED STRESS PROBLEMS

Various combinations of stress may occur in structural and machine members. One possible combination will be discussed in the following example.

EXAMPLE 15.13

The machine member shown in Fig. 15.24 is acted on by an axial force of 11 kN and a vertical force of 2.4 kN. Determine the maximum tensile, compressive, and shear stresses. Maximum stresses will occur at points A and B.

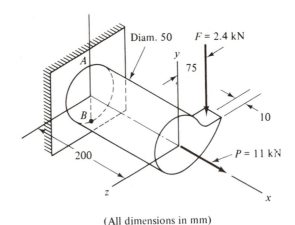

(All dimensions in mm)

FIGURE 15.24

Solution

This problem involves axial, bending, and torsion loads.

Axial Load

The cross section of the bar through A and B has an area $A = \pi d^2/4 = \pi(50)^2/4 = 1.964 \times 10^3$ mm^2. Therefore, the axial stress, given by Eq. (9.3),

$$\sigma_x = \frac{P}{A} = \frac{11}{1.964 \times 10^3} = 5.60 \times 10^{-3} \ \frac{kN}{mm^2}$$

$$= 5.60 \ MPa \tag{a}$$

Bending Loads

The moment of inertia of the bar about the z axis,

$$I_z = \frac{\pi d^4}{64} = \frac{\pi (50)^4}{64} = 306.8 \times 10^3 \ mm^4$$

and the bending moment at the cross section is $M_z = -Pd = -2.4\,(190) = 456 \ kN\cdot mm$. Therefore, the bending stress, given by Eq. (12.6),

$$\sigma_x = \pm \frac{M_z c}{I_z} = \pm \frac{456\,(25)}{306.8 \times 10^3} = \pm 37.2 \times 10^{-3} \ \frac{kN}{mm^2}$$

$$= \pm 37.2 \ MPa \tag{b}$$

The stress is tensile for point A and compressive for point B.

Torsion Load

The polar moments of inertia of the bar about the x axis, $J = 2I_z = 613.6 \times 10^3 \ mm^4$, and the torque about the x axis, $T = Fr = 2.4\,(75) = 180 \ kN\cdot mm$. Therefore, the torsional stress, given by Eq. (11.13),

$$\tau_{xy} = \frac{Tc}{J} = \frac{180\,(25)}{613.6 \times 10^3} = 7.33 \times 10^{-3} \ \frac{kN}{mm^2}$$

$$= 7.3 \ MPa \tag{c}$$

Point A

At point A the normal stresses from Eqs. (a) and (b) add to give $\sigma_x = 5.6 + 37.2 = 42.8$ MPa. The shear stress from Eq. (c) is $\tau_{xy} = 7.3$ MPa. An element at point A is shown in Fig. 15.25(a) and the corresponding Mohr's circle in Fig. 15.25(b). From the Mohr's circle,

$$\sigma_{max} = \sigma_1 = \frac{42.8}{2} + \left[\left(\frac{42.8}{2} \right)^2 + (7.3)^2 \right]^{1/2}$$

$$\sigma_1 = 21.4 + 22.6 = +40.0 \ MPa \qquad \qquad \text{Answer}$$

$$\sigma_{min} = \sigma_2 = 21.4 - 22.6 = -1.2 \ MPa \qquad \qquad \text{Answer}$$

$$\tau_{max} = 22.6 \ MPa \qquad \qquad \text{Answer}$$

Point B

At point B the normal stresses from Eqs. (a) and (b) subtract to give $\sigma_x = 5.6 - 37.2 = -31.6$ MPa and the shear stress from Eq. (c) is $\tau_{xy} = 7.3$ MPa. An element at point B

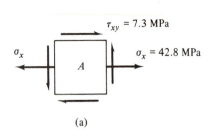

$\tau_{xy} = 7.3$ MPa

σ_x

$\sigma_x = 42.8$ MPa

A

(a)

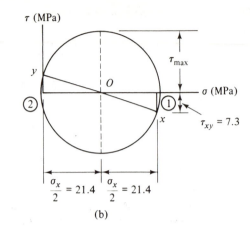

(b)

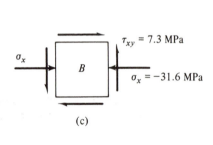

$\tau_{xy} = 7.3$ MPa

σ_x

B

$\sigma_x = -31.6$ MPa

(c)

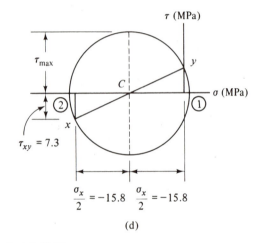

(d)

FIGURE 15.25

is shown in Fig. 15.25(c) and the corresponding Mohr's circle in Fig. 15.25(d). From the Mohr's circle,

$$\sigma_{max} = \sigma_1 = \frac{-31.6}{2} + \left[\left(\frac{31.6}{2}\right)^2 + (7.3)^2\right]^{1/2}$$

$$\sigma_1 = -15.8 + 17.4 = +1.6 \text{ MPa} \qquad \text{Answer}$$

$$\sigma_{min} = \sigma_2 = -15.8 - 17.4 = -33.2 \text{ MPa} \qquad \text{Answer}$$

$$\tau_{max} = 17.4 \text{ MPa} \qquad \text{Answer}$$

PROBLEMS

15.46 For the cantilever beam shown, use Mohr's circle to determine the principal stresses and maximum shear stress at A, B, and C for the section in the figure.

15.47 For the simply supported beam shown, use Mohr's circle to find the principal stresses and planes of principal stress at A, B, and C for the section in the figure.

15.48 A cantilever beam is loaded as shown in the figure. Use Mohr's circle to find the principal stresses at A, B, and C of the cross section.

15.49 A beam with overhang, load, and cross section is shown in the figure. Use Mohr's circle to determine the principal stresses at A, B, and C of the cross section.

15.50 A simply supported beam with loads and cross section is shown in the figure. The moment of inertia with respect to the neutral axis I_{NA} = 260.8 in.⁴. Use Mohr's circle to determine the principal stresses at A, B, and C of the cross section.

15.51 An axially loaded steel rod with a diameter of 20 mm has a tensile stress of 80 MPa on a plane that makes an angle of 35° with the axis of the rod. What is the load on the rod?

15.52 A steel bar with a cross section 1.25 in. by 0.375 in. is tested in tension. The bar has a tensile stress of 20 ksi on a plane that makes an angle of 50° with the axis of the bar. What is the load on the bar?

15.53 Two bars measuring 2 in. by 2 in. are glued together to form the member shown. If the allowable stresses in the glue are 900 psi in tension and 500 psi in shear, what is the allowable axial load on the member? Assume that the load is based on the strength of the glue.

15.54 A concrete test cylinder 6 in. in diameter and 12 in. high failed when

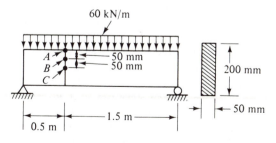

PROB. 15.46

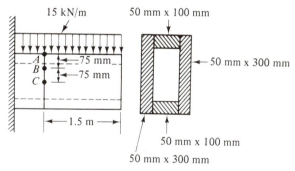

B. 15.48

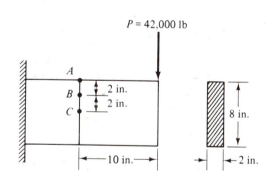

PROB. 15.47

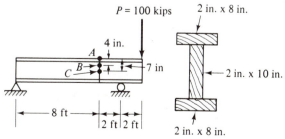

PROB. 15.49

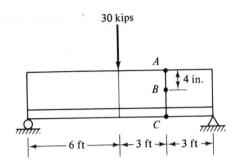

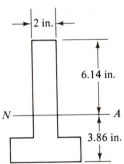

PROB. 15.50

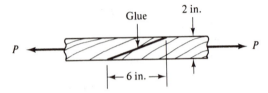

PROB. 15.53

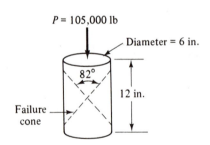

PROB. 15.54

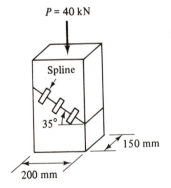

PROB. 15.55

subjected to a compressive load of 105,000 lb. The surface of failure formed a right circular cone whose angle was 82°. Determine (a) the compressive and shearing stresses on the plane of failure, and (b) the maximum compressive and shearing stresses in the cylinder at failure. (*Hint:* The plane of failure forms an angle of 41° with the axis of the cylinder.)

15.55 Two blocks are joined along a diagonal plane by three splines (keys) as shown. (a) If an axial load of $P = 40$ kN is applied to the blocks, what is the shear force on each spline? Neglect friction. (b) If each spline measures 5 mm by 20 mm by 150 mm, what is the shear stress in the splines?

15.56 A steel boiler 2.5 m in diameter has a wall thickness of 25 mm. The internal pressure is 1.6 MPa. Determine (a) the circumferential stress, and (b) the longitudinal stress. Use Mohr's circle to find (c) the maximum shearing stress in a plane parallel to the radial direction, and (d) the maximum shearing stress for any plane at a point on the outside of the cylinder.

15.57 A spherical steel boiler 8 ft in diameter has a wall thickness of 1 in. The internal pressure is 250 psi. Determine (a) the circumferential stress and (b) the longitudinal stress in the boiler due to the internal pressure. Use

Mohr's circle to find (c) the maximum shearing stress in a plane parallel to the radial direction, and (d) the maximum shearing stress for any plane at a point on the outside of the sphere.

15.58 The cylindrical tank shown has an outside diameter d_o = 225 mm and a wall thickness t = 15 mm. The internal gauge pressure in the tank is 2.5 MPa, the end torque is T = 30 kN · m, and the axial load P = 200 kN. Determine (a) the circumferential and longitudinal stress due to the inside pressure, (b) the shear stress due to the torque, and (c) the tensile stress due to the axial load. Use Mohr's circle to find (d) the principal stresses and planes of principal stress for an element at B on the front surface of the tank as shown.

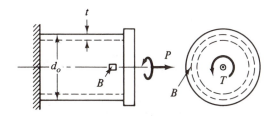

PROB. 15.58 and PROB. 15.59

15.59 The cylindrical tank shown has an outside diameter of d_o = 10 in. and a wall thickness of t = 0.625 in. The internal gauge pressure in the tank is 500 psi. The end torque is T = 350 kip-in. and the axial load P = 60 kips. Determine (a) the circumferential and longitudinal stress due to the internal pressure, (b) the tensile stress due to the axial load, and (c) the shear stress due to the torque. Use Mohr's circle to find (d) the principal stresses and planes of principal stress for an element at B on the front surface of the tank as shown.

15.60 For the bracket shown, the load P = 19.8 kN, L = 100 mm, and the cross section is b = 25 mm by h = 75 mm. Determine (a) the normal stresses on

the cross section at A, B, and C due to P, and (b) the shear stresses on the cross section at A, B, and C due to P. Use Mohr's circle to find (c) the principal stresses and maximum shear stresses at A, B, and C from the stresses found in parts (a) and (b).

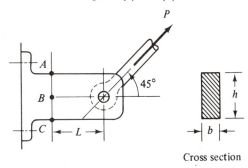

Cross section

PROB. 15.60 and PROB. 15.61

15.61 For the bracket shown, the load P = 4.3 kips, L = 4 in., and the cross section is b = 1 in. by h = 3 in. Determine (a) the normal stresses on the cross section at A, B, and C due to P, and (b) the shear stresses on the cross section at A, B, and C due to P. Use Mohr's circle to find (c) the principal stresses and maximum shear stresses at A, B, and C from the stresses found in parts (a) and (b).

15.62 For the machine part shown, the load P = 1500 lb, the diameter d = 2 in., L = 4 in., and w = 1 in. Determine the following stresses on the cross section at D, E, F, and G: (a) the normal stress due to bending, (b) the shear stress due to torsion, and (c) the shear stress due to the vertical shear force. Use Mohr's circle to find (d) the principal stresses and the maximum shear stresses at D, E, F, and G from the results in parts (a), (b), and (c).

15.63 The load P = 6.6 kN, the diameter d = 50 mm, L = 100 mm, and w = 25 mm for the machine part shown. Determine the following stresses on

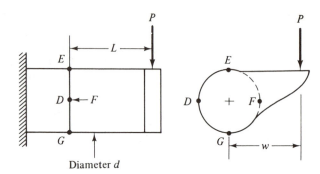

PROB. 15.62 and PROB. 15.63

the cross section at D, E, F, and G: (a) the normal stress due to the bending produced by P, (b) the shear stresses due to the torque produced by P, and (c) the shear stresses due to the vertical shear force $V = P$. Use Mohr's circle to find (d) the principal stresses and maximum shear stresses at D, E, F, and G from the results in parts (a), (b), and (c).

15.64 The bracket shown is loaded with a force $P = 1$ kip. The dimensions of the bracket are $L = 6$ in., $R = 3$ in., $w = 4$ in., and $d = 1.5$ in. Determine the following stresses on the cross section at A, B, C, and D: (a) the normal stress due to bending, (b) the shear stress due to torsion, and (c) the shear stress due to the vertical shear force.

Use Mohr's circle to find (d) the principal stresses and maximum shear stresses at A, B, C, and D from the results in parts (a), (b), and (c).

15.65 The bracket shown is loaded with a force $P = 4.4$ kN. The dimensions of the bracket are $L = 150$ mm, $R = 75$ mm, $w = 100$ mm, and $d = 40$ mm. Determine the following stresses on the cross section at A, B, C, and D: (a) the normal stress due to bending, (b) the shear stress due to torsion, and (c) the shear stress due to the vertical shear force. Use Mohr's circle to find (d) the principal stresses and maximum shear stresses at A, B, C, and D from the results in parts (a), (b), and (c).

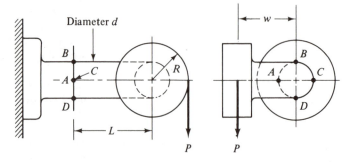

PROB. 15.64 and PROB. 15.65

15.14 PLANE STRAIN

A state of plane strain in the xy plane can be described in terms of three components of strain ϵ_x, ϵ_y, and γ_{xy}. The shear strains γ_{xy} and γ_{yx} are equal. This is similar to shear stress where $\tau_{xy} = \tau_{yx}$. To represent strain we introduce a special notation. Arrows directed outward or inward from an element represent normal strains which are either tensile or compressive [Fig. 15.26(a) and (b)]. Arrows directed along the sides of the element represent either positive or negative shear strains, depending on their

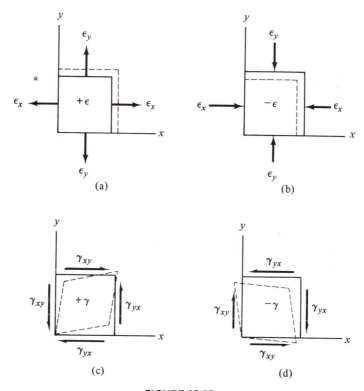

FIGURE 15.26

orientation [Fig. 15.26(c) and (d)]. Notice that the arrows follow the same sign convention as for stress. See Fig. 15.9, where stresses are positive in the directions shown.

It can be shown that strains can be found along oblique axes from strains measured with respect to the x and y axes by methods similar to those for stress. One such method is Mohr's circle of strain.

15.15 MOHR'S CIRCLE OF STRAIN

For Mohr's circle of strain, normal strains are plotted along a horizontal axis with tension positive and compression negative. Shear strains divided by *two* are plotted along a vertical axis. A *special* sign convention is required. If the arrows representing strain tend to rotate the element clockwise, they are positive and are plotted up on the vertical axis. If the arrows tend to rotate the element counterclockwise, they are negative and are plotted down on the vertical axis. (Sign conventions are similar to those for shear.)

EXAMPLE 15.14

The plane strain components in a body are $\epsilon_x = -200 \times 10^{-6}$, $\epsilon_y = 1000 \times 10^{-6}$, and $\gamma_{xy} = 800 \times 10^{6}$. Determine the principal strains and the planes of principal strain. Also find the strains on planes that form angles of $20°$ and $110°$ with the x axis.

Solution

The strains are shown on an element in Fig. 15.27(a). In Fig. 15.27(b) Mohr's circle is drawn on the basis of the strains. (Notice that the vertical axis represents $\gamma/2$). From the circle

$$\tan 2\theta_1 = \frac{400}{600}$$

$$2\theta_1 = 33.7° \qquad \theta_1 = 16.8°$$

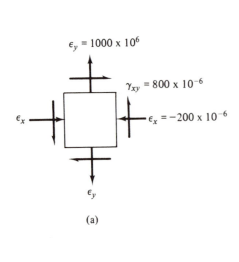

(a)

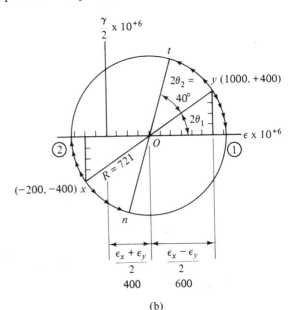

(b)

FIGURE 15.27

444

The center of the circle 0 is at

$$\epsilon = \frac{\epsilon_x + \epsilon_y}{2} = \frac{-200 + 1000}{2} = 400$$

and the radius of the circle

$$R = \left[\left(\frac{\epsilon_x - \epsilon_y}{2}\right)^2 + \left(\frac{\gamma_{xy}}{2}\right)^2\right]^{1/2} = [(-600)^2 + (400)^2]^{1/2}$$

$$= 721$$

The principal strains

$$\epsilon_1 = (400 + 721) \times 10^{-6} = 1121 \times 10^{-6}$$

$$\epsilon_2 = (400 - 721) \times 10^{-6} = -321 \times 10^{-6}$$

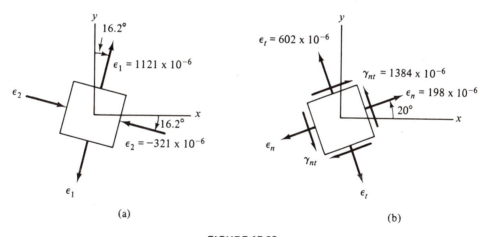

FIGURE 15.28

and the directions are shown in Fig. 15.28(a). For the strains of $20°$ and $110°$ with the x axis,

$$\epsilon_n = (400 \times 10^{-6}) - (721 \times 10^{-6}) \cos(40° + 33.7°)$$

$$= 198 \times 10^{-6}$$

$$\epsilon_t = (400 \times 10^{-6}) + (721 \times 10^{-6}) \cos(40° + 33.7°)$$

$$= 602 \times 10^{-6}$$

The shear strain coordinate for point n is negative and for point t positive; therefore, the shear strain γ_{nt} is positive.

$$\gamma_{nt} = 2(721 \times 10^{-6}) \sin(40° + 33.7°)$$
$$= 1384 \times 10^{-6}$$

The directions are shown in Fig. 15.28(b).

16

Columns ─────────────────────────

16.1 INTRODUCTION

In previous chapters we have studied the relationship between the loads applied to a member and the resultant stresses and strains and deflections. In each problem the member was in *stable* equilibrium.

Let us now study the column—a long slender member subject to an axial load. For small values of the axial load, the column will remain essentially straight and the stresses can be determined from the usual formulas for axial loads. However, as the axial load increases, the column bends and is subject to *small* lateral deflections. The stresses now depend not only on the axial loads but also on the deflection. Continued small increases in the load after it is close to a certain critical value results in sudden *large* deflections and buckling or collapse without warning. This collapse is not due to failure of the column material but to passage of the column from *stable* to *unstable* equilibrium.

Stable and Unstable Equilibrium

As a first example, consider the three smooth surfaces of Fig. 16.1. The surfaces are all horizontal at point B. A sphere will be in equilibrium at B because the weight W and normal reaction N_R will both be vertical, equal in magnitude, and opposite in direction. The sphere is now moved to C. Three possibilities exist, depending on the unbalanced force. In Fig. 16.1(a) the unbalanced force moves the sphere back to its original position B. This is called *stable* equilibrium. In Fig. 16.1(b) no unbalanced force develops—the sphere is still in equilibrium—and therefore no tendency for it to move away from the new position. This is called *neutral* equilibrium. In Fig. 16.1(c) the unbalanced force moves the sphere away from its original position B. This is called *unstable* equilibrium.

For the second example, consider the hinged bar in Fig. 16.2. The load directed up in Fig. 16.2(a) produces stable equilibrium since a small rotation produces a moment that restores the bar to the vertical equilibrium position. The downward load in Fig. 16.2(b) produces un-

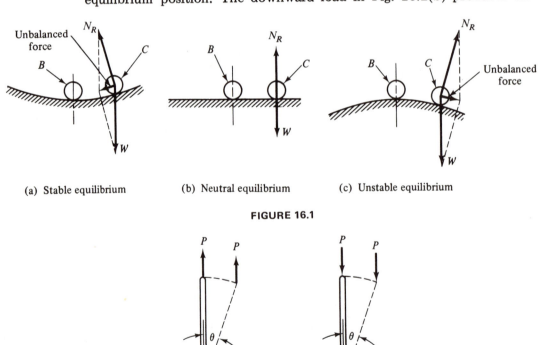

(a) Stable equilibrium   (b) Neutral equilibrium   (c) Unstable equilibrium

FIGURE 16.1

(a) Stable equilibrium   (b) Unstable equilibrium

FIGURE 16.2

stable equilibrium, since a small rotation produces a moment that moves the bar away from the vertical equilibrium position.

Consider a system in equilibrium. Any possible small displacement away from the equilibrium position produces forces. If the forces move the system back to the equilibrium position, the system is stable. If the forces do not—the system is unstable.

16.2 THE EULER COLUMN FORMULA

The long member or column shown in Fig. 16.3(a) has pin supports on both ends and is acted on by a compressive load. The smallest column load that produces buckling or failure is called the *critical* or *Euler load*. For a load less than the critical load, the column remains straight and in stable equilibrium. For a load equal to or greater than the critical load, the column is unstable and fails. The bent shape shown represents buckling or failure. Our problem is to find the smallest column load that will produce buckling or failure.

From the free-body diagram for a section at a distance x from the top of the column [Fig. 16.3(b)] and the rotational equation of equilibrium, the bending moment $M = -P\delta$. The slope of the deflection curve $\theta = \Delta\delta/\Delta x$. The slope and bending moment are related by Eq.

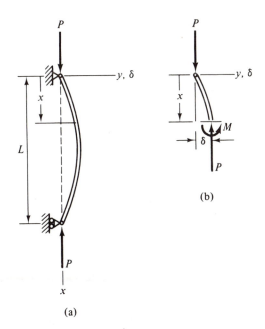

FIGURE 16.3 Deflection curve for column with pin ends.

(14.3), $\Delta\theta/\Delta x = M/EI$. Therefore, we write

$$\frac{\Delta(\Delta\delta/\Delta x)}{\Delta x} = \frac{M}{EI} \tag{16.1}$$

Substituting for the bending moment $M = -P\delta$, we have

$$\frac{\Delta(\Delta\delta/\Delta x)}{\Delta x} + \frac{P\delta}{EI} = 0 \tag{16.2}$$

The solution of this equation expresses the deflection δ as a function of x.

If we let the element of length along the column Δx decrease until it approaches a point, Eq. (16.2) becomes a differential equation. The differential equation has known solutions. Guided by these solutions, the deflective curve shown in Fig. 16.3(a) looks like a sine curve from $\theta = 0$ to $\theta = \pi$. We modify the "scale" of the sine curve to make the curve from $\theta = 0$ to $\theta = \pi$ fit the deflection curve from $x = 0$ to $x = L$. To modify the sine curve, we let

$$\delta = A \sin \frac{\pi x}{L} \tag{a}$$

as shown in Fig. 16.4(a). The slope of the deflection curve $\Delta\delta/\Delta x$, where the length Δx decreases until it approaches a point, is given by

$$\frac{\Delta\delta}{\Delta x} = A \frac{\pi}{L} \cos \frac{\pi x}{L} \tag{b}$$

as shown in Fig. 16.4(b).

The slope of the slope of the deflection curve $\Delta(\Delta\delta/\Delta x)/\Delta x$, when the length Δx decreases until it approaches a point, is given by

$$\frac{\Delta(\Delta\delta/\Delta x)}{\Delta x} = -A \frac{\pi^2}{L^2} \sin \frac{\pi x}{L} \tag{c}$$

and is shown in Fig. 16.4(c).

Substituting Eq. (a) and (c) into Eq. (16.2), we write

$$-A \frac{\pi^2}{L^2} \sin \frac{\pi x}{L} + \frac{P}{EI} A \sin \frac{\pi x}{L} = 0 \tag{d}$$

Equation (d) is satisfied if $A = 0$. This represents no buckling—zero deflection—as can be seen from Eq. (a). The other possible solution re-

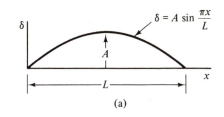

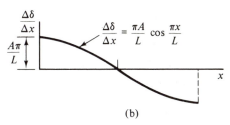

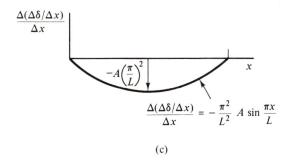

FIGURE 16.4

quires that $\pi^2/L^2 = P/EI$. The value of the load P that makes the curved shape possible, the critical or Euler load, is, therefore, given by

$$P_C = \frac{\pi^2 EI}{L^2} \tag{16.3}$$

This is the *Euler column formula* for a column that has pin supports at both ends.

16.3 EFFECTIVE LENGTH OF COLUMNS

For end conditions other than pin supports, the shapes of the deflection curves can be used to modify the Euler column formula. In Fig. 16.5 we show the deflection curve for different end conditions. In each case, the length of a single loop of a sine curve represents the effective length L_e of the column. The extent of a single loop is marked by a dot on the deflection curves. The effective length L_e can be expressed in the form

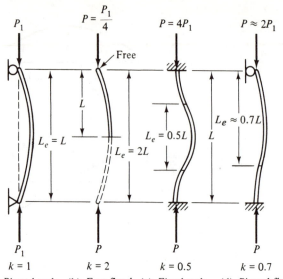

$$P = \frac{P_1}{4}$$
P_1 Free $P = 4P_1$ $P \approx 2P_1$

$L_e = L$ $L_e = 2L$ $L_e = 0.5L$ $L_e \approx 0.7L$

L L

P_1 P P P
$k = 1$ $k = 2$ $k = 0.5$ $k = 0.7$
(a) Pinned ends (b) Free-fixed (c) Fixed ends (d) Pinned-fixed

FIGURE 16.5 Various column end constraints.

$$L_e = kL$$

Substituting the effective length into Eq. (16.3), the critical load becomes

$$P_C = \frac{\pi^2 EI}{(kL)^2} \tag{16.4}$$

where the value of k depends on the end conditions. For example, with both ends fixed as shown in Fig. 16.5(c), $k = 0.5$. The critical load $P_C = 4\pi^2 EI/L^2$. Thus we see that the critical load is four times the value for an identical column with both ends pinned.

16.4 FURTHER COMMENTS ON THE EULER COLUMN FORMULA

The Euler column formula is usually expressed in terms of the radius of gyration. From Sec. 8.10 the radius of gyration $r = \sqrt{I/A}$ or $I = Ar^2$, where A is the cross-sectional area of the column and I is the minimum moment of inertia of the cross-sectional area. Substituting for the moment of inertia in Eq. (16.4), we have

$$P_C = \frac{\pi^2 EAr^2}{(kL)^2}$$

or in terms of critical stress, $\sigma_C = P_C/A$, we write

$$\sigma_C = \frac{\pi^2 E}{(kL/r)^2} \tag{16.5}$$

The critical stress σ_C is the average value of the stress over the cross-sectional area A at the critical load P_C. The effective length of the column is given by kL. Buckling will take place about the axis of minimum moment of inertia or equivalently about the axis with a minimum radius of gyration unless prevented by braces. Therefore, r is the minimum radius of gyration. The ratio of the effective length kL to the minimum radius of gyration r is called the *slenderness ratio*.

In Fig. 16.6 we show the relation between critical stress and the slenderness ratio for various materials [Eq. (16.5)]. Because the Euler column formula is based on Hooke's law, the curves are not valid for stresses above the proportional limit of the material.

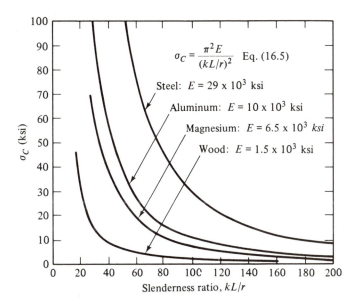

FIGURE 16.6   Euler curves.

We are now able to define what is meant by a long column. A *long column* may be defined as one in which the critical stress σ_C is less than the proportional limit σ_P for the material.

EXAMPLE 16.1

A W14 × 74 steel column has a length of 35 ft. The steel has a proportional limit of 33,000 psi. Use a modulus $E = 30 \times 10^6$ psi and a factor of safety, F.S. = 2.0. Determine the critical or Euler load and the allowable or working load if (a) the ends of the column are pinned, (b) one end is pinned and the other is fixed, and (c) both ends are fixed.

Solution

The minimum radius of gyration for the W14 X 74 column is $r = 2.48$ in. and the cross-sectional area $A = 21.8$ in.^{2}

(a) Ends pinned: for pinned ends, $k = 1$ [Fig. 16.5(a)] and the slenderness ratio $kL/r = 1(35)(12)/2.48 = 169$. The critical Euler stress

$$\sigma_C = \frac{P_C}{A} = \frac{\pi^2 E}{(kL/r)^2} = \frac{\pi^2 (30)(10^6)}{(169)^2} = 10{,}370 \text{ psi}$$

Because the critical stress is less than the proportional limit stress, Euler's formula applies and we write

$$P_C = \sigma_C A = 10{,}370 (21.8) = 226{,}000 \text{ lb}$$

Therefore, the allowable load

$$P_a = \frac{P_C}{\text{F.S.}} = \frac{226{,}000}{2.0} = 113{,}000 \text{ lb}$$

(b) One end pinned, the other fixed: for the end conditions, $k = 0.7$ [Fig. 16.5(d)] and the slenderness ratio $kL/r = 0.7(35)(12)/2.48 = 119$. The critical or Euler stress

$$\sigma_C = \frac{P_C}{A} = \frac{\pi^2 E}{(kL/r)^2} = \frac{\pi^2 (30)(10^6)}{(119)^2} = 20{,}900 \text{ psi}$$

The critical stress is less than the proportional limit and Euler's formula applies.

$$P_C = \sigma_C A = 20{,}900 (21.8) = 456{,}000 \text{ lb}$$

Therefore, the allowable load

$$P_a = \frac{P_C}{\text{F.S.}} = \frac{456{,}000}{2.0} = 228{,}000 \text{ lb}$$

(c) Ends fixed: for both ends fixed, $k = 0.5$ [Fig. 16.5(c)] and the slenderness ratio $kL/r = 0.5(35)(12)/2.48 = 85$. The critical or Euler stress

$$\sigma_C = \frac{P_C}{A} = \frac{\pi^2 E}{(kL/r)^2} = \frac{\pi^2 (30)(10^6)}{(85)^2} = 41{,}000 \text{ psi}$$

Because the critical stress is greater than the proportional limit, Euler's column formula does not apply. An allowable load based on the Euler formula would be meaningless.

16.1 A W14 × 74 steel column has a length of 30 ft. The steel has a modulus $E = 29 \times 10^3$ ksi and a proportional limit $\sigma_P = 33$ ksi. Use a factor of safety of 2. Determine the critical or Euler load and the allowable load if (a) the ends are pinned, (b) one end is pinned and the other is fixed, and (c) both ends are fixed. For section properties, see Table A.5 of the Appendix.

16.2 A W360 × 110 rolled steel column has a length of 9 m. The steel has a modulus $E = 200 \times 10^6$ kPa and a proportional limit $\sigma_P = 230 \times 10^3$ kPa. Use a factor of safety of 2. Determine the critical or Euler load and the allowable load if (a) the ends are pinned, (b) one end is pinned and the other is fixed, and (c) both ends are fixed. For section properties, see Table A.5 (SI Units) of the Appendix.

16.3 A boom consists of a 100-mm-diameter steel pipe AB and a cable BC as shown. The modulus of steel $E = 200 \times 10^6$ kPa. Determine the capacity P of the boom. Use the Euler formula with a factor of safety of 3. Neglect the weight of the pipe. Section properties of pipe: $A = 2.05 \times 10^3$ mm^2 and the radius of gyration $r = 38.4$ mm.

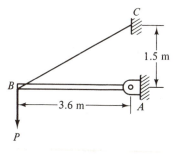

PROB. 16.3

16.4 A boom consists of a 4 in. diameter steel pipe AB and a cable as shown. The modulus of steel $E = 29 \times 10^3$ ksi. Determine the capacity P of the boom. Use the Euler formula with a factor of safety of 2.5. Neglect the weight of the pipe. Section properties of pipe: $A = 3.174$ in.^{2} and the radius of gyration $r = 1.51$ in.

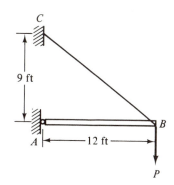

PROB. 16.4

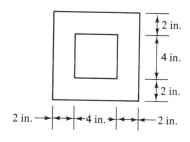

PROB. 16.5

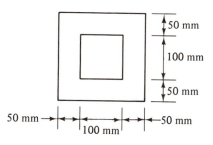

PROB. 16.6

16.5 The cross section of an 18-ft-long column is shown in the figure on page 455. Determine the slenderness ratio if (a) the ends are pinned, and (b) one end is fixed and the other end is free.

16.6 The cross section of a 5.5-m-long column is shown in the figure on page 455. Determine the slenderness ratio if (a) the ends are pinned, and (b) one end is fixed and the other end is free.

16.5 TANGENT MODULUS THEORY

The range of the Euler column formula can be extended if we define a new modulus called the *tangent modulus* E_T. The tangent modulus is the slope of the stress–strain curve at any point. It represents the instantaneous relationship between stress and strain for a particular value of stress. The stress–strain curve for a ductile material, together with the stress vs. the tangent modulus E_T curve for the same material, is shown in Fig. 16.7.

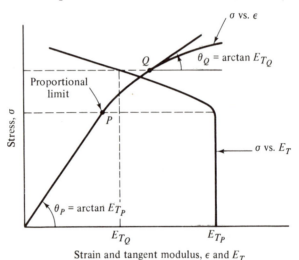

Strain and tangent modulus, ϵ and E_T

FIGURE 16.7

The slope of the stress–strain curve is a constant up to the proportional limit. Therefore, the tangent modulus is also a constant—shown in the figure as E_{T_P}. Above the proportional limit the slope drops rapidly as the stress and strain increase. This is shown in the figure for one value as E_{T_Q}. Substituting the tangent modulus E_T for the elastic modulus E in Eq. (16.5), we write

$$\sigma_C = \frac{\pi^2 E_T}{(kL/r)^2} \tag{16.6}$$

This equation is the *tangent modulus* or *Engesser formula*. Engesser first proposed the formula in 1889.

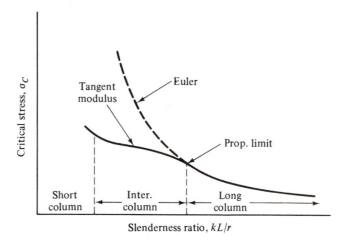

FIGURE 16.8

The Euler and tangent modulus formulas, Eqs. (16.5) and (16.6), have been plotted for the same ductile material in Fig. 16.8. Since the tangent modulus and the elastic modulus are identical below the proportional limit, the two curves for critical stress vs. the slenderness ratio coincide. Extensive test results indicate close agreement with the Euler column formula for long columns in which the critical stress falls below the proportional limit of the material and close agreement with the tangent modulus formula for intermediate columns in which the stress falls above the proportional limit.

Long columns buckle elastically and intermediate columns buckle inelastically (see Fig. 16.8). Short columns or compression blocks do not buckle. They fail by yielding for ductile materials and fracturing for brittle materials.

The tangent modulus formula requires a trial-and-error solution. Therefore, as is customary in practice, we will use empirical column formulas to find the allowable load or design load.

16.6 EMPIRICAL COLUMN FORMULAS—DESIGN FORMULAS

Many empirical formulas have been proposed and some are widely used. In general, these formulas are used to determine the allowable or working stress, and the allowable or working load is found by multiplying the stress by the cross-sectional area. Design formulas for structural steel, aluminum alloy, and wood are given in the following discussions.

Structural Steel Columns

The American Institute of Steel Construction (*AISC Steel Construction Manual*, New York: AISC, 1970) recommends a parabolic formula for short and intermediate columns and the Euler formula for long columns

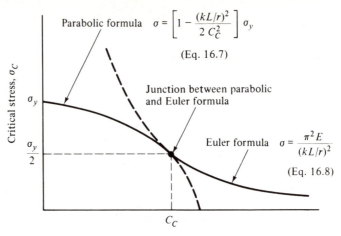

FIGURE 16.9 AISC column formulas. (Formulas do not include factors of safety.)

as shown in Fig. 16.9. Factors of safety are applied to the formulas to find the allowable load. The parabola extends from a stress equal to the yield stresses σ_y when $kL/r = 0$ to a stress equal to one-half the yield stress when $kL/r = C_c$. Because it joins the Euler curve at $kL/r = C_c$, we can evaluate C_c by equating the stress—one-half the yield stress—to the Euler formula, where $kL/r = C_c$. That is,

$$\frac{\sigma_y}{2} = \frac{\pi^2 E}{(C_c)^2} \quad \text{or} \quad C_c = \sqrt{\frac{2\pi^2 E}{\sigma_y}} \tag{a}$$

The parabolic equation is given by

$$\sigma_C = \frac{P}{A} = \left[1 - \frac{(kL/r)^2}{2C_c^2}\right]\sigma_y \quad \text{where } 0 \leqslant \frac{kL}{r} \leqslant C_c \tag{16.7}$$

and the Euler equation is given by

$$\sigma_C = \frac{P}{A} = \frac{\pi^2 E}{(kL/r)^2} \quad \text{where } C_c \leqslant \frac{kL}{r} \leqslant 200 \tag{16.8}$$

The modulus of elasticity for structural steel is taken as 29×10^3 ksi. Equation (16.7) is divided by a variable factor of safety, defined by

$$\text{F.S.} = \frac{5}{3} + \frac{3kL/r}{8C_c} - \frac{(kL/r)^3}{8C_c^3} \tag{16.9}$$

to give the allowable stress. The factor of safety has a value of 1.67 at $kL/r = 0$ and a value of 1.92 at $kL/r = C_c$. Equation (16.8) is divided by a factor of safety of $23/12 \approx 1.92$ to give the allowable stress. Therefore, the design equation becomes

$$\sigma_a = \frac{\left[1 - \frac{1}{2}\left(\frac{kL/r}{C_c}\right)^2\right]\sigma_y}{\text{F.S.}} \qquad \text{where } 0 \leqslant \frac{kL}{r} \leqslant C_c \qquad (16.10)$$

and

$$\sigma_a = \frac{149{,}000}{(kL/r)^2} \text{ (ksi)} \qquad \text{where } C_c \leqslant \frac{kL}{r} \leqslant 200 \qquad (16.11)$$

The parabolic equation [Eq. (16.7)] is used in machine design and is sometimes called the *J. B. Johnson formula*.

Aluminum Alloy Columns

The Aluminum Company of America (*ALCOA Structural Handbook*, Pittsburgh, PA: ALCOA, 1960) gives a straight-line formula for short intermediate columns and the Euler column formula for long columns as shown in Fig. 16.10. As an example, consider structural aluminum alloy 2014-T6 as given in the ALCOA handbook. The straight-line formula is

$$\sigma_C = 64.7 - 0.543 \frac{kL}{r} \text{ (ksi)} \qquad \text{where } \frac{kL}{r} < 54 \qquad (16.12)$$

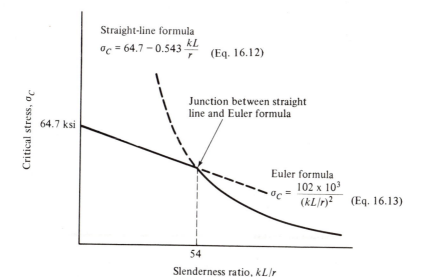

FIGURE 16.10 Alcoa column formulas for aluminum alloy 2014-T6. (Formulas do not include factor of safety.)

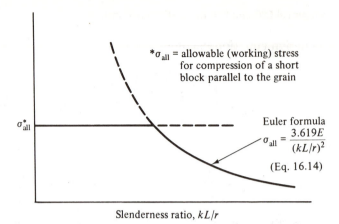

FIGURE 16.11 National Lumber Manufacturing Association formulas for wood. (Formulas include factor of safety.)

and the Euler formula is

$$\sigma_C = \frac{102,000}{(kL/r)^2} \text{ (ksi)} \qquad \text{where } \frac{kL}{r} \geqslant 54 \qquad (16.13)$$

Factors of safety must be used with Eqs. (16.12) and (16.13).

Wood Columns

The National Lumber Manufacturers Association (*NLMA National Design Specification*, Washington, D.C.: NLMA, 1960) suggests the use of the Euler column formula for long solid wood columns with a factor of safety F.S. = 2.727 and a straight horizontal line σ_a equal to allowable stress for compression of a short block parallel to the grain for the particular species of wood. The curves are shown in Fig. 16.11. The Euler formula is given by

$$\sigma_a = \frac{\pi^2 E}{2.727(kL/r)^2} = \frac{3.619E}{(kL/r)^2} \qquad (16.14)$$

EXAMPLE 16.2

Using the AISC column formulas, determine the allowable axial load for a W12 × 72 pin-ended column (a) 20 ft long, and (b) 40 ft long. The A36 structural steel used has a yield stress of 36 ksi.

Solution

The minimum radius of gyration for the W12 × 72 is r = 3.04 in. and the cross-sectional area A = 21.2 in.^{2}.

(a) The slenderness ratio for a column 20 ft long is

$$\frac{kL}{r} = \frac{1(20)(12)}{3.04} = 78.9$$

Because the slenderness ratio is less than

$$C_c = \sqrt{\frac{2\pi^2 E}{\sigma_y}} = \sqrt{\frac{2\pi^2 (29)(10)^3}{36}} = 126.1$$

we use the parabolic formula for stress. The ratio of the slenderness ratio to C_c,

$$\frac{kL/r}{C_c} = \frac{78.9}{126.1} = 0.626$$

From Eq. (16.9), the factor of safety

$$\text{F.S.} = \frac{5}{3} + \frac{3}{8}(0.626) - \frac{(0.626)^3}{8} = 1.87$$

The allowable stress is given by Eq. (16.10):

$$\sigma_a = \frac{\left[1 - \frac{1}{2}\left(\frac{kL/r}{C_c}\right)^2\right]\sigma_y}{\text{F.S.}} = \frac{[1 - \frac{1}{2}(0.626)^2]36}{1.87} = 15.5 \text{ ksi}$$

The allowable load

$$P_a = \sigma_a A = 15.5(21.2) = 329 \text{ kips} \qquad \text{Answer}$$

(b) The slenderness ratio for a column 40 ft long is

$$\frac{kL}{r} = \frac{1(40)(12)}{3.04} = 157.9$$

Because the slenderness ratio is greater than $C_c = 126.1$, we use the Euler formula for stress given by Eq. (16.11).

$$\sigma_a = \frac{149,000}{(kL/r)^2} = \frac{149,000}{(157.9)^2} = 5.98 \text{ ksi}$$

The allowable load

$$P_a = \sigma_a A = 5.98(21.2) = 126.7 \text{ kips} \qquad \text{Answer}$$

EXAMPLE 16.3

Determine the allowable axial load for a pin-ended aluminum alloy 2014-T6 angle L3 × 3 × $\frac{1}{4}$ (a) 4 ft long, and (b) 2.5 ft long. Use a factor of safety of 2.2.

Solution

We will assume that the section properties of aluminum angles are the same as those of steel. The minimum radius of gyration for the angle is $r = 0.592$ in. and the cross-sectional area $A = 1.44$ in.^{2}

(a) The slenderness ratio for a column 4 ft long is

$$\frac{kL}{r} = \frac{1(4)(12)}{0.592} = 81.1$$

Because the slenderness ratio is greater than 54 (see Fig. 16.9), we use the Euler column formula given by Eq. (16.13) divided by the factor of safety.

$$\sigma_a = \frac{102{,}000}{\text{F.S.}(kL/r)^2} = \frac{102{,}000}{2.2(81.1)^2} = 7.05 \text{ ksi}$$

The allowable load

$$P_a = \sigma_a A = 7.05(1.44) = 10.15 \text{ kips} \qquad\qquad \text{Answer}$$

(b) The slenderness ratio for a column 2.5 ft long is

$$\frac{kL}{r} = \frac{1(2.5)(12)}{0.592} = 50.7$$

Because the slenderness ratio is less than 54, we use the straight-line formula given by Eq. (16.12) divided by the factor of safety. That is,

$$\sigma_a = \frac{64.7 - 0.543(kL/r)}{\text{F.S.}} = \frac{64.7 - 0.543(50.7)}{2.2} = 16.89 \text{ ksi}$$

The allowable load

$$P_a = \sigma_a A = 16.89(1.44) = 24.3 \text{ kips} \qquad\qquad \text{Answer}$$

EXAMPLE 16.4

Find the allowable load for a pin-ended pine column 4 in. by 6 in. (a) 10 ft long, and (b) 7 ft long. The modulus $E = 1.6 \times 10^6$ psi and the allowable stress parallel to the grain for a short compression block $\sigma_a = 880$ psi.

Solution

The minimum moment of inertia $I = bh^3/12 = 6(4)^3/12$ or $I = 32$ in.^{4}. Therefore, the minimum radius of gyration $r = \sqrt{I/A} = \sqrt{32/24} = 1.155$ in.

(a) The slenderness ratio for a length of 10 ft is

$$\frac{kL}{r} = \frac{1(10)(12)}{1.155} = 103.9$$

The allowable stress from the Euler formula is given by Eq. (16.14).

$$\sigma_a = \frac{3.619E}{(kL/r)^2} = \frac{3.619(1.6)(10^6)}{(103.9)^2} = 536 \text{ psi}$$

which is less than $\sigma_{\text{all}} = 880$ psi. Therefore,

$$P_a = \sigma_a A = 536(24) = 12{,}870 \text{ lb.} \qquad \text{Answer}$$

(b) The slenderness ratio for a length of 7 ft is

$$\frac{kL}{r} = \frac{1(7)(12)}{1.155} = 72.73$$

The allowable stress, from Eq. (16.14),

$$\sigma_a = \frac{3.619E}{(kL/r)^2} = \frac{3.619(1.6)(10^6)}{(72.73)^2} = 1095 \text{ psi}$$

which is greater than $\sigma_{\text{all}} = 880$ psi. Therefore, we use a stress of 880 psi and the allowable load

$$P_a = \sigma_a A = 880(24) = 21{,}120 \text{ lb} \qquad \text{Answer}$$

PROBLEMS

16.7 The bracket shown supports a force P. Assume that both members have their ends pinned in the plane of the bracket ABC and fixed in a perpendicular plane. The modulus $E = 200 \times 10^6$ kPa and the yield stress $\sigma_y = 240 \times 10^3$ kPa. (a) Determine the load P that causes a critical stress in one of the members. Use either Eq. (16.7) or Eq. (16.8). (b) Determine the allowable load from part (a) if a factor of safety of 2 is used.

16.8 In the mechanism shown on page 464, called a toggle joint, a small force P can be used to exert a much larger force F. Assume that both members have their ends pinned in the plane of the mechanism ABC and fixed in a perpendicular plane. The modulus $E = 29 \times 10^3$ ksi and the yield stress $\sigma_y = 36$ ksi. (a) Determine the load P that will cause a critical stress in the member. Use either Eq. (16.7) or Eq.

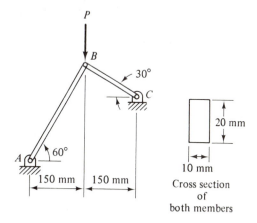

PROB. 16.7

(16.8). (b) Determine the allowable load from part (a) if a factor of safety of 2.5 is used.

16.9 What is the allowable axial load for a W10 × 45 steel section used as a column 14 ft long if the member has pin ends and is braced at the midpoint in

Cross section
of
both members

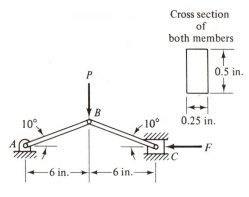

0.5 in.

0.25 in.

PROB. 16.8

the weak direction? $E = 29 \times 10^3$ ksi and $\sigma_y = 36$ ksi. Use AISC formulas—Eq. (16.10) or Eq. (16.11). For section properties, see Table A.5 of the Appendix.

16.10 What is the allowable axial load for a W250 × 67 section used as a column 4.5 m long if the member has pin ends and is braced at the midpoint in the weak direction? $E = 200 \times 10^6$ kPa and $\sigma_y = 250 \times 10^3$ kPa. Use AISC formulas—Eq. (16.10) or Eq. (16.8) together with a factor of safety of 23/12. For section properties, see Table A.5 (SI Units) of the Appendix.

16.11 Determine the allowable axial compressive load that can be supported by a steel column 10 m long with the cross section shown. $E = 200 \times 10^6$ kPa and $\sigma_y = 250 \times 10^3$ kPa. Use AISC formulas—Eq. (16.10) or Eq.

(16.8) together with a factor of safety of 23/12. For section properties, see Table A.8 (SI Units) of the Appendix.

16.12 Determine the allowable axial compressive load that can be supported by a steel column 40 ft long with the cross section shown. $E = 29 \times 10^3$ ksi and $\sigma_y = 36$ ksi. Use AISC formulas—Eq. (16.10) or Eq. (16.11). For section properties, see Table A.8 of the Appendix.

16.13 Solve Prob. 16.12 if the column is aluminum alloy 2014-T6. Use a factor of safety of 2.0 and the ALCOA formulas—Eq. (16.12) or Eq. (16.13).

16.14 Solve Prob. 16.11 if the column is aluminum alloy 2014-T6. Use a factor of safety of 2.0 and the ALCOA formulas in SI units.

$$\sigma_C = 446 - 3.74 \; \frac{kL}{r} \quad \text{(MPa)}$$

$$\text{where} \; \frac{kL}{r} < 54 \quad \text{(SI 16.12)}$$

$$\sigma_C = \frac{703 \times 10^3}{(kL/r)^2} \quad \text{(MPa)}$$

$$\text{where} \; \frac{kL}{r} \geqslant 54 \quad \text{(SI 16.13)}$$

16.15 Member *BC* of the pin-connected truss has a cross section as shown. $E = 200 \times 10^6$ kPa and $\sigma_y = 290 \times 10^3$ kPa. Determine the load *P* that will cause the allowable load in the member. Use AISC formulas—Eq. (16.10) or Eq. (16.8)—together with a factor of safety of 23/12.

16.16 Member *BC* of the pin-connected truss has a cross section as shown. $E = 29 \times 10^3$ ksi and $\sigma_y = 42$ ksi. Determine the load *P* that will cause the allowable load in the member. Use AISC formulas—Eq. (16.10) or Eq. (16.11).

16.17 Determine the allowable axial compressive load that can be supported

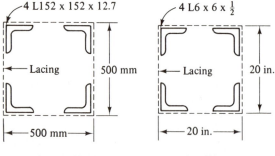

4 L152 x 152 x 12.7

Lacing   500 mm

500 mm

PROB. 16.11

4 L6 x 6 x $\frac{1}{2}$

Lacing   20 in.

20 in.

PROB. 16.12

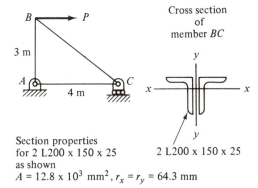

Section properties
for 2 L200 x 150 x 25
as shown
$A = 12.8 \times 10^3$ mm^2, $r_x = r_y = 64.3$ mm

PROB. 16.15

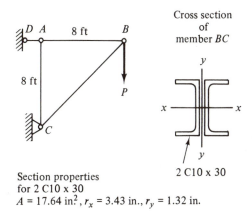

Section properties
for 2 C10 x 30
$A = 17.64$ in.^{2}, $r_x = 3.43$ in., $r_y = 1.32$ in.

PROB. 16.16

by a 10-in.-diameter wooden column
20 ft long. The modulus $E = 1.5 \times 10^6$
psi and the allowable stress for compression of a short block parallel
to the grain $\sigma_a = 1500$ psi. Use Eq.
(16.14).

16.18 Determine the allowable axial compressive load that can be supported
by a 250-mm-diameter wooden column 6.5 m long. The modulus $E =
10 \times 10^6$ kPa and the allowable stress
for compression of a short block
parallel to the grain $\sigma_a = 10 \times 10^3$
kPa. Use Eq. (16.14).

16.19 A wooden bar BD is loaded by cables
as shown. The bar has a cross section
75 mm by 75 mm. Determine the allowable load P based on buckling of
the bar BD. The modulus $E = 10 \times
10^6$ kPa and the allowable stress for
compression of a short block parallel
to the grain $\sigma_a = 10 \times 10^3$ kPa. Use
Eq. (16.14).

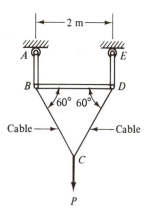

PROB. 16.19

16.20 A pin-ended wooden boom BC is supported by a cable as shown. The boom
has a cross section 5 in. by 5 in. Determine the allowable load P on the
boom. The modulus $E = 1.5 \times 10^6$ psi
and the allowable stress for compression of a short block parallel to the
grain $\sigma_a = 1500$ psi. Use Eq. (16.14).

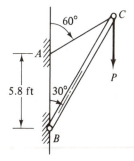

PROB. 16.20

Appendix _____

TABLES

Credits: Tables A.5 through A.10 are taken from or based on data in *AISC Steel Construction Manual*, New York: AISC, 1970. Publication is made by permission of the American Institute of Steel Construction, Inc. Tables in SI units are rounded conversions from common units by the author.

TABLE A.1
SI Prefixes

Multiplication Factor	Prefix	Symbol
$1\ 000\ 000\ 000 = 10^9$	giga	G
$1\ 000\ 000 = 10^6$	mega	M
$1\ 000 = 10^3$	kilo	k
$0.1 = 10^{-1}$	deci*	d
$0.01 = 10^{-2}$	centi*	c
$0.001 = 10^{-3}$	milli	m
$0.000\ 001 = 10^{-6}$	micro	μ (mu)

*These prefixes are not recommended except for decimeter in the measurement of areas and volumes and centimeter for nontechnical length measurements.

TABLE A.2
Conversion Factors

To Convert	To	Multiply by
From Common to SI Units		
Length		
inch (in.)	millimeter (mm)	25.40
inch (in.)	meter (m)	0.02540
foot (ft)	meter (m)	0.3048
Force		
pound (lb)	newton (N)	4.448
kilopound (kip)	kilonewton (kN)	4.448
Stress, pressure		
pound/inch^2 (lb/in.^{2})	newton/meter^2 (N/m^2)	6.895×10^3
pound/foot^2 (lb/ft^2)	newton/meter^2 (N/m^2)	47.88
From SI to Common Units		
Length		
millimeter (mm)	inch (in.)	0.03937
meter (m)	inch (in.)	39.37
meter (m)	foot (ft)	3.281
Force		
newton (N)	pound (lb)	0.2248
kilonewton (kN)	kilopound (kip)	0.2248
Stress, pressure		
newton/meter^2 (N/m^2)	pound/inch^2 (lb/in.^{2})	0.1450×10^{-3}
newton/meter^2 (N/m^2)	pound/foot^2 (lb/ft^2)	0.02089

TABLE A.3
Areas and Centroids of Simple Shapes

Shapes		Area	$\bar{x}$	$\bar{y}$
Rectangle		bh	$\dfrac{b}{2}$	$\dfrac{h}{2}$
Triangle		$\dfrac{bh}{2}$	$\dfrac{b}{3}$	$\dfrac{h}{3}$
Circle		$\dfrac{\pi d^2}{4}$	$\dfrac{d}{2}$	$\dfrac{d}{2}$
Semi-circle		$\dfrac{\pi d^2}{8}$	$\dfrac{d}{2}$	$\dfrac{4r}{3\pi}$
Quarter-circle		$\dfrac{\pi d^2}{16}$	$\dfrac{4r}{3\pi}$	$\dfrac{4r}{3\pi}$

TABLE A.4
Moments of Inertia of Simple Shapes

Shapes		I_x	I_{x_c}
Rectangle		$\dfrac{bh^3}{3}$	$\dfrac{bh^3}{12}$
Triangle		$\dfrac{bh^3}{12}$	$\dfrac{bh^3}{36}$
Circle		$\dfrac{5\pi r^4}{4}$	$\dfrac{\pi r^4}{4}$
Semicircle		$\dfrac{\pi r^4}{8}$	$0.0349\pi r^4$
Quarter-circle		$\dfrac{\pi r^4}{16}$	$0.01747\pi r^4$

TABLE A.5 (U.S. Common Units)
W Shapes: Properties for Designing
(Selected Listing)

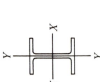

Designation	Area in.²	Depth in.	Flange Width in.	Flange Thickness in.	Web thickness in.	Axis X-X I in.⁴	Axis X-X S in.³	Axis X-X r in.	Axis Y-Y I in.⁴	Axis Y-Y S in.³	Axis Y-Y r in.
W36 × 260	76.5	36.24	16.551	1.440	0.841	17300	952	15.0	1090	132	3.77
× 160	47.1	36.00	12.000	1.020	0.653	9760	542	14.4	295	49.1	2.50
W33 × 220	64.8	33.25	15.810	1.275	0.775	12300	742	13.8	841	106	3.60
× 141	41.6	33.31	11.535	0.960	0.605	7460	448	13.4	246	42.7	2.43
W30 × 190	56.0	30.12	15.040	1.185	0.710	8850	587	12.6	673	89.5	3.47
× 116	34.2	30.00	10.500	0.850	0.564	4930	329	12.0	164	31.3	2.19
W27 × 160	47.1	27.08	14.023	1.075	0.658	6030	446	11.3	495	70.6	3.24
× 102	30.0	27.07	10.018	0.827	0.518	3610	267	11.0	139	27.7	2.15
W24 × 145	42.7	24.49	14.043	1.020	0.608	4570	373	10.3	471	67.1	3.32
× 110	32.5	24.16	12.042	0.855	1.510	3330	276	10.1	249	41.4	2.77
× 84	24.7	24.09	9.015	0.772	0.470	2370	197	9.79	94.5	21.0	1.95
W21 × 127	37.4	21.24	13.061	0.985	0.588	3020	284	8.99	366	56.1	3.13
× 96	28.3	21.14	9.038	0.935	0.575	2100	198	8.61	115	25.5	2.02
× 68	20.0	21.13	8.270	0.685	0.430	1480	140	8.60	64.7	15.7	1.80
× 55	16.2	20.80	8.215	0.522	0.375	1140	110	8.40	48.3	11.8	1.73
W18 × 114	33.5	18.48	11.833	0.991	0.595	2040	220	7.79	274	46.3	2.86
× 105	30.9	18.32	11.792	0.911	0.554	1850	202	7.75	249	42.3	2.84
× 70	20.6	18.00	8.750	0.751	0.438	1160	129	7.50	84.0	19.2	2.02
× 55	16.2	18.12	7.532	0.630	0.390	891	98.4	7.42	45.0	11.9	1.67
× 35	10.3	17.71	6.000	0.429	0.298	513	57.9	7.05	15.5	5.16	1.23

Elastic properties

TABLE A.5 (U.S. Common Units) (*cont.*)

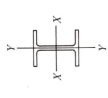

Designation	Area in.²	Depth in.	Flange Width in.	Flange Thickness in.	Web thickness in.	Axis X-X I in.⁴	Axis X-X S in.³	Axis X-X r in.	Axis Y-Y I in.⁴	Axis Y-Y S in.³	Axis Y-Y r in.
W16 × 96	28.2	16.32	11.533	0.875	0.535	1360	166	6.93	224	38.8	2.82
× 78	23.0	16.32	8.586	0.875	0.529	1050	128	6.75	92.5	21.6	2.01
× 64	18.8	16.00	8.500	0.715	0.443	836	104	6.66	73.3	17.3	1.97
× 50	14.7	16.25	7.073	0.628	0.380	657	80.8	6.68	37.1	10.5	1.59
× 36	10.6	15.85	6.992	0.428	0.299	447	56.5	6.50	24.4	6.99	1.52
× 26	7.67	15.65	5.500	0.345	0.250	300	38.3	6.25	9.59	3.49	1.12
W14 × 184	54.1	15.38	15.660	1.378	0.840	2270	296	6.49	883	113	4.04
× 167	49.1	15.12	15.600	1.248	0.780	2020	267	6.42	790	101	4.01
× 150	44.1	14.88	15.515	1.128	0.695	1790	240	6.37	703	90.6	3.99
× 136	40.0	14.75	14.740	1.063	0.660	1590	216	6.31	568	77.0	3.77
× 95	27.9	14.12	14.545	0.748	0.465	1060	151	6.17	384	52.8	3.71
× 74	21.8	14.19	10.072	0.783	0.450	797	112	6.05	133	26.5	2.48
× 68	20.0	14.06	10.040	0.718	0.418	724	103	6.02	121	24.1	2.46
× 61	17.9	13.91	10.000	0.643	0.378	641	92.2	5.98	107	21.5	2.45
× 53	15.6	13.94	8.062	0.658	0.370	542	77.8	5.90	57.5	14.3	1.92
× 43	12.6	13.68	8.000	0.528	0.308	429	62.7	5.82	45.1	11.3	1.89
× 38	11.2	14.12	6.776	0.513	0.313	386	54.7	5.88	26.6	7.86	1.54
× 34	10.0	14.00	6.750	0.453	0.287	340	48.6	5.83	23.3	6.89	1.52
× 30	8.83	13.86	6.733	0.383	0.270	290	41.9	5.74	19.5	5.80	1.49
W12 × 92	27.1	12.62	12.155	0.856	0.545	789	125	5.40	256	42.2	3.08
× 72	21.2	12.25	12.040	0.671	0.430	597	97.5	5.31	195	32.4	3.04
× 58	17.1	12.19	10.014	0.641	0.359	476	78.1	5.28	107	21.4	2.51
× 45	13.2	12.06	8.042	0.576	0.336	351	58.2	5.15	50.0	12.4	1.94
× 31	9.13	12.09	6.525	0.465	0.265	239	39.5	5.12	21.6	6.61	1.54
× 22	6.47	12.31	4.030	0.424	0.260	156	25.3	4.91	4.64	2.31	0.847
× 16.5	4.87	12.00	4.000	0.269	0.230	105	17.6	4.65	2.88	1.44	0.770

Elastic properties

TABLE A.5 (U.S. Common Units) (cont.)

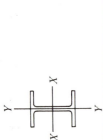

Designation	Area	Depth	Flange		Web thick-ness	Elastic properties					
			Width	Thick-ness		Axis X-X			Axis Y-Y		
						I	S	r	I	S	r
	in.²	in.	in.	in.	in.	in.⁴	in.³	in.	in.⁴	in.³	in.
W10 × 112	32.9	11.38	10.415	1.248	0.755	719	126	4.67	235	45.2	2.67
× 100	29.4	11.12	10.345	1.118	0.685	625	112	4.61	207	39.9	2.65
× 89	26.2	10.88	10.275	0.998	0.615	542	99.7	4.55	181	35.2	2.63
× 72	21.2	10.50	10.170	0.808	0.510	421	80.1	4.46	142	27.9	2.59
× 54	15.9	10.12	10.028	0.618	0.368	306	60.4	4.39	104	20.7	2.56
× 45	13.2	10.12	8.022	0.618	0.350	249	49.1	4.33	53.2	13.3	2.00
× 33	9.71	9.75	7.964	0.433	0.292	171	35.0	4.20	36.5	9.16	1.94
× 25	7.36	10.08	5.762	0.430	0.252	133	26.5	4.26	13.7	4.76	1.37
× 21	6.20	9.90	5.750	0.340	0.240	107	21.5	4.15	10.8	3.75	1.32
× 19	5.61	10.25	4.020	0.394	0.250	96.3	18.8	4.14	4.28	2.13	0.874
× 11.5	3.39	9.87	3.950	0.204	0.180	52.0	10.5	3.92	2.10	1.06	0.787
W8 × 67	19.7	9.00	8.287	0.933	0.575	272	60.4	3.71	88.6	21.4	2.12
× 58	17.1	8.75	8.222	0.808	0.510	227	52.0	3.65	74.9	18.2	2.10
× 48	14.1	8.50	8.117	0.683	0.405	184	43.2	3.61	60.9	15.0	2.08
× 40	11.8	8.25	8.077	0.558	0.365	146	35.5	3.53	49.0	12.1	2.04
× 35	10.3	8.12	8.027	0.493	0.315	126	31.1	3.50	42.5	10.6	2.03
× 31	9.12	8.00	8.000	0.433	0.288	110	27.4	3.47	37.0	9.24	2.01
× 28	8.23	8.06	6.540	0.463	0.285	97.8	24.3	3.45	21.6	6.61	1.62
× 24	7.06	7.93	6.500	0.398	0.245	82.5	20.8	3.42	18.2	5.61	1.61
× 20	5.89	8.14	5.268	0.378	0.248	69.4	17.0	3.43	9.22	3.50	1.25
× 15	4.43	8.12	4.015	0.314	0.245	48.1	11.8	3.29	3.40	1.69	0.876

TABLE A.5 (SI Units)
W Shapes: Properties for Designing
(Selected Listing)

Designation	Area	Depth	Flange Width	Flange Thickness	Web thickness	Axis X-X I	Axis X-X S	Axis X-X r	Axis Y-Y I	Axis Y-Y S	Axis Y-Y r
	mm^2	mm	mm	mm	mm	$\times 10^6\ mm^4$	$\times 10^3\ mm^3$	mm	$\times 10^6\ mm^4$	$\times 10^3\ mm^3$	mm
W920 × 387	49 350	920.5	420.4	36.6	21.4	7200	15 600	381	453	2160	95.8
× 238	30 390	914.4	304.8	25.9	16.6	4060	8 880	366	123	805	63.5
W840 × 327	41 800	844.6	401.6	32.4	19.7	5120	12 200	351	350	1740	91.4
× 208	26 840	846.1	293.0	24.4	15.4	3110	7 340	340	102	700	61.7
W760 × 283	36 130	765.0	382.0	30.1	18.0	3680	9 520	320	280	1470	88.1
× 173	22 060	762.0	266.7	21.6	14.3	2050	5 390	305	68.3	513	55.6
W690 × 238	30 390	687.8	356.2	27.3	16.7	2510	7 310	287	206	1160	82.3
× 152	19 350	687.6	254.5	21.0	13.2	1500	4 380	279	57.9	454	54.6
W610 × 216	27 550	622.0	356.7	28.0	15.4	1900	6 110	262	196	1100	84.3
× 164	20 970	613.7	305.9	21.7	13.0	1390	4 520	257	104	678	70.4
× 125	15 940	611.9	229.0	19.6	11.9	986	3 230	249	39.3	344	49.5
W530 × 189	24 130	539.5	331.7	25.0	14.9	1260	4 650	228	152	919	79.5
× 143	18 260	537.0	229.6	23.7	14.6	874	3 240	219	47.9	418	51.3
× 101	12 900	536.7	210.1	17.4	10.9	616	2 290	218	26.9	257	45.7
× 82	10 450	528.3	208.7	13.3	9.5	475	1 800	213	20.1	193	43.9
W460 × 170	21 610	469.4	300.6	25.2	15.1	849	3 610	198	114	759	72.6
× 156	19 940	465.3	299.5	25.2	14.1	770	3 310	197	104	693	72.1
× 104	13 290	457.2	222.3	19.1	11.1	483	2 110	190	35.0	315	51.3
× 82	10 450	460.2	191.3	16.0	9.9	371	1 610	188	18.7	195	42.4
× 52	6 650	450.0	152.4	10.9	7.6	214	949	179	6.45	84.6	31.2

Elastic properties

TABLE A.5 (SI Units) (cont.)

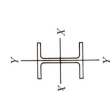

Designation	Area	Depth	Flange		Web thick-ness	Elastic properties					
			Width	Thick-ness		Axis X–X			Axis Y–Y		
						I	S	r	I	S	r
	mm^2	mm	mm	mm	mm	$\times 10^6$ mm^4	$\times 10^3$ mm^3	mm	$\times 10^6$ mm^4	$\times 10^3$ mm^3	mm
W410 × 143	18 190	414.5	292.9	22.2	13.6	566	2 720	176	93.2	636	71.6
× 116	14 840	414.5	218.1	22.2	13.4	437	2 100	171	38.5	354	51.1
× 95	12 130	406.4	215.9	18.7	11.3	348	1 700	169	30.5	283	50.0
× 74	9 480	412.8	179.7	16.0	9.7	273	1 320	170	15.4	172	40.4
× 54	6 840	402.6	177.6	10.9	7.6	186	926	165	10.2	115	38.6
× 39	4 950	397.5	139.7	8.8	6.4	125	628	159	3.99	57.2	28.4
W360 × 274	34 900	390.7	397.8	35.0	21.3	945	4 850	165	368	1850	103
× 249	31 680	384.0	396.2	31.7	19.8	841	4 380	163	329	1660	102
× 223	28 450	378.0	394.1	28.7	17.7	745	3 930	162	293	1480	101
× 202	25 800	374.7	374.4	27.0	16.8	662	3 540	160	236	1260	95.8
× 141	18 000	358.6	369.4	19.0	11.8	441	2 470	157	160	865	94.2
× 110	14 060	360.4	255.8	19.9	11.4	332	1 840	154	55.4	433	63.0
× 101	12 900	357.1	255.0	18.2	10.6	301	1 690	153	50.4	395	62.5
× 91	11 550	353.3	254.0	16.3	9.6	267	1 510	152	44.5	351	62.2
× 79	10 060	354.1	204.8	16.7	9.4	226	1 270	150	23.9	233	48.8
× 64	8 130	347.5	203.2	13.4	7.8	179	1 030	148	18.8	185	48.0
× 57	7 230	358.6	172.1	13.0	8.0	161	896	149	11.1	129	39.1
× 51	6 450	355.6	171.5	11.5	7.3	142	796	148	9.70	113	38.6
× 45	5 700	352.0	171.0	9.7	6.9	120	687	146	8.12	95.0	37.8
W310 × 137	17 480	320.5	308.7	21.7	13.8	328	2 050	137	107	692	78.2
× 107	13 680	311.1	305.8	17.0	10.9	248	1 600	135	81.2	531	77.2
× 86	11 030	309.6	254.4	16.3	9.1	198	1 280	134	44.5	351	63.8
× 67	8 520	306.3	204.3	14.6	8.5	146	954	131	20.8	203	49.3
× 46	5 890	307.1	165.7	11.8	6.7	99.5	647	130	8.99	108	39.1
× 33	4 170	312.7	102.4	10.8	6.6	64.9	415	125	1.93	37.9	21.5
× 25	3 140	304.8	101.6	6.8	5.8	43.7	288	118	1.20	23.6	19.6

TABLE A.5 (SI Units) (cont.)

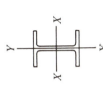

Designation	Area	Depth	Flange Width	Flange Thickness	Web thickness	Axis X-X			Axis Y-Y		
	mm^2	mm	mm	mm	mm	I $\times 10^6$ mm^4	S $\times 10^3$ mm^3	r mm	I $\times 10^6$ mm^4	S $\times 10^3$ mm^3	r mm
W250 × 167	21 230	289.1	264.5	31.7	19.2	299	2 060	119	97.8	741	67.8
× 149	18 970	282.4	262.8	28.4	17.4	260	1 840	117	86.2	654	67.3
× 132	15 900	276.4	261.0	25.3	15.6	226	1 630	116	75.3	577	66.8
× 107	13 680	266.7	258.3	20.5	13.0	175	1 310	113	59.1	457	65.8
× 80	10 260	257.0	254.7	15.7	9.3	127	990	112	43.2	339	65.0
× 67	8 520	257.0	203.8	15.7	8.9	104	805	110	22.1	218	50.8
× 49	6 260	247.6	202.3	11.0	7.4	71.2	574	107	15.2	150	49.3
× 37	4 750	256.0	146.4	10.9	6.4	55.4	434	108	5.70	78.0	34.8
× 31	4 000	251.5	146.0	8.6	6.1	44.5	352	105	4.50	61.5	33.5
× 28	3 620	260.4	102.1	10.0	6.4	40.1	308	105	1.78	34.9	22.2
× 17	2 190	250.7	100.3	5.2	4.6	21.6	172	99.6	0.874	17.4	20.0
W200 × 100	12 710	228.6	210.5	23.7	14.6	113	990	94.2	36.9	351	53.8
× 86	11 030	222.3	208.8	20.5	13.0	94.5	852	92.7	31.2	298	53.3
× 71	9 100	215.9	206.2	17.3	10.3	76.6	708	91.7	25.3	246	52.8
× 60	7 610	209.6	205.2	14.2	9.3	60.8	582	89.7	20.4	198	51.8
× 52	6 650	206.2	203.9	12.5	8.0	52.4	510	88.9	17.7	174	51.6
× 46	5 880	203.2	203.2	11.0	7.3	45.8	450	88.1	15.4	151	51.1
× 42	5 310	204.7	166.1	11.8	7.2	40.7	398	87.6	8.99	108	41.1
× 36	4 560	201.4	165.1	10.1	6.2	34.3	341	86.9	7.58	91.9	40.9
× 30	3 800	206.8	133.8	9.6	6.3	28.9	279	87.1	3.84	57.4	31.8
× 22	2 860	206.2	102.0	8.0	6.2	20.0	193	83.6	1.42	27.7	22.3

TABLE A.6 (U.S. Common Units)
S Shapes: Properties for Designing

Designation	Area in.²	Depth in.	Flange Width in.	Flange Thickness in.	Web thickness in.	Axis X-X I in.⁴	Axis X-X S in.³	Axis X-X r in.	Axis Y-Y I in.⁴	Axis Y-Y S in.³	Axis Y-Y r in.
S24 × 120	35.3	24.00	8.048	1.102	0.798	3030	252	9.26	84.2	20.9	1.54
× 105.9	31.1	24.00	7.875	1.102	0.625	2830	236	9.53	78.2	19.8	1.58
× 100	29.4	24.00	7.247	0.871	0.747	2390	199	9.01	47.8	13.2	1.27
× 90	26.5	24.00	7.124	0.871	0.624	2250	187	9.22	44.9	12.6	1.30
× 79.9	23.5	24.00	7.001	0.871	0.501	2110	175	9.47	42.3	12.1	1.34
S20 × 95	27.9	20.00	7.200	0.916	0.800	1610	161	7.60	49.7	13.8	1.33
× 85	25.0	20.00	7.053	0.916	0.653	1520	152	7.79	46.2	13.1	1.36
× 75	22.1	20.00	6.391	0.789	0.641	1280	128	7.60	29.6	9.28	1.16
× 65.4	19.2	20.00	6.250	0.789	0.500	1180	118	7.84	27.4	8.77	1.19
S18 × 70	20.6	18.00	6.251	0.691	0.711	926	103	6.71	24.1	7.72	1.08
× 54.7	16.1	18.00	6.001	0.691	0.461	804	89.4	7.07	20.8	6.94	1.14
S15 × 50	14.7	15.00	5.640	0.622	0.550	486	64.8	5.75	15.7	5.57	1.03
× 42.9	12.6	15.00	5.501	0.622	0.411	447	59.6	5.95	14.4	5.23	1.07
S12 × 50	14.7	12.00	5.477	0.659	0.687	305	50.8	4.55	15.7	5.74	1.03
× 40.8	12.0	12.00	5.252	0.659	0.472	272	45.4	4.77	13.6	5.16	1.06
× 35	10.3	12.00	5.078	0.544	0.428	229	38.2	4.72	9.87	3.89	0.980
× 31.8	9.35	12.00	5.000	0.544	0.350	218	36.4	4.83	9.36	3.74	1.00

Elastic properties

TABLE A.6 (U.S. Common Units) *(cont.)*

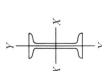

Designation	Area in.²	Depth in.	Flange Width in.	Flange Thickness in.	Web thickness in.	Axis X-X I in.⁴	Axis X-X S in.³	Axis X-X r in.	Axis Y-Y I in.⁴	Axis Y-Y S in.³	Axis Y-Y r in.
S10 X 35	10.3	10.00	4.944	0.491	0.594	147	29.4	3.78	8.36	3.38	0.901
X 25.4	7.46	10.00	4.661	0.491	0.311	124	24.7	4.07	6.79	2.91	0.954
S8 X 23	6.77	8.00	4.171	0.425	0.441	64.9	16.2	3.10	4.31	2.07	0.798
X 18.4	5.41	8.00	4.001	0.425	0.271	57.6	14.4	3.26	3.73	1.86	0.831
S7 X 20	5.88	7.00	3.860	0.392	0.450	42.4	12.1	2.69	3.17	1.64	0.734
X 15.3	4.50	7.00	3.662	0.392	0.252	36.7	10.5	2.86	2.64	1.44	0.766
S6 X 17.25	5.07	6.00	3.565	0.359	0.465	26.3	8.77	2.28	2.31	1.30	0.675
X 12.5	3.67	6.00	3.332	0.359	0.232	22.1	7.37	2.45	1.82	1.09	0.705
S5 X 14.75	4.34	5.00	3.284	0.326	0.494	15.2	6.09	1.87	1.67	1.01	0.620
X 10	2.94	5.00	3.004	0.326	0.214	12.3	4.92	2.05	1.22	0.809	0.643
S4 X 9.5	2.79	4.00	2.796	0.293	0.326	6.79	3.39	1.56	0.903	0.646	0.569
X 7.7	2.26	4.00	2.663	0.293	0.193	6.08	3.04	1.64	0.764	0.574	0.581
S3 X 7.5	2.21	3.00	2.509	0.260	0.349	2.93	1.95	1.15	0.586	0.468	0.516
X 5.7	1.67	3.00	2.330	0.260	0.170	2.52	1.68	1.23	0.455	0.390	0.522

Elastic properties

TABLE A.6 (SI Units)
S Shapes: Properties for Designing

| Designation | Area | Depth | Flange | | Web thickness | Elastic properties | | | | | |
| | | | Width | Thickness | | Axis X-X | | | Axis Y-Y | | |
	mm^2	mm	mm	mm	mm	I $\times 10^6$ mm^4	S $\times 10^3$ mm^3	r mm	I $\times 10^6$ mm^4	S $\times 10^3$ mm^3	r mm
S610 × 179	22 770	609.6	204.4	28.0	20.3	1260	4130	235	35.0	342	39.1
× 157.6	20 060	609.6	200.0	28.0	15.9	1180	3870	242	32.5	324	40.1
× 149	18 970	609.6	184.1	22.1	19.0	995	3260	229	19.9	216	32.3
× 134	17 100	609.6	180.9	22.1	15.8	937	3060	234	18.7	206	33.0
× 118.9	15 160	609.6	177.8	22.1	12.7	878	2870	241	17.6	198	34.0
S510 × 141	18 000	508.0	182.9	23.3	20.3	670	2640	193	20.7	226	33.8
× 126	16 130	508.0	179.1	23.3	16.6	633	2490	198	19.2	215	34.5
× 112	14 260	508.0	162.3	20.0	16.3	533	2100	193	12.3	152	29.5
× 97.3	12 390	508.0	158.8	20.0	12.7	491	1930	199	11.4	144	30.2
S460 × 104	13 290	457.2	158.8	17.6	18.1	385	1690	170	10.0	127	27.4
× 81.4	10 390	457.2	152.4	17.6	11.7	335	1470	180	8.66	114	29.0
S380 × 74	9 480	381.0	143.3	15.8	14.0	202	1060	146	6.53	91.3	26.2
× 63.8	8 130	381.0	139.7	15.8	10.4	186	977	151	5.99	85.7	27.2
S300 × 74	9 480	304.8	139.1	16.7	17.4	127	832	116	6.53	94.1	26.2
× 60.7	7 740	304.8	133.4	16.7	12.0	113	744	121	5.66	84.6	26.9
× 62	6 650	304.8	129.0	13.8	10.9	95.3	626	120	4.11	63.7	24.9
× 47.3	6 030	304.8	127.0	13.8	8.9	90.7	596	123	3.90	61.3	25.4

TABLE A.6 (SI Units) (cont.)

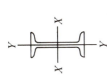

Designation	Area	Depth	Flange		Web thickness	Elastic properties					
			Width	Thickness		Axis X-X			Axis Y-Y		
						I	S	r	I	S	r
	mm²	mm	mm	mm	mm	×10⁶ mm⁴	×10³ mm³	mm	×10⁶ mm⁴	×10³ mm³	mm
S250 × 52	6 650	254.0	125.6	12.5	15.1	61.2	482	96.0	3.48	55.4	22.9
× 37.8	4 810	254.0	118.4	12.5	7.9	51.6	405	103	2.83	47.7	24.2
S200 × 34	4 370	203.2	105.9	10.8	11.2	27.0	265	78.7	1.79	33.9	20.3
× 27.4	3 490	203.2	101.6	10.8	6.9	24.0	236	82.8	1.55	30.5	21.1
S180 × 30	3 790	177.8	98.0	10.0	11.4	17.6	198	68.3	1.32	26.9	18.6
× 22.8	2 900	177.8	93.0	10.0	6.4	15.3	172	72.6	1.10	23.6	19.5
S150 × 25.7	3 270	152.4	90.6	9.1	11.8	10.9	144	57.9	0.961	21.3	17.1
× 18.6	2 370	152.4	84.6	9.1	5.9	9.20	121	62.2	0.758	17.9	17.9
S130 × 22.0	2 800	127.0	83.4	8.3	12.6	6.33	99.8	47.5	0.695	16.6	15.7
× 14.9	1 900	127.0	76.3	8.3	5.4	5.12	80.6	52.1	0.508	13.3	16.3
S100 × 14.1	1 800	101.6	71.0	7.4	8.3	2.83	55.6	39.6	0.376	10.6	14.5
× 11.5	1 460	101.6	67.6	7.4	4.9	2.53	49.8	41.7	0.318	9.41	14.8
S75 × 11.2	1 430	76.2	63.7	6.6	8.9	1.22	32.0	29.2	0.244	7.67	13.1
× 8.5	1 080	76.2	59.2	6.6	4.3	1.05	27.5	31.2	0.189	6.39	13.3

TABLE A.7 (U.S. Common Units)
Channels American Standard: Properties for Designing

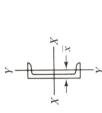

			Flange		Web thick-ness	Elastic properties						
						Axis X-X			Axis Y-Y			
Designation	Area	Depth	Width	Thickness		I	S	r	I	S	r	$\bar{x}$
	in.²	in.	in.	in.	in.	in.⁴	in.³	in.	in.⁴	in.³	in.	in.
C15 × 50	14.7	15.00	3.716	0.650	0.716	404	53.8	5.24	11.0	3.78	0.867	0.799
× 40	11.8	15.00	3.520	0.650	0.520	349	46.5	5.44	9.23	3.36	0.886	0.778
× 33.9	9.96	15.00	3.400	0.650	0.400	315	42.0	5.62	8.13	3.11	0.904	0.787
C12 × 30	8.82	12.00	3.170	0.501	0.510	162	27.0	4.29	5.14	2.06	0.763	0.674
× 25	7.35	12.00	3.047	0.501	0.387	144	24.1	4.43	4.47	1.88	0.780	0.674
× 20.7	6.09	12.00	2.942	0.501	0.282	129	21.5	4.61	3.88	1.73	0.799	0.698
C10 × 30	8.82	10.00	3.033	0.436	0.673	103	20.7	3.42	3.94	1.65	0.669	0.649
× 25	7.35	10.00	2.886	0.436	0.526	91.2	18.2	3.52	3.36	1.48	0.676	0.617
× 20	5.88	10.00	2.739	0.436	0.379	78.9	15.8	3.66	2.81	1.32	0.691	0.606
× 15.3	4.49	10.00	2.600	0.436	0.240	67.4	13.5	3.87	2.28	1.16	0.713	0.634
C9 × 20	5.88	9.00	2.648	0.413	0.448	60.9	13.5	3.22	2.42	1.17	0.642	0.583
× 15	4.41	9.00	2.485	0.413	0.285	51.0	11.3	3.40	1.93	1.01	0.661	0.586
× 13.4	3.94	9.00	2.433	0.413	0.233	47.9	10.6	3.48	1.76	0.962	0.668	0.601
C8 × 18.75	5.51	8.00	2.527	0.390	0.487	44.0	11.0	2.82	1.98	1.01	0.599	0.565
× 13.75	4.04	8.00	2.343	0.390	0.303	36.1	9.03	2.99	1.53	0.853	0.615	0.553
× 11.5	3.38	8.00	2.260	0.390	0.220	32.6	8.14	3.11	1.32	0.781	0.625	0.571

TABLE A.7 (U.S. Common Units) *(cont.)*

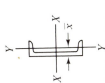

Designation	Area in.²	Depth in.	Flange Width in.	Flange Thickness in.	Web thickness in.	Axis X-X I in.⁴	Axis X-X S in.³	Axis X-X r in.	Axis Y-Y I in.⁴	Axis Y-Y S in.³	Axis Y-Y r in.	$\bar{x}$ in.
C7 X 14.75	4.33	7.00	2.299	0.366	0.419	27.2	7.78	2.51	1.38	0.779	0.564	0.532
X 12.25	3.60	7.00	2.194	0.366	0.314	24.2	6.93	2.60	1.17	0.702	0.571	0.525
X 9.8	2.87	7.00	2.090	0.366	0.210	21.3	6.08	2.72	0.968	0.625	0.581	0.541
C6 X 13	3.83	6.00	2.157	0.343	0.437	17.4	5.80	2.13	1.05	0.642	0.525	0.514
X 10.5	3.09	6.00	2.034	0.343	0.314	15.2	5.06	2.22	0.865	0.564	0.529	0.500
X 8.2	2.40	6.00	1.920	0.343	0.200	13.1	4.38	2.34	0.692	0.492	0.537	0.512
C5 X 9	2.64	5.00	1.885	0.320	0.325	8.90	3.56	1.83	0.632	0.449	0.489	0.478
X 6.7	1.97	5.00	1.750	0.320	0.190	7.49	3.00	1.95	0.478	0.378	0.493	0.484
C4 X 7.25	2.13	4.00	1.721	0.296	0.321	4.59	2.29	1.47	0.432	0.343	0.450	0.459
X 5.4	1.59	4.00	1.584	0.296	0.184	3.85	1.93	1.56	0.319	0.283	0.449	0.458
C3 X 6	1.76	3.00	1.596	0.273	0.356	2.07	1.38	1.08	0.305	0.268	0.416	0.455
X 5	1.47	3.00	1.498	0.273	0.258	1.85	1.24	1.12	0.247	0.233	0.410	0.438
X 4.1	1.21	3.00	1.410	0.273	0.170	1.66	1.10	1.17	0.197	0.202	0.404	0.437

Elastic properties

TABLE A.7 (SI Units)
Channels American Standard: Properties for Designing

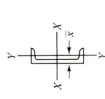

Designation	Area	Depth	Flange Width	Thickness	Web thickness	Axis X-X I	S	r	Axis Y-Y I	S	r	$\bar{x}$
	mm^2	mm	mm	mm	mm	$\times 10^6$ mm^4	$\times 10^3$ mm^3	mm	$\times 10^6$ mm^4	$\times 10^3$ mm^3	mm	mm
C375 × 74.4	9480	381.0	94.4	16.5	18.2	168	882	133	4.58	61.9	22.0	20.3
× 59.5	7610	381.0	89.4	16.5	13.2	145	762	138	3.84	55.1	22.5	19.8
× 50.4	6430	381.0	86.4	16.5	10.2	131	688	143	3.38	51.0	23.0	20.0
C300 × 44.6	5690	304.8	80.5	12.7	13.0	67.4	442	109	2.14	33.8	19.4	17.1
× 37.2	4740	304.8	77.4	12.7	9.8	59.9	395	113	1.86	30.8	19.8	17.1
× 30.8	3930	304.8	74.7	12.7	7.2	53.7	352	117	1.61	28.3	20.3	17.7
C250 × 44.6	5690	254.0	77.0	11.1	17.1	42.9	339	86.9	1.64	27.0	17.0	16.5
× 37.2	4740	254.0	73.3	11.1	13.4	38.0	298	89.4	1.40	24.3	17.2	15.7
× 29.8	3790	254.0	69.6	11.1	9.6	32.8	259	93.0	1.17	21.6	17.6	15.4
× 22.8	2900	254.0	66.0	11.1	6.1	28.1	221	98.3	0.949	19.0	18.1	16.1
C225 × 29.8	3790	228.6	67.3	10.5	11.4	25.3	221	81.8	1.01	19.2	16.3	14.8
× 22.3	2850	228.6	63.1	10.5	7.2	21.2	185	86.4	0.803	16.6	16.8	14.9
× 19.9	2540	228.6	61.8	10.5	5.9	19.9	174	88.4	0.733	15.8	17.0	15.3
C200 × 27.9	3550	203.2	64.2	9.9	12.4	18.3	180	71.6	0.824	16.6	15.2	14.4
× 20.5	2610	203.2	59.5	9.9	7.7	15.0	148	75.9	0.637	14.0	15.6	14.0
× 17.1	2180	203.2	57.4	9.9	5.6	13.6	133	79.0	0.549	12.8	15.9	14.5

Elastic properties

TABLE A.7 (SI Units) (cont.)

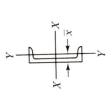

Designation	Area mm²	Depth mm	Flange Width mm	Flange Thickness mm	Web thickness mm	Axis X-X I ×10⁶ mm⁴	Axis X-X S ×10³ mm³	Axis X-X r mm	Axis Y-Y I ×10⁶ mm⁴	Axis Y-Y S ×10³ mm³	Axis Y-Y r mm	Axis Y-Y $\bar{x}$ mm
C175 × 22.0	2790	177.8	58.4	9.3	10.6	11.3	127	63.8	0.574	12.8	14.3	13.5
× 18.2	2320	177.8	55.7	9.3	8.0	10.1	114	66.0	0.487	11.5	14.5	13.3
× 14.6	1850	177.8	53.1	9.3	5.3	8.87	99.6	69.1	0.403	10.2	14.8	13.7
C150 × 19.5	2470	152.4	54.8	8.7	11.1	7.24	95.0	54.1	0.437	10.5	13.3	13.1
× 15.6	1990	152.4	51.7	8.7	8.0	6.33	82.9	56.4	0.360	9.24	13.4	12.7
× 12.2	1550	152.4	48.8	8.7	5.1	5.45	71.8	59.4	0.288	8.06	13.6	13.0
C125 × 13.4	1700	127.0	47.9	8.1	8.3	3.70	58.3	46.5	0.263	7.36	12.4	12.1
× 10.0	1270	127.0	44.4	8.1	4.8	3.12	49.2	49.5	0.199	6.19	12.5	12.3
C100 × 10.8	1370	101.6	43.7	7.5	8.2	1.91	37.5	37.3	0.180	5.62	11.4	11.7
× 8.03	1030	101.6	40.2	7.5	4.7	1.60	31.6	39.6	0.133	4.64	11.4	11.6
C75 × 8.93	1140	76.2	40.5	6.9	9.0	0.862	22.6	27.4	0.127	4.39	10.6	11.6
× 7.44	948	76.2	38.0	6.9	6.6	0.770	20.3	28.4	0.103	3.82	10.4	11.1
× 6.10	781	76.2	35.8	6.9	4.3	0.691	18.0	29.7	0.0820	3.31	10.3	11.1

Elastic properties

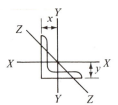

TABLE A.8 (U.S. Common Units)
Angles, Equal Legs: Properties for Designing

Size and thickness	Weight per foot	Area	Elastic properties				Axis Z–Z
			Axis X–X or Y–Y				
			I	S	r	x or y	r
in.	lb	in.^{2}	in.^{4}	in.^{4}	in.	in.	in.
L8 × 8 × $1\frac{1}{8}$	56.9	16.7	98.0	17.5	2.42	2.41	1.56
× 1	51.0	15.0	89.0	15.8	2.44	2.37	1.56
× $\frac{7}{8}$	45.0	13.2	79.6	14.0	2.45	2.32	1.57
× $\frac{3}{4}$	38.9	11.4	69.7	12.2	2.47	2.28	1.58
× $\frac{5}{8}$	32.7	9.61	59.4	10.3	2.49	2.23	1.58
× $\frac{9}{16}$	29.6	8.68	54.1	9.34	2.50	2.21	1.59
× $\frac{1}{2}$	26.4	7.75	48.6	8.36	2.50	2.19	1.59
L6 × 6 × 1	37.4	11.0	35.5	8.57	1.80	1.86	1.17
× $\frac{7}{8}$	33.1	9.73	31.9	7.63	1.81	1.82	1.17
× $\frac{3}{4}$	28.7	8.44	28.2	6.66	1.83	1.78	1.17
× $\frac{5}{8}$	24.2	7.11	24.2	5.66	1.84	1.73	1.18
× $\frac{9}{16}$	21.9	6.43	22.1	5.14	1.85	1.71	1.18
× $\frac{1}{2}$	19.6	5.75	19.9	4.61	1.86	1.68	1.18
× $\frac{7}{16}$	17.2	5.06	17.7	4.08	1.87	1.66	1.19
× $\frac{3}{8}$	14.9	4.36	15.4	3.53	1.88	1.64	1.19
× $\frac{5}{16}$	12.4	3.65	13.0	2.97	1.89	1.62	1.20
L5 × 5 × $\frac{7}{8}$	27.2	7.98	17.8	5.17	1.49	1.57	0.973
× $\frac{3}{4}$	23.6	6.94	15.7	4.53	1.51	1.52	0.975
× $\frac{5}{8}$	20.0	5.86	13.6	3.86	1.52	1.48	0.978
× $\frac{1}{2}$	16.2	4.75	11.3	3.16	1.54	1.43	0.983
× $\frac{7}{16}$	14.3	4.18	10.0	2.79	1.55	1.41	0.986
× $\frac{3}{8}$	12.3	3.61	8.74	2.42	1.56	1.39	0.990
× $\frac{5}{16}$	10.3	3.03	7.42	2.04	1.57	1.37	0.994

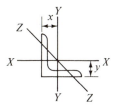

TABLE A.8 (U.S. Common Units) (*cont.*)

Size and thickness	Weight per foot	Area	Elastic properties Axis X-X or Y-Y I	S	r	x or y	Axis Z-Z r
in.	lb	in.^{2}	in.^{4}	in.^{4}	in.	in.	in.
L4 × 4 × $\frac{3}{4}$	18.5	5.44	7.67	2.81	1.19	1.27	0.778
× $\frac{5}{8}$	15.7	4.61	6.66	2.40	1.20	1.23	0.779
× $\frac{1}{2}$	12.8	3.75	5.56	1.97	1.22	1.18	0.782
× $\frac{7}{16}$	11.3	3.31	4.97	1.75	1.23	1.16	0.785
× $\frac{3}{8}$	9.8	2.86	4.36	1.52	1.23	1.14	0.788
× $\frac{5}{16}$	8.2	2.40	3.71	1.29	1.24	1.12	0.791
× $\frac{1}{4}$	6.6	1.94	3.04	1.05	1.25	1.09	0.795
L3$\frac{1}{2}$ × 3$\frac{1}{2}$ × $\frac{1}{2}$	11.1	3.25	3.64	1.49	1.06	1.06	0.683
× $\frac{7}{16}$	9.8	2.87	3.26	1.32	1.07	1.04	0.684
× $\frac{3}{8}$	8.5	2.48	2.87	1.15	1.07	1.01	0.687
× $\frac{5}{16}$	7.2	2.09	2.45	0.976	1.08	0.990	0.690
× $\frac{1}{4}$	5.8	1.69	2.01	0.794	1.09	0.968	0.694
L3 × 3 × $\frac{1}{2}$	9.4	2.75	2.22	1.07	0.898	0.932	0.584
× $\frac{7}{16}$	8.3	2.43	1.99	0.954	0.905	0.910	0.585
× $\frac{3}{8}$	7.2	2.11	1.76	0.833	0.913	0.888	0.587
× $\frac{5}{16}$	6.1	1.78	1.51	0.707	0.922	0.869	0.589
× $\frac{1}{4}$	4.9	1.44	1.24	0.577	0.930	0.842	0.592
× $\frac{3}{16}$	3.71	1.09	0.962	0.441	0.939	0.820	0.596
L2$\frac{1}{2}$ × 2$\frac{1}{2}$ × $\frac{1}{2}$	7.7	2.25	1.23	0.724	0.739	0.806	0.487
× $\frac{3}{8}$	5.9	1.73	0.984	0.566	0.753	0.762	0.487
× $\frac{5}{16}$	5.0	1.46	0.849	0.482	0.761	0.740	0.489
× $\frac{1}{4}$	4.1	1.19	0.703	0.394	0.769	0.717	0.491
× $\frac{3}{16}$	3.07	0.92	0.547	0.303	0.778	0.694	0.495

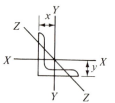

TABLE A.8 (SI Units)
Angles, Equal Legs: Properties for Designing

Size and thickness	Mass per meter	Area	Elastic properties					Axis Z–Z
			Axis X–X or Y–Y					
			I	S	r	x or y		r
mm	kg	mm^2	$\times 10^6$ mm^4	$\times 10^3$ mm^3	mm	mm		mm
L203 × 203 × 28.6	84.6	10 770	40.8	287	61.5	61.2		39.6
× 25.4	75.9	9 680	37.0	259	62.0	60.2		39.6
× 22.2	67.0	8 520	33.1	229	62.2	58.9		39.9
× 19.0	57.9	7 350	29.0	200	62.7	57.9		40.1
× 15.9	48.7	6 200	24.7	169	63.2	56.6		40.1
× 14.3	44.0	5 600	22.5	153	63.5	56.1		40.4
× 12.7	39.3	5 000	20.2	137	63.5	55.6		40.4
L152 × 152 × 25.4	55.7	7 100	14.8	140	45.7	47.2		29.7
× 22.2	49.3	6 280	13.3	125	46.0	46.2		29.7
× 19.0	42.7	5 450	11.7	109	46.5	45.2		29.7
× 15.9	36.0	4 590	10.1	92.8	46.7	43.9		30.0
× 14.3	32.6	4 150	9.20	84.2	47.0	43.4		30.0
× 12.7	29.2	3 710	8.28	75.5	47.2	42.7		30.0
× 11.1	25.6	3 260	7.37	66.9	47.5	42.2		30.2
× 9.52	22.2	2 810	6.41	57.8	47.8	41.7		30.2
× 7.94	18.5	2 350	5.41	48.7	48.0	41.1		30.5
L127 × 127 × 22.2	40.5	5 150	7.41	84.7	37.8	39.9		24.7
× 19.0	35.1	4 480	6.53	74.2	38.4	38.6		24.8
× 15.9	29.8	3 780	5.66	63.3	38.6	37.6		24.8
× 12.7	24.1	3 060	4.70	51.8	39.1	36.3		25.0
× 11.1	21.3	2 700	4.16	45.7	39.4	35.8		25.0
× 9.52	18.3	2 330	3.64	39.7	39.6	35.3		25.1
× 7.94	15.3	1 950	3.09	33.4	39.9	34.8		25.3

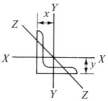

TABLE A.8 (SI Units) *(cont.)*

Size and thickness	Mass per meter	Area	Elastic properties				Axis Z-Z
			Axis X-X or Y-Y				
			I	S	r	x or y	r
mm	*kg*	*mm²*	$\times 10^6$ *mm⁴*	$\times 10^3$ *mm³*	*mm*	*mm*	*mm*
L102 × 102 × 19.0	27.5	3 510	3.19	46.0	30.2	32.3	19.8
× 15.9	23.4	2 970	2.77	39.3	30.5	31.2	19.8
× 12.7	19.0	2 420	2.31	32.3	31.0	30.0	19.9
× 11.1	16.8	2 140	2.07	28.7	31.2	29.5	19.9
× 9.52	14.6	1 850	1.81	24.9	31.2	29.0	20.0
× 7.94	12.2	1 550	1.54	21.1	31.5	28.5	20.1
× 6.35	9.8	1 250	1.27	17.2	31.8	27.7	20.2
L89 × 89 × 12.7	16.5	2 100	1.52	24.4	26.9	26.9	17.3
× 11.1	14.6	1 850	1.36	21.6	27.2	26.4	17.4
× 9.52	12.6	1 600	1.19	18.8	27.2	25.7	17.4
× 7.94	10.7	1 350	1.02	16.0	27.4	25.1	17.5
× 6.35	8.6	1 090	0.837	13.0	27.7	24.6	17.6
L76 × 76 × 12.7	14.0	1 770	0.924	17.5	22.8	23.7	14.8
× 11.1	12.4	1 570	0.828	15.6	23.0	23.1	14.9
× 9.52	10.7	1 360	0.733	13.7	23.2	22.6	14.9
× 7.94	9.1	1 150	0.629	11.6	23.4	22.1	15.0
× 6.35	7.3	929	0.516	9.46	23.6	21.4	15.0
× 4.76	5.52	703	0.400	7.23	23.9	20.8	15.1
L64 × 64 × 12.7	11.4	1 450	0.512	11.9	18.8	20.5	12.4
× 9.52	8.9	1 120	0.410	9.38	19.1	19.4	12.4
× 7.94	7.4	942	0.353	7.90	19.3	18.8	12.4
× 6.35	6.1	768	0.293	6.46	19.5	18.2	12.5
× 4.76	4.57	594	0.228	4.97	19.8	17.6	12.6

TABLE A.9 (U.S. Common Units)
Angles, Unequal Legs: Properties for Designing

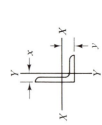

Size and thickness in.	Weight per foot lb	Area in.²	Elastic properties							
			Axis X-X				Axis Y-Y			
			I in.⁴	S in.³	r in.	y in.	I in.⁴	S in.³	r in.	x in.
L9 × 4 × 1	40.8	12.0	97.0	17.6	2.84	3.50	12.0	4.00	1.00	1.00
× $\frac{3}{4}$	31.3	9.19	76.1	13.6	2.88	3.41	9.63	3.11	1.02	0.906
× $\frac{1}{2}$	21.3	6.25	53.2	9.34	2.92	3.31	6.92	2.17	1.05	0.810
L8 × 6 × 1	44.2	13.0	80.8	15.1	2.49	2.65	38.8	8.92	1.73	1.65
× $\frac{3}{4}$	33.8	9.94	63.4	11.7	2.53	2.56	30.7	6.92	1.76	1.56
× $\frac{1}{2}$	23.0	6.75	44.3	8.02	2.56	2.47	21.7	4.79	1.79	1.47
L7 × 4 × $\frac{7}{8}$	30.2	8.86	42.9	9.65	2.20	2.55	10.2	3.46	1.07	1.05
× $\frac{5}{8}$	22.1	6.48	32.4	7.14	2.24	2.46	7.84	2.58	1.10	0.963
× $\frac{3}{8}$	13.6	3.98	20.6	4.44	2.27	2.37	5.10	1.63	1.13	0.870
L6 × 4 × $\frac{7}{8}$	27.2	7.98	27.7	7.15	1.86	2.12	9.75	3.39	1.11	1.12
× $\frac{5}{8}$	20.0	5.86	21.1	5.31	1.90	2.03	7.52	2.54	1.13	1.03
× $\frac{3}{8}$	12.3	3.61	13.5	3.32	1.93	1.94	4.90	1.60	1.17	0.941
L6 × 3$\frac{1}{2}$ × $\frac{1}{2}$	15.3	4.50	16.6	4.24	1.92	2.08	4.25	1.59	0.972	0.833
× $\frac{3}{8}$	11.7	3.42	12.9	3.24	1.94	2.04	3.34	1.23	0.988	0.787
× $\frac{1}{4}$	7.9	2.31	8.86	2.21	1.96	1.99	2.34	0.847	1.01	0.740

TABLE A.9 (U.S. Common Units) (cont.)

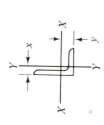

| Size and thickness | Weight per foot | Area | Elastic properties | | | | | | | |
| | | | Axis X–X | | | | Axis Y–Y | | | |
in.	lb	in.²	I in.⁴	S in.³	r in.	y in.	I in.⁴	S in.³	r in.	x in.
L5 × 3 × $\frac{1}{2}$	12.8	3.75	9.45	2.91	1.59	1.75	2.58	1.15	0.829	0.750
× $\frac{3}{8}$	9.8	2.86	7.37	2.24	1.61	1.70	2.04	0.888	0.845	0.704
× $\frac{1}{4}$	6.6	1.94	5.11	1.53	1.62	1.66	1.44	0.614	0.861	0.657
L4 × 3 × $\frac{1}{2}$	11.1	3.25	5.05	1.89	1.25	1.33	2.42	1.12	0.864	0.827
× $\frac{3}{8}$	8.5	2.48	3.96	1.46	1.26	1.28	1.92	0.866	0.879	0.782
× $\frac{1}{4}$	5.8	1.69	2.77	1.00	1.28	1.24	1.36	0.599	0.896	0.736
L3$\frac{1}{2}$ × 2$\frac{1}{2}$ × $\frac{1}{2}$	9.4	2.75	3.24	1.41	1.09	1.20	1.36	0.760	0.704	0.705
× $\frac{3}{8}$	7.2	2.11	2.56	1.09	1.10	1.16	1.09	0.592	0.719	0.660
× $\frac{1}{4}$	4.9	1.44	1.80	0.755	1.12	1.11	0.777	0.412	0.735	0.614
L3 × 2 × $\frac{1}{2}$	7.7	2.25	1.92	1.00	0.924	1.08	0.672	0.474	0.546	0.583
× $\frac{3}{8}$	5.9	1.73	1.53	0.781	0.940	1.04	0.543	0.371	0.559	0.539
× $\frac{1}{4}$	4.1	1.19	1.09	0.542	0.957	0.993	0.392	0.260	0.574	0.493
L2$\frac{1}{2}$ × 2 × $\frac{3}{8}$	5.3	1.55	0.912	0.547	0.768	0.831	0.514	0.363	0.577	0.581
× $\frac{1}{4}$	3.62	1.06	0.654	0.381	0.784	0.787	0.372	0.254	0.592	0.537

TABLE A.9 (SI Units)
Angles, Unequal Legs: Properties for Designing

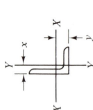

Size and thickness	Mass per meter	Area	Elastic properties							
			Axis X-X				Axis Y-Y			
			I	S	r	y	I	S	r	x
mm	kg	mm^2	$\times 10^6$ mm^4	$\times 10^3$ mm^3	mm	mm	$\times 10^6$ mm^4	$\times 10^3$ mm^3	mm	mm
L229 × 102 × 25.4	60.7	7740	40.4	288	72.1	88.9	4.99	65.5	25.4	25.4
× 19.0	46.6	5930	31.7	223	73.1	86.6	4.01	51.0	25.9	23.0
× 12.7	31.7	4030	22.1	153	74.2	84.1	2.88	35.6	26.7	20.6
L203 × 152 × 25.4	65.8	8390	33.6	247	63.2	67.3	15.9	146	43.9	41.9
× 19.0	50.3	6410	26.4	192	64.3	65.0	12.8	113	44.7	39.6
× 12.7	34.2	4350	18.4	131	65.0	62.7	9.03	78.5	45.5	37.3
L178 × 102 × 22.2	44.9	5720	17.9	158	55.9	64.8	4.25	56.7	27.2	26.7
× 15.9	32.9	4180	13.5	117	56.9	62.5	3.26	42.3	27.9	24.5
× 9.52	20.2	2570	8.57	72.8	57.7	60.2	2.12	26.7	28.7	22.1
L152 × 102 × 22.2	40.5	5150	11.5	117	47.2	53.8	4.06	55.6	28.2	28.5
× 15.9	29.8	3780	8.78	87.0	48.3	51.6	3.13	41.6	28.7	26.2
× 9.52	18.3	2330	5.62	54.4	49.0	49.3	2.04	26.2	29.7	23.9
L152 × 89 × 12.7	22.7	2900	6.91	69.5	48.8	52.8	1.77	26.1	24.7	21.2
× 9.52	17.4	2200	5.37	53.1	49.3	51.8	1.39	20.2	25.1	20.0
× 6.35	11.8	1490	3.69	36.2	49.8	50.5	0.974	13.9	25.7	18.8

TABLE A.9 (SI Units) (cont.)

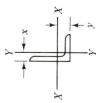

Size and thickness	Mass per meter	Area	Elastic properties							
			Axis X-X				Axis Y-Y			
			I	S	r	y	I	S	r	x
mm	kg	mm²	$\times 10^6$ mm⁴	$\times 10^3$ mm³	mm	mm	$\times 10^6$ mm⁴	$\times 10^3$ mm³	mm	mm
L127 × 76 × 12.7	19.0	2420	3.93	47.7	40.4	44.4	1.07	18.8	21.1	19.0
× 9.52	14.6	1850	3.06	36.7	40.9	43.2	0.849	14.6	21.5	17.9
× 6.35	9.8	1250	2.13	25.1	41.1	42.2	0.599	10.1	21.9	16.7
L102 × 76 × 12.7	16.5	2100	2.10	31.0	31.8	33.8	1.01	18.4	21.9	21.0
× 9.52	12.6	1600	1.65	23.9	32.0	32.5	0.799	14.2	22.3	19.9
× 6.35	8.6	1090	1.15	16.4	32.5	31.5	0.566	9.82	22.8	18.7
L89 × 64 × 12.7	14.0	1770	1.35	23.1	27.7	30.5	0.566	12.5	17.9	17.9
× 9.52	10.7	1360	1.07	17.9	27.9	29.5	0.454	9.70	18.3	16.8
× 6.35	7.3	929	0.749	12.4	28.4	28.2	0.323	6.75	18.7	15.6
L76 × 51 × 12.7	11.5	1450	0.799	16.4	23.5	27.4	0.280	7.77	13.9	14.8
× 9.52	8.8	1120	0.637	12.8	23.9	26.4	0.226	6.08	14.2	13.7
× 6.35	6.1	768	0.454	8.88	24.3	25.2	0.163	4.26	14.6	12.5
L64 × 51 × 9.52	7.9	1000	0.380	8.96	19.5	21.1	0.214	5.95	14.7	14.8
× 6.35	5.39	684	0.272	6.24	19.9	20.0	0.155	4.16	15.0	13.6

TABLE A.10 (U.S. Common Units)
American Standard Timber Sizes: Properties for Designing[*]
(Selected Listing)

Nominal size	American Standard dressed size	Area of section	Weight per foot	Moment of inertia	Section modulus
in.	*in.*	$in.^2$	*lb*	$in.^4$	$in.^3$
2 X 4	1.5 X 3.5	5.25	1.46	5.36	3.06
X 6	5.5	8.25	2.29	20.8	7.56
X 8	7.25	10.9	3.02	47.6	13.1
X 10	9.25	13.9	3.85	89.9	21.4
X 12	11.25	16.9	4.69	178	31.6
4 X 4	3.5 X 3.5	12.2	3.40	12.5	7.15
X 6	5.5	19.2	5.35	48.5	17.6
X 8	7.25	25.4	7.05	111	30.7
X 10	9.25	32.4	8.99	231	49.9
X 12	11.25	39.4	10.9	415	73.8
6 X 6	5.5 X 5.5	30.2	8.40	76.3	27.7
X 8	7.5	41.2	11.5	193	51.6
X 10	9.5	52.2	14.5	393	82.7
X 12	11.5	63.2	17.6	697	121
X 14	13.5	74.2	20.6	1128	167
8 X 8	7.5 X 7.5	56.2	15.6	263	70.3
X 10	9.5	71.2	19.8	536	117
X 12	11.5	86.2	24.0	951	165
X 14	13.5	101	28.1	1538	228
X 16	15.5	116	32.3	2327	300
10 X 10	9.5 X 9.5	90.2	25.1	679	143
X 12	11.5	109	30.3	1204	209
X 14	13.5	128	35.6	1948	289
X 16	15.5	147	40.9	2948	380
X 18	17.5	166	46.2	4243	485
X 20	19.5	185	51.5	5870	602
12 X 12	11.5 X 11.5	132	36.7	1458	253
X 14	13.5	155	43.1	2358	349
X 16	15.5	178	49.5	3569	460
X 18	17.5	201	55.9	5136	587
X 20	19.5	224	62.3	7106	729
X 22	21.5	247	68.7	9524	886
X 24	23.5	270	75.1	12440	1058

[*]All weights and properties calculated for dressed sizes. Weights based on 40 lb/ft^3. Moments of inertia and section modulus calculated from formulas.

TABLE A.10 (SI Units)
American Standard Timber Sizes: Properties for Designing*
(Selected Listing)

Nominal size	American Standard dressed size	Area of section	Mass per meter	Moment of inertia	Section modulus
mm	mm	$\times 10^3 mm^2$	kg	$\times 10^6 mm^4$	$\times 10^3 mm^3$
50 × 100	38 × 89	3.38	2.03	2.23	50.2
× 150	140	5.32	3.19	8.69	124
× 200	184	6.99	4.20	19.7	214
× 250	235	8.93	5.36	41.1	350
× 300	286	10.87	6.52	74.1	518
100 × 100	89 × 89	7.92	4.75	5.23	117
× 150	140	12.5	7.48	20.4	291
× 200	184	16.4	9.83	46.2	502
× 250	235	20.9	12.5	96.3	819
× 300	286	25.4	15.3	174	1 213
150 × 150	140 × 140	19.6	11.8	32.0	457
× 200	191	26.7	16.0	81.3	851
× 250	241	33.7	20.2	163	1 355
× 300	292	40.9	24.6	290	1 989
× 350	343	48.0	28.8	471	2 745
200 × 200	191 × 191	36.5	21.9	111	1 161
× 250	241	46.0	27.6	223	1 849
× 300	292	55.8	33.5	396	2 714
× 350	343	65.5	39.3	642	3 745
× 400	394	75.2	45.2	974	4 942
250 × 250	241 × 241	58.1	34.8	281	2 333
× 300	292	70.4	42.2	500	3 425
× 350	343	82.7	49.6	810	4 726
× 400	394	95.0	57.0	1 228	6 235
× 450	445	107	64.3	1 770	7 954
× 500	495	119	71.6	2 436	9 842
300 × 300	292 × 292	85.3	51.2	606	4 150
× 350	343	100	60.1	982	5 726
× 400	394	115	69.0	1 488	7 555
× 450	445	130	78.0	2 144	9 637
× 500	495	144	86.7	2 951	11 920
× 550	546	159	95.7	3 961	14 510
× 600	597	174	105	5 178.	17 350

*All masses and properties calculated for dressed sizes. Masses based on 600 kg/m³. Moments of inertia and section modulus calculated from formulas.

TABLE A.11

Average Mechanical Properties of Selected Engineering Materials
(Values in SI Units Given in Parentheses)

Material	Specific weight, lb/ft^3 (Density) (Mg/m^3)	Coefficient of thermal expansion, $10^{-6}/F°$ ($10^{-6}/C°$)	Elastic moduli, 10^3 ksi (GN/m^2)	
			E	G
Aluminum				
2014-T6[a]	175 (2.80)	12.8 (23.0)	10.6 (73)	4.00 (28)
6061-T6[a]	169 (2.71)	13.0 (23.4)	10.0 (70)	3.75 (26)
6062-T6[a]	169 (2.71)	13.0 (23.4)	10.0 (70)	3.75 (26)
Cast iron				
Gray, class 20[b]	440 (7.05)	6.0 (10.8)	12 (83)	4.8 (33)
Malleable, ASTM 32510[b]	460 (7.37)	6.5 (11.7)	26 (180)	10 (69)
Steel				
0.2% C, hot-rolled	490 (7.85)	6.5 (11.5)	30 (210)	12 (83)
0.2% C, cold-rolled	490 (7.85)	6.5 (11.5)	30 (210)	12 (83)
0.4% C, hot-rolled	490 (7.85)	6.5 (11.5)	30 (210)	12 (83)
0.4% C, water quenched and drawn	490 (7.85)	6.5 (11.5)	30 (210)	12 (83)
0.4% C, cast and annealed	490 (7.85)	6.5 (11.5)	30 (210)	12 (83)
3.5% Ni, 0.4% C, oil quenched and drawn	490 (7.85)	6.5 (11.5)	30 (210)	12 (83)
Magnesium				
AZ 31B-H24[b]	108 (1.73)	14.0 (25.0)	6.5 (45)	2.4 (17)
AZ 92A-T6[b]	108 (1.73)	14.0 (25.0)	6.5 (45)	2.4 (17)
Titanium				
Pure Ti (Gr 1)[b]	280 (4.50)	4.9 (8.8)	16.5 (114)	6.0 (41)
Ti-6A1-4V (Gr 5)[b]	280 (4.50)	4.9 (8.8)	16.5 (114)	6.0 (41)
Concrete				
Low strength	150 (2.40)	6.0 (10.8)	3 (21)	—
High strength	150 (2.40)	6.0 (10.8)	5 (34)	—
Wood				
Douglas fir	34 (0.54)	—	1.9 (13)	—
Southern pine	34 (0.54)	—	1.6 (11)	—

[a]Aluminum Association designation.
[b]American Society of Testing Materials designation.

Tensile strength, 10³ psi (MN/m²)		Compression strength, 10³ psi (MN/m²)		Shear strength, 10³ psi (MN/m²)		Poisson's ratio
Yield	Ultimate	Yield	Ultimate	Yield	Ultimate	
53 (365)	60 (410)	65 (450)	—	31 (210)	35 (240)	0.33
35 (240)	38 (260)	35 (240)	—	20 (140)	24 (170)	0.33
16 (110)	26 (180)	14 (96)	—	9 (60)	16 (110)	0.33
—	20 (140)	—	95 (660)	—	25 (170)	0.25
32 (220)	50 (350)	—	208 (1430)	24 (170)	62 (430)	0.25
36 (250)	60 (410)	36 (250)	—	21 (150)	40 (280)	0.29
60 (410)	80 (550)	60 (410)	—	36 (250)	60 (410)	—
52 (360)	90 (620)	52 (360)	—	36 (250)	80 (550)	—
87 (600)	102 (702)	87 (700)	—	52 (360)	90 (620)	—
32 (220)	65 (450)	30 (210)	—	18 (120)	55 (380)	—
150 (1030)	170 (1170)	150 (1030)	—	90 (620)	140 (970)	—
32 (220)	42 (290)	—	—	—	—	0.35
23 (160)	40 (280)	—	—	—	21 (150)	0.35
30 (210)	40 (280)	—	—	—	—	0.34
120 (830)	130 (900)	—	—	72 (500)	76 (520)	0.34
—	—	1.2 (8)	3 (21)	—	—	0.15
—	—	2 (14)	5 (34)	—	—	0.15
8.1 (56)	—	6.4[c] (44)	7.4 (51)	—	1.1[c] (7.6)	—
5.1 (35)	—	4.0[c] (28)	5.8 (40)	—	0.8[c] (5.5)	—

[c]Parallel to the grain of the wood.

Index

Answers
to Even-Numbered Problems _____

1.2   (a) 402 in., 33.5 ft
     (b) 1772 in., 147.6 ft
     (c) 8.03 in., 0.669 ft
     (d) 181.1 in., 15.09 ft

1.4   (a) 11.89 lb, 0.01189 kip
     (b) 1540 lb, 1.540 kips
     (c) 270 lb, 0.270 kip
     (d) 4680 lb, 4.68 kips

1.6   (a) 3.75 m^2 (b) 375 dm^2

1.8   (a) 125 000 mm^2 (b) 12.5 dm^2

1.10 (a) 0.540 ft^2 (b) 5.02 dm^2

1.12 (a) 1.200 m^2 (b) 120.0 dm^2

1.14 (a) 152 100 mm^2 (b) 15.21 dm^2

1.16 (a) 132.7 $in.^2$ (b) 8.56 dm^2

1.18 (a) 1.386 m^3 (b) 1386 dm^3 (liters)

1.20 (a) 1.800×10^6 mm^3
      (b) 1.800 dm^3 (liters)

1.26 $c = 5$ ft, $\sin \theta = 0.6$, $\cos \theta = 0.8$

1.28 $c = 25$ in., $\sin \theta = 0.960$, $\cos \theta = 0.280$

1.30 $b = 428$ m, $c = 128.6$ m

1.32 21.0 m

1.34 55.4 m

1.36 97.0 m

1.38 2630 ft

1.40 (a) $c = 3.72$ in., $b = 4.79$ in., $A = 110°$
      (b) $c = 4.57$ m, $A = 29.2°$, $B = 102.8°$
      (c) $c = 8.72$ ft, $B = 49.3°$, $C = 70.7°$
      (d) $A = 25.2°$, $B = 48.2°$, $C = 106.6°$

1.42 5.26 ft, $12.4°\angle$

1.44 $\angle GCB = 76.0°$, $\angle AHJ = 67.1°$,
      $\angle GFC = 35.0°$

CHAPTER 2

2.2   330 N ⦨ 27.5°
2.4   100 N ⦨ 58°
2.6   11.6 kN ∠ 80°
2.8 Q = 26.8 kN, R = 35.6 kN
2.10 P = 4.2 kN, Q = 7.0 kN
2.12   1430 N ⦨ 70°
2.14   13.2 kips ⦨ 35°
2.16   2.1 kips ∠ 22°
2.18   (a) 22.9 kN ←, 16.06 kN ↓
      (b) 1449 lb →, 388 lb ↑
      (c) 2.64 kips ←, 9.85 kips ↑
      (d) 289 N ←, 345 N ↓
2.20   (a) 27.6 kN ⦫ 45°, 4.86 kN ⦪ 45°
      (b) 1299 lb ∠ 45°, 750 lb ⦨ 45°
      (c) 5.10 kips ∠ 45°, 8.83 kips ⦪ 45°
      (d) 448 N ⦫ 45°, 39.2 N ⦨ 45°
2.22   (a) 2830 lb perpendicular and 2830 lb
       parallel to *m–n*
      (b) 3860 lb perpendicular and 1035 lb
       parallel to *r–s*
2.24   539 N ⦨ 21.8°
2.26   7.62 kips ∠ 6.9°
2.28   19.05 kN ∠ 65.7°
2.30   1403 lb ⦨ 43.6°
2.32   333 lb ⦨ 27.3°
2.34   101.0 N ⦨ 57.9°
2.36   11.60 N ∠ 80.0°
2.38 R = 35.6 kN, Q = 26.8 kN
2.40 P = 4.18 kN, Q = 7.00 kN
2.42   1433 N ⦨ 69.9°
2.44   13.22 kips ⦨ 35.2°
2.46   2.12 kips ∠ 22.4°
2.48   14.63 kN ⦪ 19.4°
2.50   2090 lb ⦨ 32.9°

CHAPTER 3

3.10 T_{BA} = 409 lb, T_{BC} = 436 lb
3.12 W_2 = 1400 N, T_{BC} = 849 N
3.14 T_{BA} = 158.0 N, T_{BC} = 273 N
3.16 T_{AB} = 2190 N, F_{AC} = -1960 N
3.18 T_{AB} = 4480 lb, F_{AC} = -5280 lb
3.20 m_A = 131 kg, N_R = 1051 N ⦪ 55°

3.22 F_1 = 1659 lb, F_2 = 318 lb
3.24 T_{BA} = 2780 N, T_{BC} = 2450 N,
      W_1 = 3430 N, m_1 = 350 kg
3.26 T_{BA} = 749 lb, W_2 = 646 lb
3.28 T_{BA} = 7.07 kN, F_{BC} = -7.07 kN
3.30 F_4 = 123.8 lb, θ_3 = 35.7°

CHAPTER 4

4.2 $M_O = M_R$ = 120 kN·m
      M_P = 420 kN·m
      M_Q = -180 kN·m
4.4 M_D = -40.0 kN·m,
      M_E = 140.0 kN·m,
      M_F = 49.4 kN·m,
      M_G = -174.0 kN·m
4.6 M_L = -6.24 kN·m,
      M_M = 6.11 kN·m,
      M_N = 2.80 kN·m,
      M_O = 2.47 kN·m
4.8   9.6 kN ↓, 2.75 m to the right of A
4.10   40 kN ↓, 6.9 m to the right of A
4.12   620 lb ↓, 3.94 ft from A along line
      connecting A and D
4.14   1360 lb ∠ 72.9°, 12.31 ft to the
      right of A
4.16   2550 lb ⦨ 78.7 °, 3.2 ft to the right
      of B
4.18   170 lb ∠ 41.7°, 5.42 in. to the right
      of A
4.20   1962 lb ∠ 55.8°, 6.25 ft to the right
      of S
4.22   38.1 kips ⦨ 23.2°, 18.6 ft above J
4.24   6.70 kN ⦨ 2.1°, 0.931 m above L
4.26   (a) d = 2.42 m   (b) F = 31.1 kN
4.28   (a) 4000 N → at B and 4000 N ← at C
      (b) 3000 N ↓ at D and 3000 N ↑ at E
4.30   (a) 80 kN ↑ at A and 28 kN·m ↻
      (b) 80 kN ↑ at B and 28 kN·m ↻
4.32   (a) 4.8 kN → at C and 0.72 kN·m ↺
      (b) 4.5 kN ↑ at A and 4.5 kN ↓ at B
4.34   55.2 kN ⦨ 11.3° at A and
      130 kN · m ↺
4.36   504 kN ⦫ 6.8° at M and 300 kN·m ↺
4.38   (a) 80 kN →, 4 m above A
      (b) 80 kN → at A and 320 kN·m ↺

4.40 (a) 15 kips ↓, 5.0 ft to the right of A
     (b) 15 kips ↓ at A and 75.0 kip-ft ↺
4.42 (a) 11,300 lb ↓, 6.40 ft to the right
     of A
     (b) 11,300 lb ↓ at A and 72,300 lb-ft ↺

CHAPTER 5

5.2 $A_x = 0$, $A_y = 44$ kN ↑,
    $M = 81.6$ kN·m ↺
5.4 $A = 70$ kN ↑, $B_x = 0$, $B_y = 66$ kN ↑
5.6 $A_x = 392$ N →, $A_y = 392$ N ↑,
    $F_{BC} = -392$ N
5.8 $A_x = 0$, $A_y = 6$ kN ↓, $M = 10$ kN·m ↻
5.10 $B_x = 0$, $B_y = 4.8$ kN ↑,
    $M = 9.6$ kN·m ↺
5.12 $A_x = 4.8$ kips ←, $A_y = 36$ kips ↑,
    $T_{HJ} = 39.7$ kips
5.14 $A_x = 0$, $A_y = 15$ lb ↓, $B = 15$ lb ↑
5.16 $A_x = 900$ N ←, $A_y = 833$ N ↓,
    $B = 1333$ N ↑
5.18 $E_x = 12$ kips ←, $E_y = 1.0$ kip ↓,
    $G = 20$ kips ∠ 53.1°
5.20 $O_x = 9.53$ kN ←, $O_y = 17.5$ kN ↑,
    $T = 11.0$ kN
5.22 $A_x = 91.5$ N →, $A_y = 64.2$ N ↑,
    $T = 111.7$ N, $N_R = 48.3$ N
5.24 $A_x = 2.74$ lb ←, $A_y = 28.6$ lb ↑,
    $B = 30.4$ lb ⦥ 53.1°
5.26 $C_x = 33.4$ kN ←, $C_y = 10.0$ kN ↓,
    $T = 86.7$ kN
5.28 $A_x = 83.5$ N ←, $A_y = 180.9$ N ↑,
    $B = 145.4$ N ∠ 55°
5.30 $T_{LC} = 466$ lb, $T_{AC} = 413$ lb
5.32 $F_{BC} = 392$ N ←, $A = 554$ N ∠ 45°
5.34 $A = 21.5$ kN ⦦ 9.46°, $F_{FG} = 21.2$ kN
5.36 $A = 747$ N ⦥ 8.2°, $B = 853$ N ⦧ 30°
5.38 $B = 61.5$ N ⦥ 12.5°, $C = 86.7$ N ↑
5.40 $A = 199.2$ N ⦥ 65.3°,
    $B = 145.4$ N ∠ 55°

CHAPTER 6

6.2 $A_x = 2.67$ kN ←, $A_y = 4.0$ kN ↑,
    $DC = -2.67$ kN, $AB = 2.67$ kN,
    $BC = -4.61$ kN, $AC = 4.0$ kN

6.4 $A_x = 0$, $A_y = 450$ N ↓, $C = 1350$ N ↑,
    $AB = 605$ N, $AC = -405$ N,
    $BC = -1410$ N
6.6 $A = 9.44$ kips ↑, $D_x = 2.5$ kips →,
    $D_y = 3.06$ kips ↑, $AB = -11.80$ kips,
    $BC = -2.29$ kips, $CD = -3.82$ kips,
    $DE = 4.79$ kips, $EF = 7.08$ kips,
    $FA = 7.08$ kips, $BF = 12.50$ kips,
    $BE = -3.82$ kips, $CE = 3.06$ kips
6.8 $C_x = 1800$ lb →, $C_y = 1000$ lb ↑,
    $E = 1800$ lb ←, $AB = 2060$ lb,
    $BC = 1800$ lb, $CD = DE = 1000$ lb,
    $EF = -2060$ lb, $FA = -1800$ lb,
    $BF = -1000$ lb, $FC = FD = 0$
6.10 $A_x = 0$, $A_y = 2$ kips ↑, $E = 3$ kips ↑,
    $AB = -2.60$ kips, $BC = -2.69$ kips,
    $CD = -2.69$ kips, $DE = -3.90$ kips,
    $EF = 2.5$ kips, $FG = 2.5$ kips,
    $GH = 1.67$ kips, $HA = 1.67$ kips,
    $BH = 0$, $BG = 1.30$ kips,
    $CG = 2.0$ kips, $DG = 0$, $DF = 2.0$ kips
6.12 $A = 4.8$ kips ↑, $D_x = 6.0$ kips →,
    $D_y = 3.2$ kips ↑, $AB = -6.0$ kips,
    $BC = -2.67$ kips, $CD = -2.67$ kips,
    $DE = 6.93$ kips, $EA = 3.6$ kips,
    $BE = 3.33$ kips, $CE = -4.0$ kips
6.14 $D_x = 5$ kips ←, $D_y = 0.5$ kip ↓,
    $GA = -3.5$ kips, $AB = -4.21$ kips,
    $BC = -4.67$ kips, $CD = -4.67$ kips,
    $DE = -0.60$ kip, $EF = 2.34$ kips,
    $FA = 2.34$ kips, $BF = 0$,
    $BE = 4.21$ kips, $CE = 0$
6.16 $BC = 500$ lb, $BF = 1000$ lb,
    $AF = -1060$ lb
6.18 (a) $AB = 57.1$ kN, $AF = 0$,
      $GF = -69.7$ kN
    (b) $BC = 57.1$ kN, $BE = 0$,
      $FE = -69.7$ kN
6.20 (a) $BC = 1.875$ kips, $GC = 3.125$ kips,
      $GF = -3.75$ kips
    (b) $CD = 3.75$ kips, $FD = 3.125$ kips,
      $FE = -5.62$ kips
6.22 $BC = -4.67$ kips, $BE = 4.21$ kips,
    $EF = 2.33$ kips
6.24 HF, FI, IE, EJ, LB
6.26 $IJ, JK, LI, HM, NG, PE, RC, BC$
6.28 Forces on AC: $A_x = 580$ lb ←,
    $A_y = 3440$ lb ↓, $B_x = 2830$ lb →,
    $B_y = 3440$ lb ↓ Forces on BD:

B_x = 2830 lb ←, B_y = 3440 lb ↑,
D_x = 2830 lb →, D_y = 2940 lb ↓
6.30 Forces on *AC*: A_x = 0, A_y = 1.45 kN ↑,
B_x = 2.40 kN ←, B_y = 2.09 kN ↓,
C_x = 1.60 kN ←, C_y = 0.64 kN ↑
Forces on *CE*: C_x = 1.60 kN →,
C_y = 0.64 kN ↓, D_x = 1.60 kN ←,
D_y = 1.91 kN ↓, E_y = 2.55 kN ↑
Forces on *BD*: B_x = 2.40 kN →,
B_y = 2.09 kN ↑, D_x = 1.60 kN →,
D_y = 1.91 kN ↑ Forces on pulley:
P_x = 4.0 kN ←, P_y = 4.0 kN ↓
6.32 Forces on *AC*: A_y = 496 N ↑,
C_x = 110 N ←, C_y = 204 N ↑
Forces on *CE*: C_x = 110 N →,
C_y = 204 N ↓, E_y = 204 N ↑
Tension in cable *BD*: T = 110 N
6.34 Forces on *FD*: C_x = 0,
C_y = 2000 N ↑, D_x = 0, D_y = 1000 N ↓
Forces on *CA*: C_x = 0, C_y = 2000 N ↓,
B_x = 0, B_y = 4000 N ↑, A_x = 0,
A_y = 2000 N ↓ Forces on *DE*: D_x = 0,
D_y = 1000 N ↑, B_x = 0, B_y = 4000 N ↓,
E_y = 3000 N ↑
6.36 Forces on *AD*:
A_x = 1.625 kN ←, A_y = 2.50 kN ↑,
B_x = 2.50 kN →, B_y = 2.50 kN ↓,
C_x = 1.625 kN →, C_y = 5.00 kN ↑,
D_x = 2.50 kN ←, D_y = 2.50 kN ↓
Forces on *EC*:
E_x = 1.625 kN →, E_y = 2.50 kN ↑,
C_x = 1.625 kN ←, C_y = 5.00 kN ↓
Force in *AE*: F_{AE} = 2.50 kN

CHAPTER 7

7.2   (a) 150.0 N   (b) 390 N   (c) 294 N
7.4   290 N, 41.3 N
7.6   135.4 lb, 20.9 lb
7.8   64.3 in.
7.10 1726 N, 453 N
7.12 4.65 ft
7.14 No
7.16 3120 N
7.18 597 lb
7.20 31.2 lb
7.22 (a) 0.110   (b) 0.166
7.24 2930 N

7.26 (a) 3630 N   (b) 1458 N
7.28 1.0 lb-ft
7.30 1404 N·m
7.32 0.136
7.34 (a) 0.171   (b) 1.71 turns
7.36 (a) 26.2 lb   (b) 0.175

CHAPTER 8

8.2 M_x = −66.0 kip-ft, M_y = 9.00 kip-ft,
M_z = 30.0 kip-ft
8.4 M_x = 22.0 kN·m, M_y = −68.0 kN·m,
M_z = −62.0 kN·m
8.6   6.44 in.
8.8   6.18 m
8.10 6.08 ft
8.12 390 mm
8.14 $\bar{x}$ = 1.781 in., $\bar{y}$ = 0.219 in.
8.16 $\bar{x}$ = 268 mm, $\bar{y}$ = 283 mm
8.22 $\bar{x}$ = 7.60 in., $\bar{y}$ = 4.20 in.
8.24 $\bar{x}$ = 33.4 mm, $\bar{y}$ = 28.9 mm
8.26 $\bar{x}$ = 3.00 in., $\bar{y}$ = 1.645 in.
8.28 $\bar{x}$ = 30.2 mm, $\bar{y}$ = 20.8 mm
8.30 $\bar{x}$ = 75.0 mm, $\bar{y}$ = 94.9 mm
8.32 $\bar{x}$ = 3.36 in., $\bar{y}$ = 2.40 in.
8.34 $\bar{x}$ = 0, $\bar{y}$ = 6.82 in.
8.36 $\bar{x}$ = 0, $\bar{y}$ = 364 mm
8.38 (a) 670 in.^{4} (b) 667 in.^{4}
(c) 0.5 percent
8.40 (a) 63.7 in.^{4} (b) 62.8 in.^{4}
(c) 1.43 percent
8.42 (a) 33.3 × 10^6 mm^4,
18.75 × 10^6 mm^4
(b) 159.4 × 10^6 mm^4,
478 × 10^6 mm^4
8.44 (a) 0.557 in.^{4}, 1.988 in.^{4}
(b) 7.85 in.^{4}, 29.7 in.^{4}
8.46 (a) 52.1 × 10^6 mm^4
(b) 638 × 10^6 mm^4
8.48 (a) 2.54 in.^{4} (b) 37.5 in.^{4}
8.50 (a) 74.5 × 10^6 mm^4
(b) 22.5 × 10^6 mm^4
8.52 (a) 4.10 in.^{4} (b) 1.024 in.^{4}
8.54 (a) 80.4 × 10^6 mm^4
(b) 45.8 × 10^6 mm^4
8.56 (a) 2.92 in.^{4} (b) 0.506 in.^{4}
8.58 (a) 19.52 × 10^6 mm^4
(b) 11.52 × 10^6 mm^4

8.60 (a) 4.05 in.^{4} (b) 3.92 in.^{4}
8.62 (a) 8470 in.^{4} (b) 8470 in.^{4}
8.64 (a) 70.7 × 10^6 mm^4
        (b) 37.4 × 10^6 mm^4
8.66 (a) 1.300 × 10^9 mm^4
        (b) 0.663 × 10^9 mm^4
8.68 (a) 2050 in.^{4} (b) 476 in.^{4}
8.70 1.032 in., 0.516 in.
8.72 76.8 mm, 55.9 mm

CHAPTER 9

For Probs. 9.1 through 9.16, answers for internal reactions are given for the bar to the left of a vertical cross section and above a horizontal cross section.

9.2   Section *M–M:* F = 500 N, V = 0, M = 0
        Section *N–N:* F = 500 N, V = 1500 N ↑,
        M = 750 N·m ↺
9.4   Section *M–M:* F = 0, V = 400 lb →,
        M = 1200 lb-in. ↺ Section *N–N:* F = 0,
        V = 50 lb ↓, M = 200 lb-in. ↺
9.6   Section *M–M:* F = 0.4 kN,
        V = 0.792 kN ↑, M = 0.0846 kN·m ↻
        Section *N–N:* F = V = M = 0
9.8   Section *M–M:* F = −2.0 kips,
        V = 0.5 kip ←, M = 21 kip-ft ↻
        Section *N–N:* F = 0.5 kip,
        V = 2.0 kips ↓, M = 6.0 kip-ft ↺
9.10 Section *R–R:* F = −3.64 kips,
        V = 3.89 kips ∡43.2°,
        M = 31.1 kip-ft ↻
9.12 Section *R–R:* F = 1600 lb, V = 0,
        M = 2800 lb-in. ↺
9.14 Section *R–R:* F = 0, V = 6.25 kN ↑,
        M = 2.5 kN·m ↺
9.16 Section *R–R:* F = 0, V = 24 N ↓,
        M = 2.4 N·m ↻, T = 0
        Section *S–S:* F = 40 N, V = 24 N ↑,
        M = 3.0 N·m ↺, T = 4.8 N·m ↺
9.18 7.23 kPa
9.20 101,900 psi
9.22 *CA:* 13.30 MPa, *CB:* 17.92 MPa,
        CD: 20.4 MPa
9.24 (a) 61,100 psi   (b) −122,200 psi

9.26 (a) *AB:* −440 psi, *BC:* −461 psi,
        AC: 299 psi
        (b) 479 psi   (c) 221 psi
9.28 20.0 MPa
9.30 (a) 6790 psi   (b) 156.3 psi
9.32 (a) −18.41 ksi   (b) −15.78 ksi
        (c) −10.52 ksi
9.34 24.3 mm
9.36 3.93 mm
9.38 65.5 mm^2, 51.5 mm^2
9.40 *AB:* 111.1 mm^2, *BC:* 389 mm^2,
        CD and *AD:* 622 mm^2,
        BD: 66.7 mm^2, *BE:* 440 mm^2
9.42 0.810 in., 3.95 in. square
9.44 21,600 lb
9.46 743 kPa, 90.2 kPa

CHAPTER 10

10.2   0.0320 mm
10.4   270 mm
10.6   (a) 230 MPa   (b) 209 × 10^3 MPa
          (c) 272 MPa, 248 MPa   (d) 379 MPa
          (e) 295 MPa   (f) 69.5 percent
          (g) 38.0 percent
10.8   (a) 100 MPa   (b) 100 × 10^3 MPa
          (c) 160 MPa   (d) 179.5 MPa
10.10 (a) 178 MPa   (b) 68.6 × 10^3 MPa
          (c) 265 MPa   (d) 448 MPa
10.12 0.462 mm
10.14 (a) 82.1 MPa   (b) 1.505 mm
10.16 144.0 kN
10.18 219 × 10^3 kPa, 109.9 × 10^6 kPa
10.20 80.0 kips, −160.0 kips; 20.0 ksi,
          −20.0 ksi; 0.667 × 10^{-3},
          −0.667 × 10^{-3}
10.22 88,400 lb, 5000 psi
10.24 (a) 5160 psi, 4130 psi   (b) 0.00206 in.
          (c) 3.10 in.
10.26 2000 lb, 8000 lb
10.28 0.286
10.30 557 kN, −0.010 mm, −0.035 mm
10.32 (a) 35,500 lb (tension)
          (b) 21,400 lb (compression)
10.34 4100 psi (tension)
10.36 Steel: 275 psi (compression)
          Concrete: 62.5 psi (tension)

CHAPTER 11

11.2 82.4×10^6 kPa
11.4 0.250
11.6 3180 psi
11.8 34.7×10^6 Pa
11.10 848 lb-in.
11.12 53.2 kN·m
11.14 2.05°
11.16 8.41°
11.18 417 lb-in.
11.20 15.11 kN·m
11.22 0.947 in., 21.8°
11.24 75.6 mm, 9.10°
11.26 3.44 kPa, 1.07°
11.28 3.23×10^3 kPa, 0.172°
11.30 1200 lb-in., 900 lb-in., 442 psi
11.32 316 N·m, 33.9 N·m,
$\quad$ 4.69×10^6 Pa
11.36 0.535 in.
11.38 71.5 hp

CHAPTER 12

12.2 (a) 29.0 kN, 9.00 kN, −31.0 kN
$\quad$ (b) 0, 29.0 kN·m, 43.4 kN·m, 0
12.4 (a) −8.00 kips, −18.00 kips
$\quad$ (b) 0, −32.0 kip-ft, −176.0 kip-ft
12.6 (a) 0, −10.00 kN $\quad$ (b) 4.00 kN·m,
$\quad$ 4.00 kN·m
12.8 (a) −8000 lb, 5600 lb, −4400 lb
$\quad$ (b) 0, −16,000 lb-ft, 17,600 lb-ft, 0
12.10 (a) −24.0 kN, −54.0 kN
$\quad$ (b) 0, −24.0 kN·m, −43.2 kN·m,
$\quad$ −108.0 kN·m

For Probs. 12.12 through 12.66 the location of the beam cross section is indicated by the distance measured from the left end of the beam. A prime (′) is used to indicate left of the section and a double prime (″) to the right of the section.

12.12 $V_0 = V'_3 = 11.00$ kips,
$\quad V''_3 = V'_6 = 3.00$ kips,
$\quad V''_6 = V_{12} = -7.00$ kips

12.14 $V_0 = V'_{1.6} = 18.00$ kN,
$\quad V''_{1.6} = V'_3 = -42.0$ kN,
$\quad V''_3 = V_4 = 30.0$ kN
12.16 $V_0 = V'_{1.2} = -5.20$ kN,
$\quad V''_{1.2} = V_{3.6} = -1.400$ kN
12.18 $V_0 = V'_4 = -9.00$ kips,
$\quad V''_4 = V'_7 = 23.5$ kips,
$\quad V''_7 = V'_{16} = -6.50$ kips,
$\quad V''_{16} = V_{18} = 12.00$ kips
12.20 $V_0 = 16.00$ kips, $V_{5.33} = 0$,
$\quad V_8 = V_{12} = -8.00$ kips
12.22 $V_0 = V'_1 = 22.0$ kN,
$\quad V''_1 = V_2 = -2.00$ kN,
$\quad V_4 = -18.00$ kN
12.24 $V_0 = V'_2 = -1.200$ kN,
$\quad V''_2 = V'_4 = 4.40$ kN,
$\quad V''_4 = 0.800$ kN, $V_{4.67} = 0$,
$\quad V_8 = -4.00$ kN
12.26 $V_0 = V_8 = -4.00$ kips,
$\quad V_{20} = -10.00$ kips
12.28 $V_0 = V'_6 = -6.00$ kips,
$\quad V''_6 = 12.33$ kips, $V_{12.16} = 0$,
$\quad V'_{18} = -11.67$ kips,
$\quad V''_{18} = V_{22} = 8.00$ kips
12.30 $V_0 = 24.0$ kN, $V_{2.4} = V_3 = 0$
12.32 $M_0 = 0$, $M_3 = 33.0$ kip-ft,
$\quad M_6 = 42.0$ kip-ft, $M_{12} = 0$
12.34 $M_0 = 0$, $M_{1.6} = 28.8$ kN·m,
$\quad M_{2.29} = 0$, $M_3 = -30.0$ kN·m, $M_4 = 0$
12.36 $M_0 = 0$, $M_{1.2} = -6.24$ kN·m,
$\quad M_{3.6} = -9.60$ kN·m
12.38 $M_0 = 0$, $M_4 = 36.0$ kip-ft, $M_{5.53} = 0$,
$\quad M_7 = 34.5$ kip-ft, $M_{12.31} = 0$,
$\quad M_{16} = -24.0$ kip-ft, $M_{18} = 0$
12.40 $M_0 = 0$, $M_{5.33} = 42.7$ kip-ft,
$\quad M_8 = 32.0$ kip-ft, $M_{12} = 0$
12.42 $M_0 = 0$, $M_1 = 22.0$ kN·m,
$\quad M_2 = 20.0$ kN·m,
$\quad M_3 = 14.00$ kN·m, $M_4 = 0$
12.44 $M_0 = 0$, $M_2 = -2.40$ kN·m, $M_{2.55} = 0$,
$\quad M_4 = 6.40$ kN·m, $M_{4.67} = 6.67$ kN·m,
$\quad M_8 = 0$
12.46 $M_0 = 0$, $M_8 = -32.0$ kip-ft,
$\quad M_{14} = -65.0$ kip-ft,
$\quad M_{20} = -116.0$ kip-ft
12.48 $M_0 = 0$, $M_6 = -36.0$ kip-ft,
$\quad M_{12.16} = 2.00$ kip-ft,
$\quad M_{18} = -32.0$ kip-ft, $M_{22} = 0$

12.50 $M_0 = -41.6$ kN·m,
      $M_{1.2} = -20.0$ kN·m,
      $M_{2.4} = M_3 = -12.80$ kN·m

12.52 $V_0 = V'_4 = 6.0$ kips,
      $V''_4 = V_{10} = -4.0$ kips, $M_0 = 0$,
      $M_4 = 24.0$ kip-ft, $M_{10} = 0$

12.54 $V_0 = V'_a = 4P/3$, $V''_a = V'_{2a} = P/3$,
      $V''_{2a} = V_{3a} = -5P/3$, $M_0 = 0$,
      $M_a = 4Pa/3$, $M_{2a} = 5Pa/3$, $M_{3a} = 0$

12.56 $V_0 = V'_2 = 64.0$ kN,
      $V''_2 = V'_4 = 16.0$ kN,
      $V''_4 = V_6 = -80.0$ kN, $M_0 = 0$,
      $M_2 = 128.0$ kN·m,
      $M_4 = 160.0$ kN·m, $M_6 = 0$

12.58 $V_0 = 12.0$ kips, $V_6 = 0$,
      $V_{12} = -12.0$ kips, $M_0 = 0$,
      $M_6 = 36.0$ kip-ft, $M_{12} = 0$

12.60 $V_0 = V_L = -P$, $M_0 = 0$, $M_L = -PL$

12.62 $V_0 = V_2 = V_4 = -25.0$ kN, $M_0 = 0$,
      $M_2 = -50.0$ kN·m, $M_4 = -100.0$ kN·m

12.64 $V_0 = 0$, $V_{7.5} = -22.5$ kips,
      $V_{15} = -45.0$ kips, $M_0 = 0$,
      $M_{7.5} = -84.4$ kip-ft,
      $M_{15} = -338$ kip-ft

12.66 $V_0 = V'_a = -P/2$, $V''_a = V'_{2a} = -3P/2$,
      $V''_{2a} = V_{3a} = P$, $M_0 = 0$, $M_a = Pa/2$,
      $M_{4a/3} = 0$, $M_{2a} = -Pa$, $M_{3a} = 0$

CHAPTER 13

13.2   (a) 34.2 MPa, -34.2 MPa
      (b) -25.7 MPa

13.4   8890 lb-ft

13.6   (a) 8.35 MPa, -8.35 MPa
      (b) -7.23 MPa

13.8   751 kip-in.

13.10   348 mm

13.12   43.8 kip-in.

13.14   (a) 21.2 MPa, -21.2 MPa
      (b) 12.73 MPa

13.16   4.67 ksi, -4.67 ksi

13.18   14.99 MPa

13.20   7.44 ksi, -7.44 ksi

13.22   51.1 MPa, -51.1 MPa

13.24   30.0 kips

13.26   261 kN/m

13.28   2.70 kips

13.30   181.4 kN/m

13.32   (a) 9.30 ksi, -9.30 ksi
      (b) 25.8 kips/ft

13.34   (a) 8.91 ksi, -8.91 ksi
      (b) 2.86 kips/ft

13.36   1200 psi

13.38   154.5 kN

13.40   230 psi

13.42   678 kN

13.44   2060 psi

13.46   (a) 22.7 MPa   (b) 22.7 MPa
      (c) 0.08 percent

13.48   6.44 ksi

13.50 $\tau_{max} = 1.826$ MPa

13.52   114.0 mm

13.54   5.82 in.

13.56   7.76 in.

13.58   5.13 in.

CHAPTER 14

14.2   (a) -523 kN·m$^3/EI$
      (b) 250 kN·m$^2/EI$

14.4   (a) -6.72×10^6 kip-in.$^3/EI$
      (b) 6.22×10^4 kip-in.$^2/EI$

14.6   (a) -2.104 MN·m$^3/EI$
      (b) 0.701 MN·m$^2/EI$

14.8   (a) -5.15×10^7 kip-in.$^3/EI$
      (b) 3.42×10^5 kip-in.$^2/EI$

14.10   (a) -0.837 MN·m$^3/EI$
      (b) 0.300 MN·m$^2/EI$

14.12   (a) -1.920 MN·m$^3/EI$
      (b) 0.800 MN·m$^2/EI$

14.14   (a) -6.07×10^5 kip-in.$^3/EI$
      (b) 8.86×10^3 kip-in.$^2/EI$

14.16   (a) -15.76 mm   (b) $0.376°$

14.18 -3.22×10^6 kip-in.$^3/EI$

14.20 -162.8 kN·m$^3/EI$

14.22 -4.52×10^6 kip-in.$^3/EI$

14.24 -30.6 kN·m$^3/EI$

14.26 -6.86×10^5 kip-in.$^3/EI$

14.28 -3.15×10^6 kip-in.$^3/EI$

14.30 -0.333 in.

14.32   (a) -523 kN·m$^3/EI$
      (b) 250 kN·m$^2/EI$

14.34   (a) -6.72×10^6 kip-in.$^3/EI$
      (b) 6.22×10^4 kip-in.$^2/EI$

14.36 (a) $-2.10 \text{ MN·m}^3/EI$
　　　(b) $0.701 \text{ MN·m}^2/EI$
14.38 (a) $-5.15 \times 10^7 \text{ kip-in.}^3/EI$
　　　(b) $3.42 \times 10^5 \text{ kip-in.}^2/EI$
14.40 (a) $-0.837 \text{ MN·m}^3/EI$
　　　(b) $0.300 \text{ MN·m}^2/EI$
14.42 $-3.22 \times 10^6 \text{ kip-in.}^3/EI$
14.44 $-162.8 \text{ kN·m}^3/EI$
14.46 $-4.52 \times 10^6 \text{ kip-in.}^3/EI$
14.48 $M_A = 27.0 \text{ kip-ft}, R_A = 11.25 \text{ kips},$
　　　$R_B = 6.75 \text{ kips}$
14.50 $M_A = 4PL/27, R_A = 13P/27,$
　　　$R_B = 14P/27$
14.52 $M_A = 53.3 \text{ kip-ft}, R_A = 9.63 \text{ kips},$
　　　$R_B = 10.37 \text{ kips}$
14.54 $M_A = 101.5 \text{ kip-ft}, R_A = 27.3 \text{ kips},$
　　　$R_B = 24.7 \text{ kips}$
14.56 $M_A = 12PL/125, M_B = 18PL/125,$
　　　$R_A = 44P/125, R_B = 81P/125$
14.58 $M_A = 24.0 \text{ kip-ft}, M_B = 36.0 \text{ kip-ft},$
　　　$R_A = 8.80 \text{ kips}, R_B = 16.20 \text{ kips}$
14.60 $M_A = M_B = 48.0 \text{ kip-ft},$
　　　$R_A = R_B = 24.0 \text{ kips}$
14.62 $M_A = 9PL/64 + wL^2/12,$
　　　$M_B = 3PL/64 + wL^2/12,$
　　　$R_A = 27P/32 + wL/2,$
　　　$R_B = 5P/32 + wL/2$
14.64 $M_A = 48.1 \text{ kN·m}, M_B = 29.4 \text{ kN·m},$
　　　$R_A = 72.2 \text{ kN}, R_B = 37.8 \text{ kN}$

CHAPTER 15

15.2　(a) $33.3 \text{ kPa}, -26.7 \text{ kPa}$
15.4　(a) $78.9 \text{ ksi}, -78.9 \text{ ksi}$
　　　(b) $72.0, -85.9 \text{ ksi}$
15.6　(a) $48.5 \text{ MPa}, -56.2 \text{ MPa}$
15.8　(a) $6040 \text{ psi}, -6590 \text{ psi}$
15.10 (a) $109.0 \text{ MPa}, -120.3 \text{ MPa}$
15.12 (a) $9500 \text{ psi}, -5500 \text{ psi}$
15.14 (a) $16.67 \text{ MPa}, -3.07 \text{ MPa}$
15.16 (a) $10,980 \text{ psi}, -11,940 \text{ psi}$
15.18 $147.9 \text{ MPa}, -147.9 \text{ MPa}$
15.20 37.1 kN
15.22 (a) A and B: zero, C and D: -0.417 ksi,
　　　　neutral axis along side AB
　　　(b) A: $-0.375 \text{ ksi}, B$: 0.375 ksi,

C: $-0.0416 \text{ ksi}, D$: 0.792 ksi,
　　neutral axis intersects side AB
　　10 in. from B and side BC 21.6 in.
　　from B.
15.24 A: $-28.1 \text{ MPa}, B$: -1.423 MPa,
　　　C: $-23.7 \text{ MPa}, D$: -50.4 MPa
15.26 $\sigma_n = 8200 \text{ psi}, \sigma_t = -4200 \text{ psi}$,
　　　$\tau_{nt} = -1270 \text{ psi}$
15.28 $\sigma_n = 12,160 \text{ psi}, \sigma_t = -14,160 \text{ psi}$,
　　　$\tau_{nt} = -2800 \text{ psi}$
15.30 $\sigma_n = 1.368 \text{ ksi}, \sigma_t = -1.368 \text{ ksi}$,
　　　$\tau_{nt} = 3.76 \text{ ksi}$
15.32 $\sigma_n = 7.20 \text{ MPa}, \sigma_t = 0.805 \text{ MPa}$,
　　　$\tau_{nt} = -6.47 \text{ MPa}$
15.34 $\sigma_n = 9.00 \text{ MPa}, \sigma_t = -1.000 \text{ MPa}$,
　　　$\tau_{nt} = 8.66 \text{ MPa}$
15.36 (a) $8320 \text{ psi}, -4320 \text{ psi}, \angle 35.8°$
　　　(b) $6325 \text{ psi}, \angle 80.8°$
15.38 (a) $12,450 \text{ psi}, -14,450 \text{ psi}, \angle 66.0°$
　　　(b) $-13,450 \text{ psi}, \angle 21.0°$
15.40 (a) $4.00 \text{ ksi}, -4.00 \text{ ksi}, \angle 45.0°$
　　　(b) $-4.00 \text{ ksi}, 0°$
15.42 (a) $11.21 \text{ MPa}, -3.21 \text{ MPa}, \angle 61.8°$
　　　(b) $-7.21 \text{ MPa}, \angle 16.8°$
15.44 (a) $14.00 \text{ MPa}, -6.00 \text{ MPa}, 0°$
　　　(b) $10.00 \text{ MPa}, \angle 45.0°$
15.46 A: $19,690 \text{ psi}, 0, 0° B$: $10,660 \text{ psi}$,
　　　$-818 \text{ psi}, \diagdown 15.5° C$: 3940 psi,
　　　$-3940 \text{ psi}, \diagdown 45.0°$
15.48 A: $6.60 \text{ MPa}, 0, 0° B$: 3.51 MPa,
　　　$-0.212 \text{ MPa}, \diagdown 13.8° C$: 1.027 MPa,
　　　$-1.027 \text{ MPa}, \diagdown 45.0°$
15.50 A: $0, -12.71 \text{ ksi } B$: 0.196 ksi,
　　　$-4.63 \text{ ksi } C$: $7.99 \text{ ksi}, 0$
15.52 15.98 kips
15.54 (a) $-1598 \text{ psi}, 1839 \text{ psi}$
　　　(b) $-3714 \text{ psi}, 1854 \text{ psi}$
15.56 (a) 80.0 MPa　(b) 40.0 MPa
　　　(c) 20.0 MPa　(d) 40.0 MPa
15.58 (a) $18.75 \text{ MPa}, 9.38 \text{ MPa}$
　　　(b) 30.8 MPa　(c) 20.2 MPa
　　　(d) $55.4 \text{ MPa}, -7.08 \text{ MPa}, \diagdown 40.0°$
15.60 (a) $-52.3 \text{ MPa}, 7.47 \text{ MPa}, 67.2 \text{ MPa}$
　　　(b) $0, 11.20 \text{ MPa}, 0$
　　　(c) A: $0, -52.3 \text{ MPa}, 26.1 \text{ MPa}$
　　　B: $15.50 \text{ MPa}, -8.07 \text{ MPa}, 11.81 \text{ MPa}$
　　　C: $67.2 \text{ MPa}, 0, 33.6 \text{ MPa}$

15.62 (a) 0, 7640 psi, 0, −7640 psi
       (b) 955 psi   (c) 637 psi, 0, 637 psi, 0
       (d) *D:* 318 psi, −318 psi, 318 psi
       E: 7760 psi, −117 psi, 3940 psi
       F: 1592 psi, −1592 psi, 1592 psi
       G: 117 psi, −7760 psi, 3940 psi
15.64 (a) 0, 27.2 ksi, 0, −27.1 ksi
       (b) 6.04 ksi
       (c) 0.755 ksi, 0, 0.755 ksi, 0
       (d) *A:* 6.78 ksi, −6.78 ksi, 6.78 ksi
       B: 28.4 ksi, −1.28 ksi, 14.86 ksi
       C: 5.28 ksi, −5.28 ksi, 5.28 ksi
       D: 1.28 ksi, −28.4 ksi, 14.86 ksi

CHAPTER 16

16.2   (a) 680 kN (b) 1387 kN (c) 1617 kN
16.4   30.0 kips
16.6   (a) 85.2 (b) 170.2
16.8   (a) 0.436 kips (b) 0.174 kips
16.10 1114 kN
16.12 409 kips
16.14 2000 kN
16.16 167.6 kips
16.18 491 kN
16.20 11,230 lb